Advances in Bridge Design and Structural Performance: 2nd Edition

Advances in Bridge Design and Structural Performance: 2nd Edition

Guest Editors

Zhongxiang Liu
Tong Guo

MDPI

Basel • Beijing • Wuhan • Barcelona • Belgrade • Novi Sad • Cluj • Manchester

Guest Editors
Zhongxiang Liu
Department of Bridge Engineering
Southeast University
Nanjing
China

Tong Guo
School of Civil Engineering
Southeast University
Nanjing
China

Editorial Office
MDPI AG
Grosspeteranlage 5
4052 Basel, Switzerland

This is a reprint of the Special Issue, published open access by the journal *Applied Sciences* (ISSN 2076-3417), freely accessible at: https://www.mdpi.com/journal/applsci/special_issues/2VB6Z72UH9.

For citation purposes, cite each article independently as indicated on the article page online and as indicated below:

Lastname, A.A.; Lastname, B.B. Article Title. *Journal Name* **Year**, *Volume Number*, Page Range.

ISBN 978-3-7258-8187-1 (Hbk)
ISBN 978-3-7258-8188-8 (PDF)
https://doi.org/10.3390/books978-3-7258-8188-8

Cover image courtesy of Zhongxiang Liu

Contents

About the Editors

Zhongxiang Liu

Zhongxiang Liu is an associate professor and PHD supervisor at Southeast University. Dr. Liu has been devoted to research in structural intelligent monitoring and condition assessment, structural fluid dynamics analysis and control, and new sensing and structure intelligence. Dr. Liu has participated in more than 10 national-level research projects, including two National Key Research and Development Project and two projects funded by the National Natural Science Foundation of China. He has published more than 70 SCI papers. His research achievements have been applied in more than ten important projects, including structural analyses and performance assessments of the Runyang Bridge in China, Hercílio Luz Bridge in Brazil, photovoltaic structures, etc. Dr. Liu has been awarded the first prize of Science and Technology Progress Award of Jiangsu, the Grand prize of the Science and Technology Progress Award from the China Highway and Transportation Society, the gold medal with congratulations of the Jury of International Exhibition of Inventions of Geneva, and the LIU Huixian Earthquake Engineering Scholarship Award.

Tong Guo

Tong Guo is a chief professor and PHD supervisor at Southeast University. Prof. Guo has long committed himself to research in structural condition assessment and performance upgrade, and has made achievements in interpreting structural damage mechanisms, simulating time-variant conditions and effective performance upgrade methods. A series of seismic-resilient structures has been developed. His research achievements have been applied in more than twenty important domestic and international projects. Prof. Guo is the principal investigator of twenty research projects, including two National Key Research and Development Projects (1.6 million USD) and five projects funded by the National Natural Science Foundation of China. Prof. Guo has published more than 200 SCI papers and was selected as a Chinese Highly Indexed Researcher. He is the lead author or co-author of fifteen technical standards and two books. Prof. Guo has been awarded the second prize of the National Science and Technology Progress Award twice as the second awardee. As the first awardee, the applicant has been awarded the first prize of the provincial and ministerial level Science and Technology Progress Award six times. Prof. Guo was elected as an ASCE fellow in 2020, was the Recipient of the National Outstanding Youth Fund, joined the Ten Thousand People Plan as a leading talent, and has been a Changjiang Youth Scholar and a Jiangsu Specially-Appointed Professor, etc. He currently serves as the editor of *Engineering in Extreme Environments*, the associate editors of two ASCE Journals (i.e. *Journal of Bridge Engineering* and *Journal of Performance of Constructed Facilities*), a member of the working group (forensic engineering) of the international association IABSE, and a committee member of the Chinese Association for Vibration Control and Structural Health Monitoring, etc. He was also appointed as a guest professor at Oslo Metropolitan University in 2019.

Preface

Bridges, as the backbone of transportation networks, are pivotal in ensuring safe, resilient, and sustainable mobility. With escalating traffic volumes and prolonged exposure to harsh environmental conditions, their structural performance inevitably deteriorates over time. This degradation is driven by a synergistic effect of fatigue, creep, shrinkage, and corrosion, which collectively erode the reliability of bridge components. In severe scenarios, the failure of critical load-bearing elements can compromise public safety, potentially triggering catastrophic progressive collapse.

This work, Advances in Bridge Design and Structural Performance, is dedicated to addressing these pressing challenges. It encompasses a comprehensive scope, covering fatigue performance assessment and life prediction, safety analysis of degraded bridges under accidental loads such as earthquakes, explosions, impacts, fires, and typhoons, as well as design and retrofit strategies for bridge members. The primary aim is to provide a robust scientific foundation for the long-term, high-quality functionality of bridges, ensuring their ability to withstand both routine operational stresses and extreme events.

The motivation for this work stems from the growing need to balance infrastructure demands with safety and sustainability. As bridge networks age and face increasing operational pressures, traditional design and maintenance approaches are no longer sufficient. This work is intended for a diverse audience, including bridge engineers, researchers, policymakers, and practitioners in the field of civil engineering. It seeks to disseminate cutting-edge research, practical methodologies, and innovative solutions to enhance the resilience and longevity of bridge infrastructures, ultimately contributing to the safety and efficiency of global transportation systems.

Zhongxiang Liu and Tong Guo
Guest Editors

applied sciences

MDPI

Editorial

Advances in Bridge Design and Structural Performance

Zhongxiang Liu [1,2,*] and Tong Guo [2,3,*]

1 School of Transportation, Southeast University, Nanjing 210096, China
2 State Key Laboratory of Safety, Durability and Healthy Operation of Long Span Bridges, Nanjing 210096, China
3 School of Civil Engineering, Southeast University, Nanjing 210096, China
* Correspondence: zhongxiang@seu.edu.cn (Z.L.); guotong@seu.edu.cn (T.G.)

1. Introduction

Bridges are critical to ensure the safe, resilient and long-term operation of the traffic network. However, the performance and reliability of bridges are bound to decline over time with the continuous increase in traffic load and long-term erosion of harsh environments [1]. Fatigue, creep, contraction, corrosion and other factors will accelerate bridge aging, and even cause structural chain collapse from the failure of load-bearing parts [2,3], posing a great threat to public safety. Under this background, evaluating the fatigue performance of bridges, predicting the fatigue life, and analyzing the safety of aging bridges under unexpected loads [4–7] have become the core research directions that need to be urgently promoted. Solving these problems can not only eliminate potential safety hazards over time, but also provide reference for the design and renovation of bridge components. The long-term stable operation of bridge infrastructure, maintaining its role as the lifeblood of society and economy, is the most basic [8]. Extensive efforts have been made in bridge design and structural performance, and relevant important progress can also reflect new trends, new solutions and future directions in the field [9,10].

The application of high-performance materials, such as ultra-high-performance concrete (UHPC) [11], high-strength steels [12] and fiber-reinforced polymers [13], has completely changed the thinking of bridge design, enabling a lighter, stronger and more durable structure. UHPC has been applied in deck panels, joints, and load-bearing components, demonstrating its ability to reduce material usage, extend service life, and enhance resistance to environmental degradation. Similarly, the use of high-strength steels allows for more efficient designs with reduced self-weight and improved fatigue performance, where welding remains the focus. Studies have focused on optimizing welding for less stress concentration and higher fatigue performance and addressing buckling challenges [14,15]. For structures and components, optimization has become a critical tool for achieving efficient and sustainable bridge designs [16]. Multi-objective optimization of cost, performance and environment has gained prominence with the progress in advanced analysis methods and computing devices [17].

Focusing on ensuring the lifecycle, safety and functionality of structures, performance-based design (PBD) and resilience-based design (RBD) have become the core development directions of modern bridge engineering [18,19]. Bridge design has been upgraded from meeting only the minimum requirements of standards such as "no collapse" to meeting graded performance goals for different load levels such as "usability, repairability, and rapid recovery", which is widely used in critical passages. The maturity of advanced computing, building information modeling (BIM), and digital twin technology enables real-time tracking of the performance of the entire design–construction–operation process [20,21],

 https://doi.org/10.3390/app16115546

supporting the development of PBD through real-time simulation and optimization capabilities. Climate change has intensified the risks of chloride ion erosion and extreme loads that bridges are exposed to, necessitating time-varying vulnerability analysis to assess long-term performance degradation [22]. RBD that prioritizes the rapid functional recovery of structures following extreme events such as earthquakes and floods has seen rapid development in recent years [23].

Industrialized and prefabricated bridge construction has entered a stage of large-scale application, with prefabrication rates exceeding 80% in key projects. Modular design strategies have been adopted during bridge design, enabling precise splitting of components, such as segmental box-girders and prefabricated piers [24]. This places greater demands on phased analysis and management throughout the construction process for the safety of prefabricated construction and the long-term performance of the overall structure. To effectively enhance the integrity and stability of assembled structures, innovations in connection technologies, such as grouting sleeves, UHPC wet joints, friction-type high-strength steel joints and module connections, and assembled joints, have been developed and improved [25]. However, challenges persist, particularly in harsh environments prone to chloride corrosion and cyclic loading. These include performance unreliability stemming from construction defects and complex processes, as well as concerns over the long-term durability of structural nodes and joints. The integration of digital fabrication and additive manufacturing technologies has unlocked new potential for industrialized bridge construction and high-performance assembled connections [26]. Extensive experiments and finite element analysis (FEA) have been carried out to assess optimal connection construction [27–29]. Automatic construction technologies such as 3D printing and robot welding are also being explored, which are expected to reduce labor costs, improve construction accuracy and realize the construction of complex modeling bridges [30]. Meanwhile, technologies such as BIM and digital twins can facilitate lifecycle management, making the bridge field develop towards intelligence.

Accurate load modeling is the basis to avoid insufficient design or over-design, and can also ensure the structural performance of the bridge in actual use environments. For example, the traditional traffic load model design stipulated in the specifications of various countries often cannot completely restore the complex situation of actual traffic and may underestimate the impact on the structure. Recent studies have begun to build load models based on real traffic data, traffic simulation and vehicle–bridge interaction analysis around major projects or special analyses [31,32]. Research on Runyang Bridge, Jiangyin Bridge and Brazil Highway Bridge shows that the design load model may greatly underestimate the local fatigue effect, traffic flow accumulation and cantilever negative bending moment brought by heavy-duty vehicles. This also shows that we need a more refined load model, taking the dynamic effect and the variability in actual traffic into account [33,34]. In addition, recent studies have also focused on the development of local non-stationary wind field models and data-driven wind load models [35,36]. These models combine high-resolution field measurement, computational fluid dynamics simulation, and ML algorithms to capture the spatiotemporal variation characteristics of wind power and improve the accuracy of structural wind response evaluation.

Dynamic loads such as vehicles, wind, and earthquakes will have a great impact on the performance and service life of bridges. Advanced numerical simulation techniques, including FEA and AI+FEA, have been employed to study dynamic responses and fatigue crack propagation [37]. For example, studies on steel bridges have analyzed the influence of the stress ratio on fatigue crack growth rate and then developed a more accurate prediction model. Some scholars put forward a multi-factor competition model (namely Liu-Guo's model), taking the coupling effect of fatigue, fretting wear, corrosion and other factors into account [38]. In addition, the Structural Health Monitoring (SHM) system has also been put into use to collect real-time data of the dynamic behavior of bridges, which can not only verify the accuracy of the numerical model, but can also help us formulate maintenance strategies based on actual conditions [39]. An intelligent monitoring system equipped with advanced sensors, data fusion algorithms and ML models can detect defects, predict structural aging, and identify potential failure risks [40]. These systems can make us more aware of the long-term performance changes of bridges, help arrange maintenance work in advance, and reduce the probability of sudden failures [37]. Digital twin modeling, which creates virtual replicas of physical bridges, has emerged as a powerful tool for simulating structural behavior, optimizing maintenance schedules, and evaluating the impact of design modifications.

In addition, bridges are now increasingly facing such extreme situations as fire, explosions, impacts and extreme weather, which will seriously threaten the safety and durability of bridges. In order to improve the disaster resistance of bridges, researchers are developing new design strategies, evaluation methods and protection technologies. According to research on refractory bridge materials and structural systems, the use of expansive coatings and composite materials to improve fire resistance has been explored [41]. Similarly, in anti-explosion and anti-impact research, energy absorption devices and retrofitting techniques have also been developed to reduce damage [42,43]. Machine learning algorithms have also been integrated into resilience assessment frameworks, which can predict the performance of structures under extreme conditions and optimize protection strategies [44].

This Special Issue series shows the frontier progress in theory, technology and application in the field of bridge design and structural performance. The Special Issue series contains 20 articles, of which 8 are from the second edition and 12 are from the first edition. The authors of these 20 articles come from different institutions in 16 countries, reflecting the global influence and extensive attention of bridge design and structural performance research. The articles included cover a wide range, involving the basic theory, innovative design technology and practical application of bridge structure performance. From the new calculation model of load analysis to the environmental protection materials that can innovate the construction mode, and from the anti-seismic and disaster-resistant strategy of long-span bridges to the intelligent monitoring system of aging infrastructure, these contents comprehensively present the latest developments in the field. The rest of this editor's note gives a brief overview of each article, focusing on their core viewpoints and contributions. I strongly recommend that readers read these articles completely. I believe this Special Issue can bring valuable ideas and inspiration to everyone.

2. Summary of the Special Issue Contents

The following is an overview of the research results collected in this Special Issue series according to the article number.

Karriqi et al. (contribution 1) took a multi-span highway bridge in Albania as the research object to evaluate the performance of a precast reinforced concrete bridge under the combined action of earthquake load and local scour caused by a flood. They combined scour depth estimation, seismic acceleration time history spectrum matching, nonlinear time history analysis, and vulnerability curves at the component and system level to quantify the ground acceleration (PGA) and flood flow scenarios with different peak values, determining the influence of foundation scour on bridge vulnerability. The value of this research lies in breaking through the limitation of single disaster assessment.

Zheng et al. (contribution 2) proposed a set of laser Doppler vibration measurement systems equipped with unmanned aerial vehicles to identify debonding damage of a hidden frame glass curtain wall through non-contact vibration detection. Combined with

UAV remote testing, wavelet packet energy decomposition and the SSD damage index, they proved that this method can not only detect the existence of debonding damage, but also intuitively locate the damage area. This research makes structural performance detection move towards safer and more efficient high-altitude damage detection.

Liu et al. (contribution 3) studied the dynamic mechanical properties of a continuous rigid frame bridge with cracks in service under the coupling vibration of vehicles and the bridge. They deduced the fracture damage stiffness reduction model, established the multi-scale finite element model, and analyzed the bridge response under different speeds, vehicle weights, and damage degrees with the Newmark-β method. The results show that bridge damage can significantly amplify stress, displacement and impact coefficient, and the damaged bridge is particularly sensitive to the vehicle speed of about 90 km/h.

Dong et al. (contribution 4) proposed a notched T-perfobond connector with an integrated flange to improve pull-out resistance and simplify rebar installation in shallow-embedded composite structures. Through pull-out tests, ABAQUS explicit finite element modeling, and 54 parametric simulations, they clarified the failure modes, load–separation behavior, and effects of rib geometry, flange size, notch width, rebar diameter, and material properties on uplift capacity. The study refines the design basis for perfobond connectors by addressing a less-studied pull-out failure mode, and it further provides a tailored calculation method for estimating the pull-out capacity of notched T-perfobond connectors.

Rajchel and Siwowski (contribution 5) assessed the structural safety of a 55-year-old thin-walled steel road bridge using the new Polish national guidelines based on prEN 1990-2. The study is valuable not only as a rigorous case-based diagnosis of an aging steel bridge, but also as one of the first practical demonstrations of a modern code-based assessment framework leading directly to a maintenance decision: bridge closure and replacement rather than repair or strengthening.

Zhou et al. (contribution 6) developed an adaptive MEC–RBF neural-network model for predicting construction-stage deflection in prestressed concrete continuous rigid-frame bridges. The study identified the cantilever-to-span length ratio and girder block length as dominant factors and achieved high prediction accuracy for the Hannan Yangtze River Bridge. The work strengthens intelligent construction control by replacing repeated time-consuming finite element recalculation with a fast, data-driven prediction tool.

Ganyukov et al. (contribution 7) optimized the supporting arrangement for the mobile overpass, which is used to maintain traffic above the urban road pipeline maintenance trench. Compared with the initial arrangement scheme, the horizontal displacement of the trench wall is reduced by more than half.

Kadyrov et al. (contribution 8) developed the undercarriage system of a mobile overpass intended to preserve traffic flow during utility and road repair works in dense urban areas. The study advances the mobile-overpass concept from structural feasibility toward practical deployment, showing that the proposed undercarriage can provide reliable positioning, transportability, and operational stability, with experimental deflection following an almost linear trend and a regression fit of $R^2 = 0.97$.

Zhang et al. (contribution 9) developed a practical method for calculating the rotational constraint stiffness of a universal hinge support in the tower–pier connection of special-shaped single-tower cable-stayed bridges. This research breaks through the assumption of simplifying the ideal hinge and captures the problems caused by friction, the actual constraint effect caused by geometry and vertical load.

Cao et al. (contribution 10) focused on the stress ratio and fatigue crack propagation law of bridge steel Q345qD. The crack propagation equation based on the energy release rate containing the stress ratio was derived, integrating into the extended finite element

(XFEM) numerical framework. It provides a more reliable tool for fatigue life assessment of steel bridge components (especially welding details affected by residual stress).

Kang et al. (contribution 11) examined the tensile performance of longitudinal joints in precast bridge deck systems. The study gives a clear experimental basis for improving the connectivity and constructability of precast bridge deck joints with both faster installation and reliable structural continuity.

Rossigali et al. [34] developed live-load models for Brazilian short-span highway bridges. Two representative load configurations for two-lane girder-type bridges were developed. They indicated that the current Brazilian design load may underestimate actual traffic effects, especially negative moments in cantilevers. The work sharpens the link between bridge design loads and measured traffic reality, offering a more rational basis for safety assessment.

Li et al. (contribution 12) investigated the dynamic behavior of a lightweight cable-supported rail system. Vehicle–rail contact forces under different spans and speeds were qualified by an improved time-domain load-identification method. The study clarifies how contact force fluctuations are governed by span length, vehicle passing time, and coupled vibration, providing a future design of this novel flexible transportation structure.

Gocál et al. (contribution 13) examined the real stiffness and fatigue resistance of stringer-to-cross-girder connections in old riveted steel railway bridges. Field measurements on an existing bridge and laboratory fatigue tests on representative specimens were conducted to investigate the non-negligible bending stiffness and local stresses. The work offers practical guidance for refined assessment of aging riveted railway bridges.

Liu et al. [2] established a corrosion–fatigue life assessment framework for bridge cables based on the equivalent initial flaw size model. This study quantified how corrosion defects and traffic-induced stress ranges jointly govern cable-wire fracture, which offers a mechanics-based basis for cable inspection, maintenance timing, and replacement decisions.

Ouyang et al. (contribution 14) developed and optimized a rhombus hanging-basket walking-track system for cantilever casting construction of long-span bridges, with the Hai-he Bridge used as the engineering background. Its value lies in shifting attention from the main truss alone to the walking mechanism and construction equipment system, offering a lighter, safer, and more efficient solution for bridge cantilever construction.

Zhu et al. (contribution 15) evaluated the residual performance of in-service tied-arch bridge hangers protected by outer steel pipes, using a bridge with severe anchor-head water damage as a case study. The study translates a difficult-to-observe hidden deterioration problem into a practical inspection and evaluation procedure.

Yang et al. (contribution 16) analyzed the stay-cable vibration of a long-span rail-cum-road cable-stayed bridge under high-speed train loading, using the Huanggang Yangtze River Bridge as the prototype. A double-end-excitation cable vibration model was derived. The work is notable for separating cable vibration from the global bridge response while retaining realistic end excitation.

Li et al. (contribution 17) proposed an idealized time-history contact-force model for AERORail structures, transforming experimentally identified vehicle–rail interaction forces into standard single-valley and double-valley load curves. By converting scattered experimental contact-force histories into a usable dynamic load model, this work provides a practical basis for the analysis, design, and future performance evaluation of lightweight cable-supported rail systems.

Xu et al. (contribution 18) studied the shear performance of a web-embedded composite shear connector that integrates concrete dowels, penetrating rebars, studs, and an embedded steel web for steel–concrete composite bridge structures. The study advances

https://doi.org/10.3390/app16115546

connector design by showing that the proposed composite connector offers higher shear stiffness and good ductility.

3. Conclusions

This collection of articles covers various insights into the evolution and transformation of bridge design and structural performance, driven by material innovation, development of design ideas and concepts, industrialization and prefabrication of construction, refinement of analysis and evaluation, advances in computing and digitization, and increasing attention to sustainability, resilience and intelligence. These articles address various types of problems from many application fields. For example, some of the problems considered involve the application of high-performance materials, structural optimization, modular construction, dynamic load analysis, and extreme condition resilience. Some advancements show that smart monitoring has significantly improved the design, construction, and maintenance of bridges. Although many of the articles focus on performance uncertainties stemming from construction defects and intricate processes, long-term durability issues at structural nodes and joints, and the risks posed by chloride corrosion and cyclic loading in harsh operating environments, the advancements generally open up new avenues for enhancing structural performance. The collection of articles spans a broad spectrum of critical themes from chloride corrosion resistance design and high-performance assembled connections to additive manufacturing, intelligent construction, data-driven load modeling, multi-factor coupling performance evaluation, and cyclic load durability enhancement, indicating the far-reaching scope and pivotal significance of bridge design and structural performance research. Furthermore, sustainable and circular design, climate resilience, digitalization, multi-hazard resilience, lifespan extension, and vibration control will be key focuses in future bridge engineering research. Pushing forward in these areas will drive innovation, enabling smarter, greener, and more durable infrastructure through integrated technological and design solutions.

Conflicts of Interest: The authors declare no conflicts of interest.

List of Contributions:

1. Karriqi, T.; Matos, J.C.; Dang, N.-S.; Xia, Y. Bridge Assessment under Earthquake and Flood-Induced Scour. *Appl. Sci.* **2024**, *14*, 5174. https://doi.org/10.3390/app14125174.
2. Zheng, H.; Guo, T.; Zhi, G.; Hu, Z. A Novel Debonding Damage Identification Approach of Hidden Frame-Supported Glass Curtain Walls Based on UAV-LDV System. *Appl. Sci.* **2024**, *14*, 5412. https://doi.org/10.3390/app14135412.
3. Liu, Y.; Wang, L.; Jiang, C.; Shu, H. Research on Mechanical Performance of In-Service Continuous Rigid-Frame Bridge Based on Vehicle-Bridge Coupling Vibration. *Appl. Sci.* **2024**, *14*, 6950. https://doi.org/10.3390/app14166950.
4. Dong, Z.; Zheng, S.; Jiao, L.; Xu, X.; Yao, Y.; Gao, Z.; Li, H. Experimental and Parametric Analysis of Pull-Out Resistance of Notched T-Perfobond Connectors. *Appl. Sci.* **2024**, *14*, 7089. https://doi.org/10.3390/app14167089.
5. Rajchel, M.; Siwowski, T. Structural Safety Assessment of an Existing Steel Bridge According to the New Polish National Guidelines Based on prEN 1990-2. *Appl. Sci.* **2025**, *15*, 10890. https://doi.org/10.3390/app152010890.
6. Zhou, C.; Gao, Q.; He, Q.; Sun, L.; Tian, D. Adaptive MEC–RBF Neural Network-Based Deflection Prediction for Prestressed Concrete Continuous Rigid Frame Bridges During Construction. *Appl. Sci.* **2025**, *15*, 11326. https://doi.org/10.3390/app152111326.
7. Ganyukov, A.; Kadyrov, A.; Kukesheva, A.; Zhumabekov, A.; Sinelnikov, K.; Amanbayev, S.; Karsakova, A. Optimization of Mobile Overpass Support Placement Considering the Nonlinear Properties of the Soil Foundation. *Appl. Sci.* **2026**, *16*, 2075. https://doi.org/10.3390/app16042075.

8. Kadyrov, A.; Kukesheva, A.; Ganyukov, A.; Aidar, Z.; Kirill, S.; Nursultan, Z.; Zhanara, Z. Development of the Undercarriage of a Mobile Overpass for Operation During Repair Works in Dense Urban Areas. *Appl. Sci.* **2026**, *16*, 3879. https://doi.org/10.3390/app16083879.
9. Zhang, Y.; Zhao, Y.; Zhou, Y.; Yang, X. Calculation Method of Rotational Constraint Stiffness for a New Tower-Pier Connected System. *Appl. Sci.* **2022**, *12*, 11221. https://doi.org/10.3390/app122111221.
10. Cao, B.; Cheng, S.; Li, A.; Deng, Y.; Fang, Z. Fatigue Crack Propagation Study of Bridge Steel Q345qD Based on XFEM Considering the Influence of the Stress Ratio. *Appl. Sci.* **2022**, *12*, 12782. https://doi.org/10.3390/app122412782.
11. Kang, S.; Cho, S.; Na, H.; Han, J.; Lee, H. Tension Performance of Precast Bridge Deck Longitudinal Joints with Different Configurations. *Appl. Sci.* **2022**, *12*, 12892. https://doi.org/10.3390/app122412892.
12. Li, F.; Guo, Z.; Cui, Y.; Wu, P. Dynamic Load Test and Contact Force Analysis of the AERORail Structure. *Appl. Sci.* **2023**, *13*, 2011. https://doi.org/10.3390/app13032011.
13. Gocál, J.; Vičan, J.; Jošt, J. Real Stiffness and Fatigue Resistance of Stringer-to-Cross-Girder Connection of Riveted Steel Railway Bridges. *Appl. Sci.* **2023**, *13*, 2278. https://doi.org/10.3390/app13042278.
14. Ouyang, Y.; Huang, J.; Song, K. Development of Rhombus Hanging Basket Walking Track Robot for Cantilever Casting Construction in Bridges. *Appl. Sci.* **2023**, *13*, 10635. https://doi.org/10.3390/app131910635.
15. Zhu, L.; Chen, T.; Chen, L.; Lu, Z.; Hu, X.; Huang, X. Experimental Testing and Residual Performance Evaluation of Existing Hangers with Steel Pipe Protection Taken from an In-Service Tied-Arch Bridge. *Appl. Sci.* **2023**, *13*, 11070. https://doi.org/10.3390/app131911070.
16. Yang, F.; Zhao, H.; Li, A.; Fang, Z. Experimental–Numerical Analysis on the Cable Vibration Behavior of a Long-Span Rail-Cum-Road Cable-Stayed Bridge under the Action of High-Speed Trains. *Appl. Sci.* **2023**, *13*, 11082. https://doi.org/10.3390/app131911082.
17. Li, F.; Guo, Z.; Wu, P.; Cui, Y. A Time-History Contact Force Model of the Dynamic Load of AERORail Structures. *Appl. Sci.* **2024**, *14*, 3055. https://doi.org/10.3390/app14073055.
18. Xu, X.; Sun, T.; Hu, S.; Tan, H.; Wang, Y.; Zeng, Y. Study on Shear Performance of Web-Embedded Composite Shear Connector. *Appl. Sci.* **2024**, *14*, 5143. https://doi.org/10.3390/app14125143.

References

1. Liu, M.; Frangopol, D.M.; Kim, S. Bridge system performance assessment from structural health monitoring: A case study. *J. Struct. Eng.* **2009**, *135*, 733–742. [CrossRef]
2. Liu, Z.; Guo, T.; Yu, X.; Niu, S.; Correia, J. Corrosion fatigue assessment of bridge cables based on equivalent initial flaw size model. *Appl. Sci.* **2023**, *13*, 10212. [CrossRef]
3. Guo, T.; Sause, R.; Frangopol, D.M.; Li, A. Time-dependent reliability of PSC box-girder bridge considering creep, shrinkage, and corrosion. *J. Bridge Eng.* **2011**, *16*, 29–43. [CrossRef]
4. Apayd, N.M. Earthquake performance assessment and retrofit investigations of two suspension bridges in Istanbul. *Soil Dyn. Earthq. Eng.* **2010**, *30*, 702–710. [CrossRef]
5. Khan, M.A.; Khan, A.A.; Domada, R.; Usmani, A. Fire hazard assessment, performance evaluation, and fire resistance enhancement of bridges. *Structures* **2021**, *34*, 4704–4714. [CrossRef]
6. Gholipour, G.; Zhang, C.; Mousavi, A.A. Nonlinear numerical analysis and progressive damage assessment of a cable-stayed bridge pier subjected to ship collision. *Mar. Struct.* **2020**, *69*, 102662. [CrossRef]
7. Li, H.; Bao, Y.Q.; Li, S.L.; Chen, W.L.; Laima, S.J.; Ou, J.P. Monitoring, Evaluation and Control for Life-Cycle Performance of Intelligent Civil Structures. *Adv. Sci. Technol.* **2013**, *83*, 105–114.
8. Zhou, X.; Zhang, X. Thoughts on the development of bridge technology in China. *Engineering* **2019**, *5*, 1120–1130. [CrossRef]
9. Chen, A.; Frangopol, D.M.; Ruan, X. Bridge design, assessment and monitoring. *Struct. Infrastruct. Eng.* **2017**, *13*, 417. [CrossRef]
10. Aldabagh, S.; Khan, S.; Hossain, F.; Alam, M.S. Performance-based seismic assessment and design of long-span concrete deck arch bridges. *J. Bridge Eng.* **2023**, *28*, 04023077. [CrossRef]
11. Shao, X.; Deng, L.; Cao, J. Innovative steel-UHPC composite bridge girders for long-span bridges. *Front. Struct. Civ. Eng.* **2019**, *13*, 981–989. [CrossRef]
12. Miki, C.; Homma, K.; Tominaga, T. High strength and high performance steels and their use in bridge structures. *J. Constr. Steel Res.* **2002**, *58*, 3–20. [CrossRef]

13. Al-Rousan, R.Z.; Alnemrawi, B.A.R. Nonlinear finite element analysis for the torsional and bending behavior of heat-damaged RC beams strengthened with CFRP composites. *Int. J. Struct. Integr.* **2025**, *16*, 1529–1556. [CrossRef]
14. Sangeetha, P.; Anbarasu, M.; Dar, M.A.; Kumar, K.A.; Phanindra, K.S.; Balraj, P.M.M.; Radheshyam, G. Tests on cold-formed steel square tubular T-joints reinforced with side plates under combined axial compression and bending. *Int. J. Struct. Integr.* **2025**, *16*, 765–792. [CrossRef]
15. Tong, G.; Aiqun, L.; Jianhui, L. Fatigue life prediction of welded joints in orthotropic steel decks considering temperature effect and increasing traffic flow. *Struct. Health Monit.* **2008**, *7*, 189–202. [CrossRef]
16. Li, Y.; Lai, Y.; Lu, G.; Yan, F.; Wei, P.; Xie, Y.M. Innovative design of long-span steel–concrete composite bridge using multi-material topology optimization. *Eng. Struct.* **2022**, *269*, 114838. [CrossRef]
17. Liu, Z.; Guo, T.; Correia, J.; Wang, L. Reliability-based maintenance strategy for gusset plate connections in steel bridges based on life-cost optimization. *J. Perform. Constr. Facil.* **2020**, *34*, 04020088. [CrossRef]
18. Mackie, K.R.; Stojadinović, B. Performance-based seismic bridge design for damage and loss limit states. *Earthq. Eng. Struct. Dyn.* **2007**, *36*, 1953–1971. [CrossRef]
19. Akiyama, M.; Frangopol, D.M.; Ishibashi, H. Toward life-cycle reliability-, risk- and resilience-based design and assessment of bridges and bridge networks under independent and interacting hazards: Emphasis on earthquake, tsunami and corrosion. *Struct. Infrastruct. Eng.* **2020**, *16*, 26–50. [CrossRef]
20. Lu, X.; Liao, W.; Zhang, Y.; Huang, Y. Intelligent structural design of shear wall residence using physics-enhanced generative adversarial networks. *Earthq. Eng. Struct. Dyn.* **2022**, *51*, 1657–1676. [CrossRef]
21. Lee, J.H.; Park, J.J.; Yoon, H. Automatic bridge design parameter extraction for scan-to-BIM. *Appl. Sci.* **2020**, *10*, 7346. [CrossRef]
22. Zucca, M.; Landi, F.; Puppio, M.L.; Mistretta, F.; Formichi, P.; Croce, P. Climate change impact on corrosion of reinforced concrete bridges and their seismic performance. *Appl. Sci.* **2023**, *14*, 60. [CrossRef]
23. Ghaffarpasand, B.; Samadian, D.; Dehkordi, M.R.; Bocchini, P.; Eghbali, M. Investigation of different resilience-based optimisation strategies for retrofitting curved bridges. *Struct. Infrastruct. Eng.* **2024**, *20*, 1925–1939. [CrossRef]
24. Tugilimana, A.; Thrall, A.P.; Coelho, R.F. Conceptual design of modular bridges including layout optimization and component reusability. *J. Bridge Eng.* **2017**, *22*, 04017094. [CrossRef]
25. Tumbeva, M.D.; Thrall, A.P.; Zoli, T.P. Modular joint for the accelerated fabrication and erection of steel bridges. *J. Bridge Eng.* **2021**, *26*, 04021022. [CrossRef]
26. Tuvayanond, W.; Prasittisopin, L. Design for manufacture and assembly of digital fabrication and additive manufacturing in construction: A review. *Buildings* **2023**, *13*, 429. [CrossRef]
27. Feng, S.; Guan, D.; Ni, L.; Lin, Y.; Liu, Z.; Guo, Z.; Li, G. Experimental study on seismic behavior of joints connecting precast H-steel reinforced concrete beams and concrete-filled steel tube columns. *J. Build. Eng.* **2022**, *45*, 103444. [CrossRef]
28. Russell, H.G.; Ralls, M.L.; Tang, B.M. Prefabricated bridge elements and systems in Japan and Europe. *Transp. Res. Rec.* **2005**, *1928*, 102–109. [CrossRef]
29. Liu, J.; Xie, J.; Xiong, G.; Wang, X.; Zou, Y.; Shu, R.; Chen, Y.F. Full-scale push-out testing of headed stud-steel block connectors in prefabricated steel-concrete composite beams. *Eng. Struct.* **2023**, *285*, 116020. [CrossRef]
30. Xia, Y.; Liu, J.; Qi, H.; Wang, L.; Li, J.; Feng, L. Graph-based BIM generation method for integrated design of steel modular buildings. *J. Build. Eng.* **2025**, *106*, 112476. [CrossRef]
31. Ma, Y.; Zhou, H.; He, Y.; Li, X.; Wang, L. Subcycle corrosion fatigue crack growth model for bridge suspender wires under random time-series loads. *Eng. Fail. Anal.* **2025**, *169*, 109183. [CrossRef]
32. Crespo-Minguillon, C.; Casas, J.R. A comprehensive traffic load model for bridge safety checking. *Struct. Saf.* **1997**, *19*, 339–359. [CrossRef]
33. Liu, Z.; Guo, T.; Chai, S. Probabilistic fatigue life prediction of bridge cables based on multiscaling and mesoscopic fracture mechanics. *Appl. Sci.* **2016**, *6*, 99. [CrossRef]
34. Rossigali, C.E.; Pfeil, M.S.; Sagrilo, L.V.S.; de Oliveira, H.M. Load models representative of Brazilian actual traffic in girder-type short-span highway bridges. *Appl. Sci.* **2023**, *13*, 1032. [CrossRef]
35. Li, S.; Laima, S.; Li, H. Data-driven modeling of vortex-induced vibration of a long-span suspension bridge using decision tree learning and support vector regression. *J. Wind Eng. Ind. Aerodyn.* **2018**, *172*, 196–211. [CrossRef]
36. Wang, H.; Guo, T.; Tao, T.-Y.; Li, A.-Q. Study on wind characteristics of runyang suspension bridge based on long-term monitored data. *Int. J. Struct. Stab. Dyn.* **2016**, *16*, 1640019. [CrossRef]
37. Liu, Y.; Qu, C.-X.; Yi, T.-H.; Li, H.-N.; Zhai, B.-Z. Modal parameters identification of highly non-orthogonal structures based on curve fitting frequency domain decomposition method. *Eng. Struct.* **2026**, *357*, 122545. [CrossRef]
38. Liu, Z.; Guo, T.; Ren, H.; Correia, J.; Liu, J.; Feng, M.Q. Corrosion-fretting fatigue analysis of bridge cables based on multifactor competition model and time-dependent reliability. *J. Bridge Eng.* **2024**, *29*, 04023106. [CrossRef]
39. Liu, Z.; Cai, H.; Guo, T.; Liu, X.; Bai, Y.; Qu, C. Monitoring-Based Evaluation of Wind-Induced Vibration and Travel Comfort of Long-Span Suspension Bridge. *Struct. Control Health Monit.* **2025**, *2025*, 9962003. [CrossRef]

https://doi.org/10.3390/app16115546

40. Zhang, Z.; Lin, D.-W.; Sun, C.; Sun, Z. Physics-Informed Neural Network--Based TMD Parameter Identification and Response Prediction. *Struct. Control Health Monit.* **2025**, *2025*, 2157493. [CrossRef]
41. Zybina, O.; Gravit, M. *Intumescent Coatings for Fire Protection of Building Structures and Materials*; Springer: Berlin/Heidelberg, Germany, 2020; Volume 216.
42. Hao, H.; Hao, Y.; Li, J.; Chen, W. Review of the current practices in blast-resistant analysis and design of concrete structures. *Adv. Struct. Eng.* **2016**, *19*, 1193–1223. [CrossRef]
43. Yang, S.; Yang, Y.; Wang, Z.; Cheng, Y.; Xu, W.; Tao, S. Damping Ratio Identification of Beam Bridge Based on Vehicle Bridge Coupling Theory Under Single Vehicle. *Chin. J. Theor. Appl. Mech.* **2023**, *56*, 1098–1109.
44. Cai, H.; Liu, Z.; Guo, T.; Cai, C.; Sun, Z.; Zong, Y.; Zhou, C. *Investigation on Nonstationary Wind Effects and Intelligent Multi-Level Warning Method for Wind-Induced Vibration of Bridges*; Elsevier: Amsterdam, The Netherlands, 2026; p. 111112.

https://doi.org/10.3390/app16115546

applied sciences

MDPI

Article

Development of the Undercarriage of a Mobile Overpass for Operation During Repair Works in Dense Urban Areas

Adil Kadyrov, Aliya Kukesheva *, Alexandr Ganyukov *, Aidar Zhumabekov, Kirill Sinelnikov, Nursultan Zharkenov and Zhanara Zhunusbekova

Department of Transport Engineering and Logistics Systems, Abylkas Saginov Karaganda Technical University, 56 Nursultan Nazarbayev Avenue, Karaganda 100027, Kazakhstan; adil.suratovich@gmail.com (A.K.); aidar7478@gmail.com (A.Z.); coolzero7777@gmail.com (K.S.); zharkenovnursultan@gmail.com (N.Z.); zhanarazhumashevna@gmail.com (Z.Z.)

* Correspondence: alewa878@gmail.com (A.K.); severchell@gmail.com (A.G.)

Abstract

The study presents an engineering solution for maintaining traffic flow during road and utility operations, such as trench excavation. The analysis of existing organizational and technical approaches, along with global experience in temporary bridge use, showed that most foreign analogs were developed for military purposes and are not fully suitable for urban conditions in Kazakhstan and CIS countries. As an alternative solution, the development of a mobile overpass adapted for operation in dense urban environments is proposed. The present study continues earlier research focused on optimizing the placement of mobile overpass supports while accounting for the nonlinear deformation behavior of the soil foundation. At the previous stage, a rational distance between the supports and the trench edge was substantiated, and horizontal soil deformations were reduced. In the current study, the primary focus is on the design of the undercarriage, which determines the mobility, stability, and operational feasibility of the structure. A morphological analysis and synthesis method is applied to select a rational configuration of the undercarriage. A 3D model and a 1:4 scale test bench were developed, followed by load tests of 50–200 kg. The maximum deflection of −1.19 mm at 200 kg demonstrated an almost linear deformation pattern. The constructed regression model ($R^2 = 0.97$) confirmed the accuracy and reliability of the design. The developed mobile overpass is versatile, cost-effective, and practical, improving the resilience of urban transport infrastructure, reducing traffic congestion during roadworks, and creating a foundation for serial production in Kazakhstan and CIS countries.

Keywords: mobile overpass; road repairs; morphological analysis; undercarriage design; experimental branch of mobile overpass; regression model; structural deflection

1. Introduction

In recent years, there has been a significant increase in the number of road vehicles in many countries, including Kazakhstan [1]. For example, in countries such as Germany [2], Japan, and the United States [3], vehicle density has risen substantially due to improved transport accessibility and rapid urbanization. This, in turn, has intensified the load on road transport infrastructure [4], increased traffic intensity, and led to chronic congestion [5]. Urban roadways are particularly vulnerable to serious traffic jams during road maintenance and repair operations [6]. Temporary reductions in the number of lanes [7] or complete closures of certain road sections [8] significantly decrease road capacity and are among the primary causes of traffic congestion. Such conditions not only increase travel delays for drivers but also result in economic losses, higher fuel consumption [9], and negative environmental impacts [10] due to elevated exhaust gas emissions [11].

Work zones are also characterized by higher accident rates, as irregular road markings, narrowed lanes [12], and unexpected driver maneuvers [13] frequently lead to traffic incidents. In dense urban areas, even short-term traffic restrictions [14] can severely disrupt traffic flows. The problem is particularly acute in the central districts of large cities, where the transport infrastructure operates near its maximum capacity and there is no spare bandwidth for redistribution of flows [15].

Various organizational and technical solutions are used to reduce traffic jams during road operations. The most effective ones include use of contra-flow traffic systems [16], temporary detour schemes [17], and the use of Intelligent Transportation Systems (ITSs) [18] for adaptive flow management and real-time optimization. Table 1 presents an analysis of these methods with their advantages and disadvantages.

Table 1. Existing organizational and technical solutions reducing traffic jams during road operations.

Method	Description	Advantages	Disadvantages and Limitations
Contra-flow traffic systems	Changing the direction of lanes depending on load.	- Flexibility in redistributing traffic flows [19]. - Increase in the capacity during rash hours [20]. - Relatively low cost of implementation with the existing infrastructure [21].	- Risk of head-on collisions in case of control errors [22]. - Need for accurate marking and constant monitoring [23]. - Restriction of passing bays for emergency services [24]. - Applicable only in case of traffic asymmetry [25].
Temporary detours	Organization of alternative routes around the areas under repair.	- Complete separation of traffic from the operations zone [26]. - Improved safety of workers and drivers [27]. - A simple and straightforward method for traffic arrangement [28].	- Increase in route time and distance [29]. - Inability to apply in dense urban areas due to lack of space [30]. - High costs for the construction of temporary roads [31]. - Poor quality of temporary pavements (unpaved roads) [32].
Intelligent Transportation Systems (ITSs)	Application of traffic control and information systems for traffic optimization.	- Reduction in the overall delay through adaptive traffic light control [33]. - Alerting drivers about traffic jams and closures in advance [34]. - Minimization of unexpected increases in the load on alternative routes [35].	- Do not address the physical limitation of road capacity [36]. - Effective only when alternative routes are available [37]. - High initial costs and requirements for infrastructure digitalization [38]. - Limited distribution in less developed cities [39].

The analysis shows that the existing methods are an adaptation measure rather than a radical elimination of the causes of traffic jams. They partially compensate for the negative impacts, but do not eliminate the traffic jam and road blockage problem itself. In this regard, the search for new promising solutions is relevant.

There is also another well-known way to keep traffic moving when repairing roads. These include temporary bridges [40] used over the areas under repair. Historically, the first such bridge was the Bailey Bridge (Figure 1a), created in the 1940s for military needs. It was assembled manually from modular metal elements, easily transported and quickly assembled [41]. Later, it was improved into the Medium Girder Bridge system (Figure 1b) (medium-span girder bridge) [42], used by the military and in the civilian sphere in more than 40 countries.

https://doi.org/10.3390/app16083879

(**a**) (**b**)

Figure 1. Prefabricated bridge used for military and civil purposes. (**a**) Bailey Bridge. (**b**) Medium Girder Bridge.

Nowadays, modern companies such as Acrow (Parsippany, NJ, USA) and Mabey Bridge (Lydney, UK) also offer modular bridges for temporary crossings. For example, in the US, a 137 m long two-lane temporary Acrow (Parsippany, NJ, USA) bridge leased for the construction period was installed during the repair of the I-295 interchange [43]. This allowed for avoiding a complete traffic stop and saved time (Figure 2).

Figure 2. Modular bridges for temporary crossings.

In Europe, the ASTRA Bridge (Federal Roads Office (ASTRA), Bern, Switzerland) developed in Switzerland became an innovative solution. This is a 257 m long portable bridge that carries traffic over the repair operations. The structure is mounted on motorized wheels and moves as the operations progress [44]. The system was successfully tested in 2022–2023 and generated interest in Germany and the UK (Figure 3).

https://doi.org/10.3390/app16083879

Figure 3. Portable ASTRA bridge (Federal Roads Office (ASTRA), Bern, Switzerland).

Similar technologies are also being developed in Asia. In Japan, an origami-based mobile folding bridge has been developed [45]; it can be deployed manually without heavy machinery, which significantly simplifies installation in constrained environments (Figure 4) [46].

Figure 4. Origami-based mobile folding bridge.

In China, ultra-fast bridge replacement is practiced [47]. For example, in Beijing, a bridge was completely replaced in 43 h using launching technology [48]. Russia is developing its own engineering solutions for temporary bridges including modules created by the Research Institute of Transport and Road Management. These structures are assembled from welded beams of box-shapes or double T-section 6–12 m long, as well as prefabricated metal or reinforced concrete supports [49]. Installation is possible with the use of hydraulic jacks and light cranes, without the need for heavy machinery. Installation is possible using hydraulic jacks and cranes with low lifting capacity, without involving heavy machinery. Articulated joints and bolted assemblies allow the bridge to be assembled on site in 1–2 shifts by a small team. All the elements are transported on vehi-

https://doi.org/10.3390/app16083879

cles such as KAMAZ (KAMAZ PJSC, Naberezhnye Chelny, Russia), Volvo FMX (Volvo Group, Gothenburg, Sweden), Mercedes-Benz (Daimler Truck AG, Stuttgart, Germany) Arocs and other analogs. Such structures are characterized by low cost, high maintainability, and the possibility of use in remote areas. However, their load-bearing capacity is limited to 30–40 tons, there is no automation of the installation process, and the maximum speed on the bridge is 40–60 km/h. Military pontoon bridges, such as the PMP-3 (Russian military engineering system, Russia), designed for passage of heavy equipment, serve as an alternative. They consist of pontoons connected by coupling units and can carry up to 60 tons. A 300 m long bridge can be assembled in 4–6 h [50]. In emergency situations, such structures are re-equipped for civilian use, but require installation of additional elements, fencing, pavement, and lighting (Figure 5).

Figure 5. Military pontoon bridge PMP-3 (Russian military engineering system, Russia).

Their advantages are high load-carrying capacity and speed of assembly, but they require specialized equipment and personnel, as well as certification for their use in civil infrastructure. Also in this sphere, there is our research on the development of a mobile overpass [51]. A mobile overpass is a single-span truss structure [52] with a reinforced metal frame, designed to organize a temporary passage of vehicles through dangerous road [53] sections or sections under repair [54] (Figure 6).

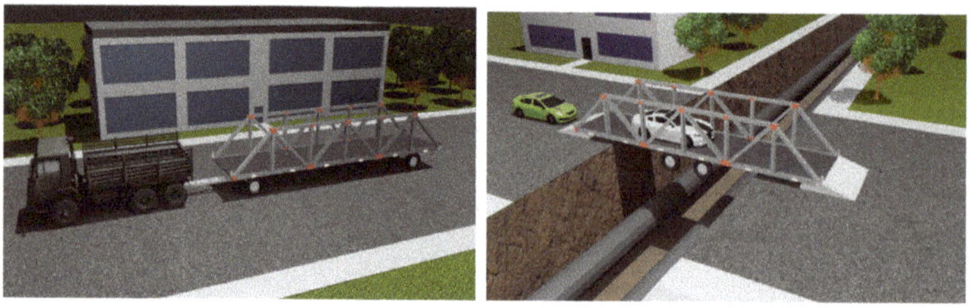

(**a**) transfer position (**b**) operational position

Figure 6. Mobile overpass.

Standard rolled metal profiles are used in the design, which ensures its availability in production and ease of adaptation for different operating conditions [55]. One of the key features of this solution is the use of wheeled bogies, which perform a dual function [56]; they provide transportation of the structure and serve as intermediate supports when it is installed on site [57]. Due to its modular structure, the overpass is easily scalable, can be

https://doi.org/10.3390/app16083879

rapidly assembled, and does not require complex machinery, which is especially important in time- or resource-constrained environments. Ease of installation, versatility of the design, and cost-effectiveness make this option attractive for mass application, especially in the regions with limited opportunity for rapid reconstruction of road infrastructure.

The conducted analysis of engineering solutions applied in global practice during roadworks shows that the existing approaches can be systematized into three main groups: organizational and technical traffic management methods, temporary detour solutions, and engineering structures such as temporary bridges and mobile overpasses. The primary purpose of the first group is the redistribution of traffic flows without changing the physical structure of the road. However, these solutions do not eliminate the key limitation, namely the reduction in road capacity during repair works, but rather adapt the system to deteriorated operating conditions. The second group, temporary detours, makes it possible to completely exclude the repair zone from the main traffic flow. Despite their high effectiveness in terms of safety, their application is significantly limited in dense urban environments due to the lack of available space and the high costs associated with the construction of temporary infrastructure. The most promising direction is represented by the engineering solutions of the third group—temporary bridges and mobile overpasses. Their development can be conditionally divided into three stages: military prefabricated bridges (Bailey bridge (UK), Medium Girder Bridge (MGB) (UK)), universal modular bridge systems (Acrow (Acrow Bridge, Parsippany, NJ, USA), Mabey (Mabey Bridge, Lydney, UK), Janson (Janson Bridging, Hank, the Netherlands)), and modern mobile and movable structures (ASTRA Bridge (Federal Roads Office (ASTRA), Bern, Switzerland), MB4.0 (China)). A comparative analysis of the considered temporary bridges and the mobile overpass is presented in Table 2.

Based on Table 2, most temporary bridge solutions developed for military applications and highway infrastructure are characterized by high load-carrying capacity and modularity. However, their application in urban environments is associated with several significant limitations:

- The need for heavy machinery and long installation time;
- Low adaptability to confined urban intersection space;
- Orientation toward linear road sections rather than complex transport nodes;
- Lack of solutions ensuring high structural mobility within the urban road network.

Even the closest analog, the ASTRA Bridge ASTRA Bridge (Federal Roads Office (ASTRA), Bern, Switzerland) , is primarily oriented toward highway applications and does not address the problem of rapid relocation and adaptability under conditions of dense urban development.

Among the considered solutions, the mobile overpass represents a relevant and practically significant engineering solution capable of effectively reducing traffic load in dense urban environments and under space constraints. Unlike most international analogs, the proposed design was developed from the outset with consideration of urban logistics. This determines its uniqueness and targeted applicability under conditions of high traffic density and constrained intersections.

However, the analysis of organizational, technical, and engineering approaches in the global context shows that existing developments are fragmented and focused on individual aspects—either traffic optimization, structural characteristics of bridge systems, or local engineering solutions. Therefore, there is a need for a comprehensive literature review aimed at generalizing existing research results, identifying the research gap, and determining insufficiently studied aspects. The results of the literature review are presented in Table 3.

https://doi.org/10.3390/app16083879

Table 2. Comparative analysis of the considered options of temporary bridges and overpasses.

Name	Country of Origin	Development Year	Purpose	Structure Material	Superstructure Length	Max. Load-Carrying Capacity	Assembly Method	Need for Machinery	Installation Speed	Standard Use	Examples of Use	Modularity and Adaptiveness	Service Life
MGB	UK	1971	Military passages, emergencies	Aluminum alloy	Up to 76 m	Up to MLC 70	Manual	Not required	Up to 1 h (31 m)	Army, emergencies	Falkland Islands, Bosnia	High	Up to 50 years
Bailey Bridge	UK	1941	Military and civil passages	Steel	Up to 60 m	~70 t	Manual/pull-out	Not required	1–2 days	Army, repair	WWII, China, Canada	Very high	Up to 70 years
Acrow Bridge	USA/Canada	After 1945	Civil construction	HDGS	Up to 100 m	Up to 250 t	Crane/pull-out	Desirable	2–5 days	Emergencies, detours	USA, Ukraine	Very high	Up to 100 years
Janson Bridging	Netherlands	1975+	Temporal/permanent	Steel slabs and beams	Up to 100 m	Up to 250 + t	Crane/mechanics	Partially	From hours to weeks	Army, construction	A1 (Amsterdam), Poland	Maximum	Up to 100 years
Mabey Bridge	UK	1970s	Temporal/permanent	High-strength galvanized steel	Up to 100 m	Up to 250 t	Manual/crane	Desirable	1–5 days	Civil construction	Iraq, UK, Australia	Up to four lanes	Up to 100 years
RDB	USA/Germany	2000s	Military, emergencies	Welded metals	Up to 20 m	Up to MLC 80	Unfolding	Minimal	<1 h	Military missions	NATO, Europe	Medium	15–30 years
ASTRA Bridge	Switzerland	2020s	Highways with repairs	Steel modules	257 m	Up to 15–20 t	Modular with engines	Required	Several hours	Road repair	A1 in Switzerland	Medium	Up to 20 years
Transportation Research Institute	Russia	2020s	Civil, emergencies, repair	Welded beams and reinforced concrete supports	Up to 24 m	Up to 30–40 t	Cranes, jacks	Desirable	1–2 days	Road repair	Developed in RF	High	Up to 30 years
PMP-3	Russia	1980s	Military and emergencies	Steel and aluminum	Up to 300 m	Up to 60 t	Pontoon, tugboats	Required	4–6 h	Army, floods	Emergency in Russia	Medium	15–20 years
Mobile overpass	Kazakhstan	2024–2025	Temporal overpass for bypassing overloaded urban crossroads	Steel (channel, I-beams, sheets, bolts, rivets)	10–24 m (depending on the crossroad)	Up to 30 t	Modular (bolted connection) with cranes and manual assembly	Desirable (rig, crane, platform)	<1 h (depending on the crossroad difficulty)	Temporary solution in transportation logistics, construction in limited traffic conditions	Development and tests in Kazakhstan, city network modeling	Adapts to city conditions, crossroads and street widths	~30 years (taking into account storage conditions and anti-corrosive treatment)

https://doi.org/10.3390/app16083879

Table 3. Results of the literature review.

No.	Article Title	Year	Research Conducted and Key Findings
1	Review of Accelerated Bridge Construction Systems for Bridge Superstructures and Their Adaptability for Cold Weather	2023	The authors analyzed more than 300 bridges assembled from prefabricated elements. Such bridges are constructed significantly faster and cause less disruption to traffic. However, in cold climates, their service life is slightly reduced and requires special materials [58].
2	Optimization of Mobile Overpass Support Placement Considering the Nonlinear Properties of the Soil Foundation	2026	Laboratory soil tests and numerical modeling were carried out. Optimal spacing between supports was determined, reducing soil deformation by more than two times and improving structural stability [53].
3	Experimental and Numerical Analysis of a Temporary Bridge Connector (Upper Bearing Block–Lower Pin)	2024	Full-scale testing showed that the connection withstands high loads and allows rapid assembly without welding [59].
4	A Comprehensive Decision Support Tool for Accelerated Bridge Construction	2025	A decision support tool was developed. Considering safety and social impact reduces accident rates by up to 60% and delays by up to 70% [60].
5	Efficiency of Modular Bridge Configurations: A Study on the Structural Stability and Capacity of Single- and Double-Story Modular Bridges	2025	Double-level structures reduce deflection by nearly two times and increase load capacity by more than 30% [61].
6	Accelerated Bridge Construction on Maui's Hana Highway	2023	A real project showed that bridge replacement can be completed in four days instead of several months [62].
7	Temporary Works in Construction of Bridges Near Third Party Assets	2021	Key risks (deformations and impact on nearby structures) were identified, and mitigation methods were proposed [63].
8	Methodology for Evaluating Behavior of Reinforced Concrete Slabs in Temporary Traffic Bridge Systems	2026	It was found that insufficient base strength leads to cracking; load limits must be controlled [64].
9	Development of Mobile Communal Overpasses Applied during Repair of Urban Utility Networks	2023	A mobile overpass was developed that can be installed quickly without heavy equipment while maintaining load capacity [65].
10	Development of Constructions of Mobile Road Overpasses	2017	The structure can be assembled in stages without stopping traffic [66].
11	Temporary Structural Health Monitoring of Historical Széchenyi Chain Bridge	2024	A sensor-based monitoring system enables early detection of critical deformations [67].
12	Improving Traffic Flow at Long-term Roadworks	2016	Road capacity decreases by 20–50% even with traffic management measures [68].
13	Study of Work Zone Traffic Safety under Adverse Driving Conditions	2020	Accident risk increases significantly under poor weather and limited visibility [69].
14	Detection Method for All Types of Traffic Conflicts in Work Zones	2022	A method for detecting hazardous traffic situations was developed [70].
15	Framework for Assessing Bridges Construction Impact on Work Zone Traffic Using BrIM	2024	A model was developed to predict congestion and improve planning [71].
16	Decision Support for Construction Traffic Management	2024	Local repair works can cause congestion across the entire urban network [72].
17	Impact of Construction Work Zone on Urban Traffic Environment	2023	A method for traffic redistribution was proposed to reduce delays [73].
18	A Decision-Making System for Traffic Management during Large-Scale Road Network Construction	2025	A real-time system allows adaptation of traffic schemes during roadworks [74].
19	Evaluation of Merge Control Strategies at Interstate Work Zones	2016	Traffic organization strategies significantly affect congestion and safety [75].

The results of the conducted analysis of scientific publications indicate the absence of a comprehensive, scientifically grounded engineering solution that ensures the conti-

https://doi.org/10.3390/app16083879

nuity of traffic flow directly within roadwork zones through the integration of structural, technological, and operational requirements. Existing studies address this problem in a fragmented manner, focusing either on traffic flow optimization, reduction in construction time, or improvement of individual structural elements of bridge systems.

At the same time, studies devoted to mobile overpasses lack a systematic consideration of these structures as mobile transport engineering systems operating simultaneously in both transport and working modes. In particular, insufficient attention has been paid to the design of the undercarriage as a key functional element responsible for structural movement, maneuverability, precise positioning, stability under load, and load transfer to the supporting base.

Thus, the identified scientific and technical gap lies in the absence of scientifically substantiated undercarriage design solutions for mobile overpasses that ensure their efficient operation in dense urban environments while simultaneously meeting the requirements for load-carrying capacity, mobility, maneuverability, rapid deployment, and operational reliability.

The key role of the undercarriage is determined by the fact that it ensures the implementation of the main operational functions of the mobile overpass, including structural movement, precise positioning at the site, stability under load, and load transfer to the supporting base. At the same time, the analysis of existing design solutions shows that current undercarriage systems do not fully meet the requirements of urban operation. Their main disadvantages include high material consumption, limited driver visibility due to elevated wheel positioning, increased complexity of axle steering operations, low maintainability, and vulnerability to complex ground conditions.

The combination of these shortcomings indicates that existing undercarriage designs do not meet modern requirements for mobility, safety, and operational reliability, which necessitates their improvement based on scientifically grounded approaches.

Therefore, further development of mobile overpasses should be focused on improving the undercarriage as the functionally determining element of the system. This will ensure the required maneuverability in confined urban conditions, reduce deployment time, improve structural stability, and provide reliable operation under various loading and operating conditions. At the same time, the undercarriage design should account for:

- Maneuverability in confined urban intersections;
- Stability and reliability under loads up to 30 t;
- Ease of assembly and maintenance under limited time and resources;
- Compatibility with tractor trucks (KAMAZ (KAMAZ PJSC, Naberezhnye Chelny, Russia), URAL (Ural Automotive Plant, Miass, Russia) and analogs);
- Operability within a wide temperature range (from −40 to +40 °C).

Based on the above, the aim of the study is to develop the undercarriage design of a mobile overpass adapted to dense urban environments and intended to ensure rapid reduction in traffic load during roadworks, accidents, and other situations limiting the capacity of the road network. To achieve this aim, the following tasks were accomplished:

- A morphological table of undercarriage design parameters was developed;
- Requirements for the undercarriage of the mobile overpass were defined, taking into account maneuverability in confined intersections, load-carrying capacity and stability, compatibility with the tractor fleet, and operation under various climatic conditions;
- Synthesis of possible undercarriage configurations was performed and the optimal configuration was selected;

https://doi.org/10.3390/app16083879

- A 3D model of the undercarriage and a scaled experimental test bench were developed to validate the design solutions;
- Experimental studies were conducted to evaluate the load and stability of the test bench;
- A regression equation describing the structural behavior under load was obtained and its convergence was assessed.

The scientific novelty of the study is as follows:

- For the first time, a systematic approach to undercarriage design based on morphological analysis and synthesis was proposed, enabling the transition from selecting individual design solutions to forming an optimal configuration based on a set of criteria (load-carrying capacity, maneuverability, weight, manufacturability, cost, and compatibility with transport vehicles);
- A new structural scheme of the undercarriage was developed, including a mechanical–pneumatic–electrical system that integrates transportation, installation, leveling, and fixation functions without the need for stationary foundations;
- A regression relationship describing structural behavior under load was obtained with high accuracy ($R^2 = 0.97$), enabling prediction of overpass performance and application in engineering calculations.

The obtained results make a significant contribution to solving the problem of ensuring traffic flow during roadworks and are expressed as follows:

- A new class of engineering solutions is proposed, focused not on traffic adaptation but on the physical preservation of road capacity directly within the work zone;
- The identified scientific and technical gap related to the lack of integrated solutions combining mobility, structural reliability, and operational adaptability has been addressed;
- The possibility of rapid deployment of a mobile overpass in urban environments without extensive preparatory work and without the need for heavy machinery has been ensured;
- An engineering basis for the design of mobile overpasses applicable in countries with high traffic density has been established;
- Prerequisites for further development of regulatory frameworks and implementation of mobile overpasses in transport planning and urban infrastructure management have been formed.

Thus, in contrast to existing studies primarily focused on span structures or organizational measures, this work establishes a new approach in which the key object of research is the mobile undercarriage system of the overpass as the basis for its practical implementation in urban environments.

2. Materials and Methods

2.1. Morphological Analysis and Synthesis of the Overpass Undercarriage

According to the goals of the research, a morphological analysis was carried out for the mobile overpass undercarriage design. Morphological analysis is a theoretical method that allows one to systematically identify all possible variants of the design, technical solutions, or processes [54]. It is based on the extraction of classification features of an object and formation of all possible combinations of them in the form of a morphological table (also called the "black morphological box") [55].

The algorithm for the morphological analysis includes:

- Determination of classification features of the design;
- Building of a morphology table;

- Analysis of the obtained combinations and selection of a rational variant [56].

The purpose of the morphological analysis was to study and systematize possible technical solutions for the design of the undercarriage of a mobile overpass used in urban environments to overcome trenches in utility networks.

To perform a morphological analysis of the mobile overpass undercarriage, key design parameters (classification features) were identified that affect its operability, mobility, compatibility with the urban environment, and technical characteristics. Fifteen such classification features were selected (Table 4).

Table 4. Classification features of the undercarriage design.

No.	Parameter	Variants
1	Undercarriage design type	Low-frame 2-axle; low-frame 3-axle; high-clearance 2-axle; modular girder with adapters; crawler track; semi-trailer with removable journal.
2	Main function	Only transportation; transportation + fixation; transportation + installation; transportation + fixation + self-positioning.
3	Chassis/frame type	Rigid welded; bolt-welded demountable; framed with girders and cross bars; open frame; cast-in modular.
4	Number of axles/wheels	1 axle (2 wheels); 2 axles (4 wheels); Tandem 2 axles (8 wheels twinned); 3 axles (6 wheels); Rocker–bogie suspension.
5	Useful load	Up to 6 t; up to 8 t; up to 10 t; up to 12 t; 14–16 t.
6	Weight of fully equipped chassis	2 t; 2.5 t; 3 t; 3.5–4 t; > 4 t.
7	Fully loaded mass of overpass	Up to 11 t; up to 13 t; up to 15 t; 16–18 t.
8	Frame dimensions	8×2.5 m; 9.5×2.5 m; 10.9×2.373 m; 12×2.5 m; telescopic 8–12 m.
9	Road clearance	250–300 mm; 320–350 mm; 380 mm; 420–450 mm; adjustable hydraulic lift.
10	Max. speed on asphalt	40 km/h; 60 km/h; 80 km/h; 90–100 km/h.
11	Max. speed on soil	20 km/h; 30 km/h; 40 km/h; 50 km/h.
12	Tire type	9.00R20; 10.00R20; 11.00R20; 385/65R22.5; 425/65R22.5.
13	Tire pressure	4.5–5.0; 5.75; 6.5; centralized pumping.
14	Tow hitch	King journal for fifth-wheel coupling; DIN loop; Loop GOST 2349 H = 750; ball 50 mm; universal replacement insert.
15	Suspension	Leaf-spring suspension; springs + shock absorbers; pneumatic suspension; torsion suspension; Bogie suspension.

The selected parameters have different origins and were defined considering both regulatory requirements and real operating conditions. First, regulatory constraints applicable in the Republic of Kazakhstan and other member states of the Eurasian Economic Union (EAEU) were taken into account, including permissible vehicle dimensions (not exceeding 2.55 m in width) [76], axle load limits, and traffic regulations [77]. The parameters of coupling devices were adopted in accordance with the requirements of GOST 2349 [78].

Operational parameters, including the temperature range (−40 to +40 °C), operating conditions (dust exposure, contamination, uneven road surfaces), and characteristics of tractor trucks (KAMAZ (KAMAZ PJSC, Naberezhnye Chelny, Russia), URAL (Ural Automotive Plant, Miass, Russia) and their analogs), were determined based on the analysis of operating conditions of transport platforms and trailer systems used in road construction and heavy structure transportation. This made it possible to account for real operating conditions of the mobile overpass and ensure the applicability of the proposed design in practical environments.

https://doi.org/10.3390/app16083879

Ranges of operating speeds, load levels, and the selection of suspension types and tires were defined based on an analysis of structural and operational characteristics of serial trailer systems and transport modules. This ensured the comparability of the developed design with existing engineering solutions and increased the validity of the adopted parameters.

Some parameters, including permissible combinations of structural elements, as well as mass and cost constraints, were defined as engineering assumptions. These assumptions are based on generalized design experience and are aimed at ensuring mobility, manufacturability, and the practical feasibility of the structure.

Thus, all selected parameters have a justified origin, are consistent with regulatory requirements, and reflect real operating conditions of the designed system, which ensures the reliability and reproducibility of the obtained results.

All features and corresponding variants were summarized in a morphological table (Table 5).

Table 5. All possible variants of the mobile overpass undercarriage.

Parameter	V1	V2	V3	V4	V5	V6
Undercarriage design type (n1)	Low-frame 2-axle	Low-frame 3-axle	High-clearance 2-axle	Modular girder with adapters	Crawler track	Semi-trailer with removable journal
Main function (n2)	Only transportation	Transportation + fixation	Transportation + installation	Transportation + fixation + self-positioning	–	–
Chassis/frame type (n3)	Rigid welded	Bolt-welded demountable	Framed with girders and cross bars	Open frame	Cast-in modular	–
Number of axles/wheels (n4)	1 axle (2 wheels)	2 axles (4 wheels)	Tandem 2 axles (8 wheels twinned)	3 axles (6 wheels)	Rocker–bogie suspension	–
Useful load (n5)	Up to 6 t	Up to 8 t	Up to 10 t	Up to 12 t	14–16 t	–
Weight of fully equipped chassis (n6)	2 t	2.5 t	3 t	3.5–4 t	>4 t	–
Fully loaded mass of overpass (n7)	Up to 11 t	Up to 13 t	Up to 15 t	16–18 t	–	–
Frame dimensions (n8)	8 × 2.5 m	9.5 × 2.5 m	10.9 × 2.373 m	12 × 2.5 m	Telescopic 8–12 m	–
Road clearance (n9)	250–300 mm	320–350 mm	380 mm	420–450 mm	Adjustable hydraulic lift	–
Max. speed on asphalt (n10)	40 km/h	60 km/h	80 km/h	90–100 km/h	–	–
Max. speed on soil (n11)	20 km/h	30 km/h	40 km/h	50 km/h	–	–
Tire type (n12)	9.00R20	10.00R20	11.00R20	385/65R22.5	425/65R22.5	–
Tire pressure/system (n13)	4.5–5.0	5.75	6.5	Centralized pumping	–	–

https://doi.org/10.3390/app16083879

Table 5. *Cont.*

Parameter	V1	V2	V3	V4	V5	V6
Tow hitch (n14)	King journal for fifth-wheel coupling	DIN loop	Loop GOST 2349 H = 750	Ball 50 mm	Universal replacement insert	–
Suspension (n15)	Leaf-spring suspension	Springs + shock absorbers	Pneumatic suspension	Torsion suspension	Bogie suspension	–

Based on the compiled morphological table, the total number of all possible configurations of the overpass undercarriage (n_1–n_{15}) was established using the following formula:

$$N_{\text{konf}} = n_1 \cdot n_2 \cdot n_3 \cdot \ldots \cdot n_{15} \quad (1)$$

We substitute the corresponding values:

$$N_{\text{konf}} = 6 \times 4 \times 5 \times 5 \times 5 \times 5 \times 4 \times 5 \times 5 \times 4 \times 4 \times 5 \times 4 \times 5 \times 5 = 12{,}000{,}000{,}000$$

In order to reduce the number of all possible configurations and to determine the optimal design of the overpass undercarriage, their synthesis was carried out. The synthesis allows for selecting one or more optimal configurations from all possible combinations of morphological table parameters, which corresponds to the goal of the research.

To perform the synthesis, the pros and cons of all the different variants were considered in terms of overpass undercarriage parameters (Table 6).

Table 6. Pros and cons of all possible variants of the design parameters.

Parameter	Variant	Pros	Cons
Undercarriage design type	Low-frame 2-axle	Low center of gravity, good stability, easier to drive in equipment	Lower load-carrying capacity compared to 3-axle trucks
	Low-frame 3-axle	High load-carrying capacity, less axle load	Harder to maneuver, more expensive
	High-clearance 2-axle	Greater ground clearance	Higher center of gravity, worse stability
	Modular girder with adapters	Versatility for different platforms	More complex design and weight
	Crawler track	Excellent floatation ability	Slow speed, high cost
	Semi-trailer with removable journal	Easy top loading	More complex design, greater weight
Main function	Only transportation	Simplicity	No fixation or setting in place
	Transportation + fixation	Universality, safety in operation	Increasing complexity
	Transportation + installation	Fast assembly	Heavier and more expensive chassis
	Transportation + fixation + self-positioning	Maximum autonomy	Very complex system, high cost
Chassis/frame type	Rigid welded	High strength, durability	Cannot be disassembled for repair
	Bolt-welded demountable	Easier to repair	Decreased rigidity
	Framed with girders and cross bars	Good load distribution	Heavier
	Open frame	Lightweight with high strength	Expensive to manufacture
	Cast-in modular	High geometric accuracy	Expensive, difficult to repair

Table 6. *Cont.*

Parameter	Variant	Pros	Cons
Number of axles/wheels	1 axle (2 wheels)	Minimal weight	Low stability and load-carrying capacity
	2 axles (4 wheels)	Balance of maneuverability and load-carrying capacity	Higher load on each axle than in 3-axle variant
	Tandem 2 axles (8 wheels twinned)	Less ground pressure	Increasing complexity
	3 axles (6 wheels)	High load-carrying capacity	Longer base, worse maneuverability
	Rocker–bogie suspension	Good floatation ability	More difficult maintenance
Useful load	Up to 6 t	Light chassis	Limitation by type of equipment
	Up to 8 t	Suitable for most lightweight vehicles	Not enough for heavy equipment
	Up to 10 t	Balance of versatility	A little heavier and more expensive
	Up to 12 t	For heavy equipment	High load on the structure
	14–16 t	Maximum of opportunities	Very heavy and expensive chassis
Weight of fully equipped chassis	~2 t	Easier transportation	Lower strength
	~2.5 t	Balance of weight and strength	Limited load
	~3 t	High strength, stability	A little heavier
	3.5–4 t	Very strong	Greatly makes heavier
	>4 t	Maximum strength	Greatly decreases mobility
Frame dimensions	~8 × 2.5 m	Compact	Smaller support zone
	~9.5 × 2.5 m	Balance of compactness and stability	Limited in length
	~10.9 × 2.373 m	Fits in with traffic regulations	Increased weight
	~12 × 2.5 m	Maximum support zone	Limited transportation
	Telescopic 8–12 m	Flexibility	Complex mechanics
Road clearance	250–300 mm	Very stable platform	Low floatation ability
	320–350 mm	Balance	Lower stability
	380 mm	Good floatation ability	A little higher center of gravity
	420–450 mm	Excellent floatation ability	Decreased stability
	Adjustable hydraulic lift	Versatility	Complexity and price
Max. speed on asphalt	40 km/h	Safe	Slow
	60 km/h	Balance of safety and speed	Longer transportation
	80 km/h	Fast enough	Requires qualitative tires
	90–100 km/h	Very fast	High design requirements
Max. speed on soil	20 km/h	Maximum safety	Very slow
	30 km/h	Balance	Slower than needed
	40 km/h	Good mobility	Requires reinforced suspension
	50 km/h	Fast	Increased wear
Tire type	9.00R20	Lighter	Lower load-carrying capacity
	10.00R20	Balance of price and strength	Not for extreme loads
	11.00R20	High load-carrying capacity	Heavier
	385/65R22.5	High stability	More expensive
	425/65R22.5	Maximum stability	Very expensive
Tire pressure/system	4.5–5.0	Soft ride	Greater wear
	5.75	Optimal load-carrying capacity	Harder ride
	6.5	High strength	Hard ride
	Centralized pumping	Versatility	Expensive

https://doi.org/10.3390/app16083879

Table 6. *Cont.*

Parameter	Variant	Pros	Cons
Tow hitch	King journal for fifth-wheel coupling DIN loop Loop GOST 2349 H = 750 Ball 50 mm Universal replacement insert	For semi-trailers Reliable Versatility Light Flexible	Not versatile Not everywhere compatible Requires specific tractor truck Limited load Expensive
Suspension	Leaf-spring suspension Springs + shock absorbers Pneumatic suspension Torsion suspension Bogie suspension	Simple Softer Very comfortable Compact Good floatation ability	Harder A little more complex Expensive and complex Limited load Difficult in repair

Based on the pros and cons of each variant, strict limitations were set to the selected configuration of the chassis, and only those variants that do not contradict the technical conditions, traffic regulations, dimensions, compatibility with KAMAZ (KAMAZ PJSC, Naberezhnye Chelny, Russia), Volvo FMX (Volvo Group, Gothenburg, Sweden), or Mercedes-Benz Arocs (Daimler Truck AG, Stuttgart, Germany), and requirements for operation in −40…+40 °C, as well as real speed and load modes, were retained (Table 7).

The final number of feasible combinations after applying the constraints was calculated according to Equation (1) and amounted to

$$N_{\text{konf}} = 4 \times 3 \times 5 \times 4 \times 3 \times 4 \times 2 \times 5 \times 4 \times 3 \times 3 \times 2 \times 4 \times 1 \times 2 = 16{,}588{,}800 \text{ configurations.}$$

The evaluation was carried out by an expert group consisting of five specialists in transport engineering and structural design. The experts performed pairwise comparisons of the criteria according to their relative importance using a standard scale from 1 to 9 (where 1 indicates equal importance and 9 indicates strong dominance of one criterion over another). Based on these comparisons, a pairwise comparison matrix was constructed, from which the final weights of the criteria were calculated. In addition, the consistency of expert judgments was assessed. The obtained value of the consistency index ($CI \ll 0.1$) indicates that the expert opinions are consistent and can be considered reliable. The following factors were taken into account:

- Operating conditions (dense urban environment, confined space, ground clearance, temperature range from −40 to +40 °C);
- Requirements for reliability and safety during transportation and installation of the structure;
- Compatibility with the existing fleet of transport vehicles and ease of maintenance;
- Constraints related to weight and cost affecting the feasibility of serial production and implementation.

The resulting criteria weights are presented in Table 8.

According to Table 8, the highest weight (0.3) was assigned to the reliability criterion, since structural failure during operation may lead to critical consequences, including accidents and disruption of traffic flow.

The criterion of compatibility with tractor trucks was assigned a weight of 0.2, as the ability to use the existing vehicle fleet is a prerequisite for the practical implementation of the mobile overpass.

The criteria of ground clearance and ease of maintenance were each assigned a weight of 0.15, as they determine the operational suitability of the structure under confined urban conditions, as well as the possibility of its use without specialized equipment and complex maintenance procedures.

The structural weight criterion was assigned a weight of 0.1, since, despite its influence on transportability and axle loads, it is considered an optimizable parameter that does not have a critical impact on system performance provided that regulatory constraints are satisfied.

Table 7. Limitations on selected chassis configuration.

Parameter	Acceptable Variants
Undercarriage design type	Low-frame 2-axle; low-frame 3-axle; high-clearance 2-axle; modular girder with adapters (excluded: crawler, low speed/cost; semi-trailer with removable journal, other concept of towbar)
Main function	Transportation + fixation; transportation + installation; transportation + fixation + self-positioning (excluded: only transportation, without safe fixation at site)
Chassis/frame type	Rigid welded; bolt-welded demountable; framed with girders and cross bars; open frame; cast-in modular (all are acceptable by "strict" criteria)
Number of axles/wheels	2 axles (4 wheels); Tandem 2 axles (8 wheels twinned); 3 axles (6 wheels); Rocker–bogie suspension (excluded: 1 axle; it is not suitable due to its load-carrying capacity/stability)
Useful load	Up to 10 t; up to 12 t; 14–16 t (excluded: up to 6 t, up to 8 t, they do not ensure the required load class)
Weight of the fully equipped chassis	∼2.5 t; ∼3 t; 3.5–4 t; >4 t (excluded: ∼2 t, risk of insufficient strength for the required loads)
Fully loaded mass of the overpass	Up to 13 t; up to 15 t (excluded: up to 11 t, does not ensure useful load of ≥10 t; 16–18 t, excessively heavy/more expensive and complex in terms of traffic regulations)
Frame dimensions	∼8 × 2.5 m; ∼9.5 × 2.5 m; ∼10.9 × 2.373 m; ∼12 × 2.5 m; telescopic 8–12 m (all are acceptable by their width ≤2.55 m)
Road clearance	320–350 mm; 380 mm; 420–450 mm; adjustable hydraulic lift (excluded: 250–300 mm—insufficient floatation ability)
Max. speed on asphalt	40 km/h; 60 km/h; 80 km/h (excluded: 90–100 km/h, increased requirements to brakes/tires for this weight)
Max. speed on soil	20 km/h; 30 km/h; 40 km/h (excluded: 50 km/h, excessive risk of wear and loss of steering response)
Tire type	10.00R20; 11.00R20 (excluded: 9.00R20, low load margin; 385/65R22.5 and 425/65R22.5, worse compatibility/cost)
Tire pressure/system	4.5–5.0; 5.75; 6.5; centralized pumping (all are acceptable, final choice depends on tires/mode)
Tow hitch	Loop GOST 2349 H = 750 (excluded: king journal for fifth-wheel coupling, ball 50 mm, other trailers category; DIN/universal insert, potential issues of compatibility/cost)
Suspension	Spring; springs + shock absorbers (excluded: pneumatic, risks in frosty weather/cost; torsion, load limitations; bogie suspension, more difficult in repair)

A similar weight (0.1) was assigned to the cost criterion, as it affects economic efficiency and implementation feasibility; however, it is not a determining factor in the selection of the design solution compared to reliability and operational characteristics.

Thus, the adopted system of weighting coefficients reflects the priority of operational reliability and functional suitability of the structure over secondary parameters such as weight and cost.

Then, 5 most suitable configurations of the overpass undercarriage were chosen and assessed according to the above criteria (Table 9).

https://doi.org/10.3390/app16083879

The integral index was calculated for each configuration according to

$$I = \sum_{i=1}^{6} W_i \cdot B_i, \qquad B_i \in [1,5], \tag{2}$$

where W_i are the criteria weights (Table 8), and B_i are the averaged scores for the criteria [57].

Table 8. Choice and weight criteria W_i.

Criterion	Weight (W_i)	Priority	Substantiation
Reliability in difficult conditions	0.3	Very high	Operation in the range of −40...+40 °C, dust, mud.
Compatibility with tractor truck fleet	0.2	High	KAMAZ, Volvo FMX, Mercedes-Benz Arocs and other foreign equivalents are used.
Simplicity in maintenance	0.15	High	Possibility of repair in the field with minimum resources.
Floatation ability	0.15	High	Operation in cross-country, trenches.
Structure weight	0.1	Medium	Balance of strength and transportability is needed.
Cost	0.1	Medium	Important for serial production.

The methodology for calculating the integral index is as follows. Each configuration parameter (type of structure, frame, axles, load, clearance, speeds, tires, coupling device, suspension, etc.) was preliminarily evaluated according to six criteria: reliability, compatibility, maintenance, ground clearance, weight, and cost. The evaluation was performed using a five-point scale, where 1 corresponds to the lowest level of compliance with the criterion and 5 to the highest. Intermediate values reflect gradations of performance: low, medium, and high.

Table 9. Optimal variants of configuration for the mobile overpass undercarriage.

Variant	Configuration
A	Low-frame 2-axle; rigid welded frame; 2 axles; useful load of up to 10 t; full load of up to 13 t; 10.9 × 2.373 m; clearance of 380 mm; 80/40 km/h; 10.00R20; 5.75; GOST-loop 750; springs + shock absorbers.
B	Low-frame 3-axle; rigid frame; 3 axles; up to 12 t; up to 15 t; the same dimensions; 380 mm; 80/40; 11.00R20; 5.75; GOST-loop; springs + shock absorbers.
C	High-clearance 2-axle; girders and cross bars; 2 axles; up to 10 t; up to 13 t; 10.9 × 2.373 m; clearance of 420–450 mm; 60/50; 11.00R20; 5.75; GOST-loop; springs.
D	Low-frame 2-axle; rigid frame; 2 axles; up to 10 t; up to 13 t; 9.5 × 2.5 m; clearance of 380 mm; 90–100/40; 385/65R22.5; centered pumping; DIN loop; pneumatic suspension.
E	Modular girder; bolt-welded; 2 axles; up to 10 t; up to 13 t; 9.5 × 2.5 m; clearance of 320–350 mm; 60/30; 10.00R20; 4.5–5.0; GOST-loop; springs.

The expert evaluation was carried out by a group of five specialists with experience in transport engineering and structural design. Each expert independently assessed all parameters according to the selected criteria, after which the average score for each indicator B_i was calculated.

To improve the objectivity of the results and reduce the influence of subjective factors, the following procedures were applied:

https://doi.org/10.3390/app16083879

- Averaging of individual expert scores;
- Control of value dispersion (the standard deviation did not exceed $\sigma \leq 0.3$);
- In the presence of significant deviations, verification and, if necessary, exclusion of extreme values.

It should be noted that differences in scores within the range of 0.1–0.2 fall within the acceptable limits of expert uncertainty and do not significantly affect the final ranking of alternatives. Nevertheless, these differences are taken into account in the calculation of the integral index, allowing for a more accurate comparison of configurations with similar characteristics.

The results of these calculations are presented in Table 10.

Table 10. Averaged criteria scores B_i.

Configuration	Reliability (av. Score)	Compatibility (av. Score)	Maintenance (av. Score)	Ground Clearance (av. Score)	Weight (av. Score)	Cost (av. Score)
A	5.00	5.00	4.62	4.08	4.46	4.31
B	4.92	5.00	4.54	4.08	4.54	4.23
C	4.85	5.00	4.46	4.15	4.15	4.08
D	4.69	4.92	4.62	3.85	4.54	4.38
E	5.00	4.92	4.23	3.92	4.00	3.92

Based on the obtained averaged criteria scores B_i, the integral index of each configuration was calculated. An example of the calculation for configuration A is shown below:

By reliability:

$$I_{rel} = 0.3 \times 5 = 1.5, \quad (3)$$

By compatibility:

$$I_{com} = 0.2 \times 5 = 1, \quad (4)$$

By maintenance:

$$I_{maint} = 0.15 \times 4.62 = 0.693, \quad (5)$$

By floatation ability:

$$I_{fl} = 0.15 \times 4.08 = 0.612, \quad (6)$$

By weight:

$$I_{weight} = 0.1 \times 4.46 = 0.446, \quad (7)$$

By cost:

$$I_{cost} = 0.1 \times 4.31 = 0.431. \quad (8)$$

The integral index for configuration A was calculated according to Equation (2):

$$I = \sum_{i=6}(W_i \times B_i) = 1.5 + 1 + 0.693 + 0.612 + 0.446 + 0.431 = 4.682.$$

Based on the above methodology, the integral index was calculated for other configurations. The calculation results are presented in Table 11.

After comparative analysis by the criteria, the best combination turned out to be configuration A, which includes:

- Type of construction: low-frame 2-axle, ensures stability, acceptable load, and maneuverability;
- Primary function: transportation + stable fixation, a simple but safe solution;
- Chassis type: rigid welded frame with cross beams, reliability, and durability with moderate weight;

https://doi.org/10.3390/app16083879

- Number of axles/wheels: 2 axles (4 wheels), balance between load-carrying capacity and maneuverability;
- Useful load: up to 10 t, suitable for passage of most vehicles and special equipment;
- Weight of the fully equipped chassis: about 3 t, strength with moderate weight;
- Fully loaded mass of the overpass: up to 13 t, permissible according to traffic regulations, compatible with standard tractor trucks;
- Dimensions: 10,900 × 2373 mm, fits into the shipping dimensions, sufficient support area;
- Road clearance: 380 mm, good ground clearance without loss of stability;
- Speed: up to 80 km/h on asphalt and 40 km/h on soil, safely and quickly;
- Tires: 10.00R20 OI-73B tire model (Russian specification)–reliable, unified for tractor trucks;
- Tire pressure: 5.75 kgf/cm^2, optimum for load-carrying capacity and durability;
- Tow hitch: loop according to GOST 2349 H=750 mm, full compatibility with traction trucks.;
- Suspension: semi-elliptic springs with shock absorption, simplicity, maintainability, vibration damping;
- Platform: movable carriages with rollers, precise positioning with minimum friction;
- Positioning drive: electromechanical cable, accuracy, reliability, operation at low temperature;
- Fixation of bridges: automatic, pneumatic cylinders with pins, fast, safe, no risk of human error;
- Pneumatic system power supply: from the tractor network, minimum additional equipment;
- Braking system: single-line pneumatic with diaphragm chambers 24, compatibility and reliability;
- Parking brake: separate circuit to the rear axle, hill holder;
- Emergency braking: auto-trigger in case of brake pipe failure, basic safety;
- Compatibility: KAMAZ, Volvo FMX or Mercedes-Benz Arocs and equivalents, ready for operation;
- Temperature range: −40…+40 °C, operation in most climatic zones of Kazakhstan and neighboring countries;

Table 11. Integral index of the configurations.

Configuration	Integral Index
A	4.682
B	4.500
C	4.519
D	4.415
E	4.554

Thus, out of the possible combinations of parameters, the chosen configuration presents the optimal balance between functionality, reliability, mobility, and cost. This is neither the maximum in complexity nor the minimum in capability undercarriage configuration, but a rational middle option that covers all the key needs of mobile overpass operation.

https://doi.org/10.3390/app16083879

2.2. Design Solution for the Suggested Overpass and Development of the Overpass Test Bench

According to the results of morphological analysis and synthesis, the design of a single-span overpass was developed, which is a prefabricated metal structure in the form of a spatial truss with a rigid frame (Figure 7).

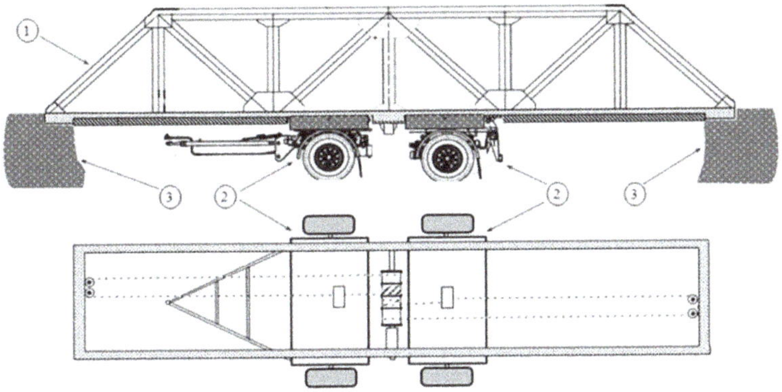

Figure 7. Construction of a single-span modular overpass in the operational position above the trench: 1—supporting structure (platform) of the mobile overpass; 2—movable suspension of the overpass; 3—side of the trench.

The design of the overpass (Figure 8a) is based on a stable triangular grid of longitudinal and diagonal links connected by vertical posts, which ensures high strength and uniform load distribution. The upper platform of the overpass is supported by two central wheeled trolleys with a biaxial undercarriage and pneumatic tires (Figure 8b), which serve as both a transport platform and an intermediate support in the working position.

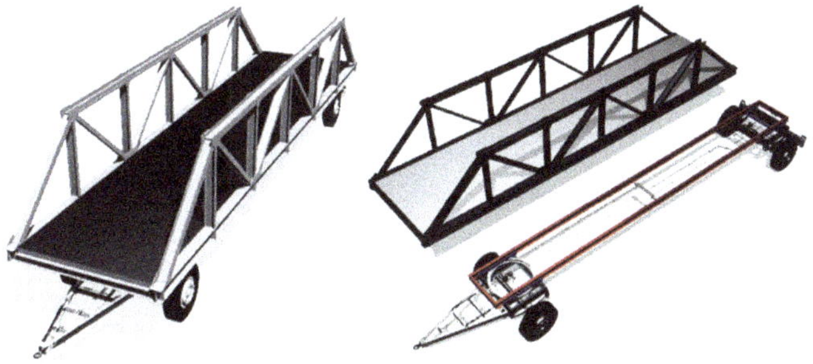

(a) (b)

Figure 8. Three -dimensional model of the modular overpass with undercarriage: (**a**) overpass framework based on a triangular grid system; (**b**) wheeled trolley with biaxial undercarriage and pneumatic tires.

The halves of the span are joined in the middle zone by means of a hinge joint, which ensures continuity and uniformity of the roadway. Installation is carried out on compacted pads with possible use of damping layers to protect against vibration and shock loads. The undercarriage with controls and movement mechanisms is a set of structures that ensure transportation, stability, and maneuvering of the overpass (Figure 9).

https://doi.org/10.3390/app16083879

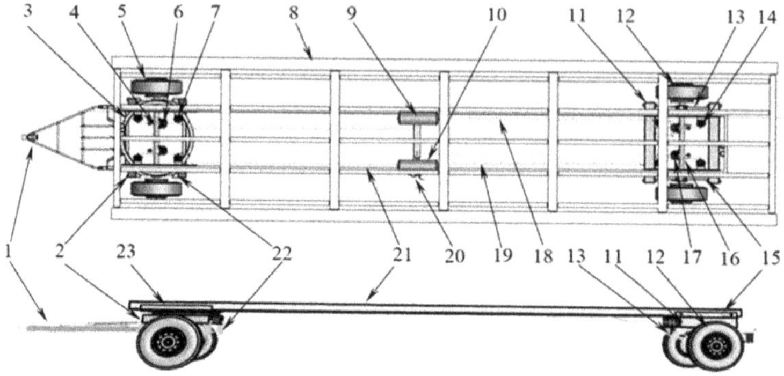

Figure 9. Structural diagram of the undercarriage of the overpass. (Explication of constructive elements: 1—drawbar of the overpass; 2—front suspension; 3—front suspension swivel; 4—front axle; 5—front driving wheel with tire; 6—front brake chamber; 7—pneumatic chamber for moving the finger of the front movable carriage; 8—upper platform; 9—compressed air receiver for finger movement of pneumatic systems; 10—compressed air receiver for pneumatic brakes; 11—rear suspension; 12—rear driving wheel with tire; 13—rear springs; 14—pneumatic chamber for finger movement of the rear movable carriage; 15—rear movable carriage; 16—rear axle; 17—rear brake chamber; 18—hose for pneumatic finger movement; 19—hose for pneumatic brakes; 20—brake valve for brake force distribution; 21—I-beam (connecting the front and rear); 22—front springs; 23—front movable carriage).

The design is based on the following:

- Wheel pairs with reinforced axles mounted on mobile carriages that move on longitudinal I-beams acting as guide rails;
- Shock absorption system—multi-leaf springs and hydraulic shock absorbers that reduce dynamic loads and ensure smooth running;
- Swivel mechanism of the front axle, which ensures maneuverability of the overpass during transportation;
- Cable actuating mechanism of transportation—the motor rotates a spool on which a cable is wound to move the carriage along the beam.
- Fixation in the desired position is ensured by a pneumatic lock, which is activated by the operator and locks the carriage with a locking pin.
- Pneumatic braking system—includes drum brakes, receiver and pneumatic valves connected to the tractor truck system by means of quick couplers.
- Electrical system—controls movement, lighting, and alarms (operates from 24 V with stabilization capability).

The adopted technical solution for the undercarriage allows for moving the structure on different types of road surfaces and positioning it accurately on the installation site without the need to build fixed foundations. This system ensures uniform distribution of loads from the span and vehicles traveling over the overpass, and reduces dynamic effects through damping. Given the high requirements for the strength, reliability, and adaptability of the structure, this engineering solution allows for the sustainable operation of the overpass throughout its temporary operation.

2.3. Development of an Experimental Overpass Test Stand

Based on the 3D modeling results, an overpass test bench was developed. The creation of the bench was the most important stage of the research, which allowed moving from theoretical design to practical realization of technical solutions. The development was carried out at a scale of 1:4 and included a full cycle of design, calculation, and production operations.

https://doi.org/10.3390/app16083879

At the preparatory stage, the dimensions of the bench adapted to laboratory conditions were determined, and a list of necessary elements was made: metal profiles, fastening joints, moving and supporting assemblies. The main load-bearing parts were manufactured by laser cutting and bending, which ensured high geometric accuracy. The assemblies were joined by welding, bolted, or riveted connections, depending on the required rigidity and the need for subsequent disassembly.

Special attention was paid to the undercarriage of the overpass to ensure its mobility. It comprises:

- wheel pairs with reinforced axles mounted on mobile carriages;
- spring suspension that reduces dynamic loads during movement;
- fixation and locking mechanisms ensuring stability of the structure in the operating position.

To increase operational reliability, the undercarriage is coated with an anti-corrosion compound, and the main components are connected by welded and bolted elements, which increases rigidity and facilitates maintenance. The rear and front axles are mounted on movable carriages. The front axle is rotary, which ensures maneuverability of the overpass (Figure 10).

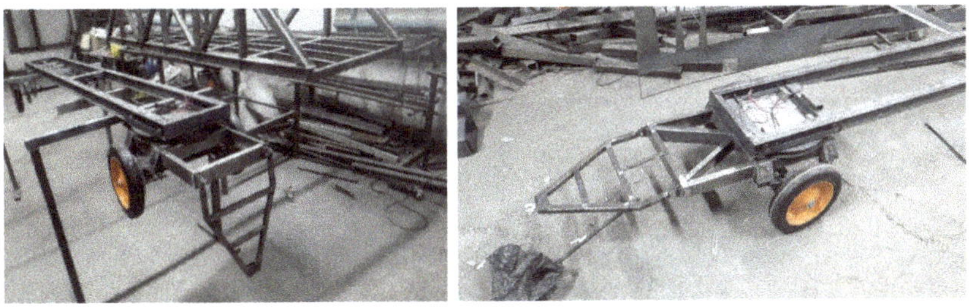

(a) (b)

Figure 10. Developed overpass undercarriage: (**a**) suspended position; (**b**) movement on wheeled undercarriage.

The front and rear chassis (Figure 11, pos. 1) are fixed on special structures: mobile carriages (Figure 11a,b, pos. 2). The carriages move left and right (horizontally) along the two longitudinal I-beam girders (Figure 11, pos. 4). Their movement is directed inward, closer to the center of the platform, which allows automatic adjustment of the axle position when the structure is in operation.

(**a**) general view (**b**) structure assembly

Figure 11. Mobile carriage of the mobile overpass: 1—front and rear chassis; 2—mobile carriage; 3—connecting frame elements; 4—longitudinal I-beam girders.

https://doi.org/10.3390/app16083879

A special mechanism is provided to facilitate the movement of the axles. Rollers are fastened on the surface of the beam, where the axles and carriages are installed (Figure 11, pos. 3)—four for each carriage. These rollers allow the structure to glide smoothly, while reducing friction and providing easy movement. Analysis of different options for mechanical solutions—rail, crawler and wheeled—showed the advantages of the latter: high maneuverability, ease of implementation, and versatility for use on different types of surfaces. The carriages move along the frame, along the longitudinal I-beams that act as guide rails (Figure 12).

(**a**) (**b**)

Figure 12. Front and rear suspensions moving along I-beam guides: (**a**) front suspension; (**b**) rear suspension.

Movement is performed by a mechanism involving an electric motor and a cable system. When power is supplied, the motor rotates the spool on which the cable is wound, moving the carriage (Figure 13). The axle position is changed by turning the special drum (Figure 13, pos. 1), to which force is applied via a cable (Figure 13, pos. 2). When the drum rotates, the rope tension changes, which causes the bridge to move (Figure 13, pos. 3) to the required position. An electric motor (Figure 13, pos. 4) actuates the movement, while ensuring accurate positioning.

The carriages are secured with a pneumatic lock: when the set position is reached, the operator activates the lock by pressing the button and the locking pin enters the hole on the overpass frame, securing the assembly.

Remarks to the Test Bench. The design of the original overpass includes pneumatic locks; however, when creating a model (scale 1:4), their application was impossible due to difficulties with the selection and installation of small-sized pneumatic chambers that would ensure correct operation of the mechanism in a reduced scale. In this regard, electronic clamps (electric pins) operating from a 12 V power supply were used for the test bench.

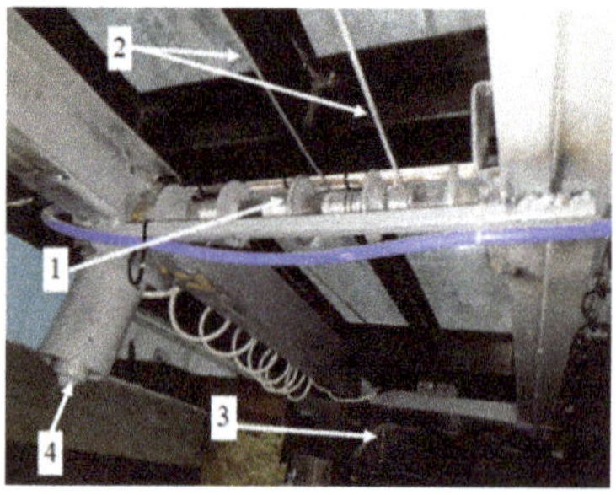

Figure 13. Mobile platform control mechanism: 1—drum; 2—cable; 3—movable carriage (bridge displacement); 4—electric motor.

https://doi.org/10.3390/app16083879

Electric pins perform the same function—insertion of the locking pin into the carriage bore—but they are controlled by an electrical signal. This preserved the functionality of the original system and ensured reliable fixation of the assemblies in laboratory conditions. The movable suspension in the stand is controlled by eight electronic clamps connected to the control panel and powered from a 12 V source.

Mobile carriages of the front and rear axles in shipment position are located at the extreme points of I-beam frame members and are rigidly fixed with electronic pins (Figure 14a,b, pos. 1).

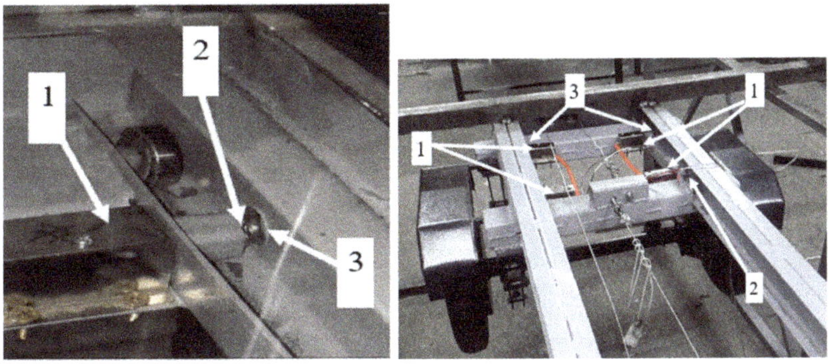

(a) (b)

Figure 14. Mobile carriage fixation mechanism: (**a**) fixation mechanism with cylinder rods acting as steel fixators; (**b**) arrangement of carriages on the front axle; 1—electronic pins; 2—cylinder rods (steel fixators); 3—holes in the carriage body and I-beam flange.

The cylinder rods act as steel fixators (Figure 14a, pos. 2) and enter through holes made in the carriage body and the upper flange of the I-beam (Figure 14a, pos. 3). Two carriages are installed at each axle:

- On the front axle—two carriages (Figure 14b, pos. 3) with four holes (Figure 14a, pos. 3);
- On the rear axle—similarly, two carriages with four matching holes.

When compressed air is supplied from the tractor's pneumatic system, the rod pin extends outward and, aligning with the holes, provides a rigid connection between the carriages and the frame, preventing longitudinal displacement of the axles. For travel, the carriages are released by switching the pneumatic valve (Figure 14a,b, pos. 1) to the "release" position: pressure is released and the rod and pin retract inside the cylinder, allowing the carriage to move freely along the girder to the next fixed position.

The choice of the pneumatic actuator in the original design is conditioned by the presence of a single tractor air system, which simplifies the design, reduces the weight of the assembly, and provides standardized maintenance. Additionally, the undercarriage is equipped with an air brake system including drum brakes, the receiver, and pneumatic valves. Pneumatic and electrical systems of the overpass are connected to the tractor systems by means of quick couplers.

The pneumatic braking system is powered by compressed air, which is supplied from the air intake valve of the towing vehicle through a special line hose. This air is accumulated in the air receivers (Figure 15a, pos. 1) and through the distributor valve (Figure 15a, pos. 2) is fed into the brake chambers (Figure 15a,b, pos. 3) of each wheel. When the driver presses the brake pedal, a valve automatically opens and compressed air flows through the hoses into the brake chambers, pressing the pistons against the jaws, which rest against the brake drum (Figure 15b, pos. 4). When the pedal is released, air is vented back into

https://doi.org/10.3390/app16083879

the atmosphere and the jaws release the brake drum, allowing the wheels to turn freely. In an emergency situation (for example, in case of a leaking line), an emergency valve is activated which completely blocks the air supply and sets the brake chambers in the "braking" position, safely keeping the trailer from moving. Therefore, the pneumatic system ensures smooth and safe deceleration of the overpass using simple and reliable compressed air control.

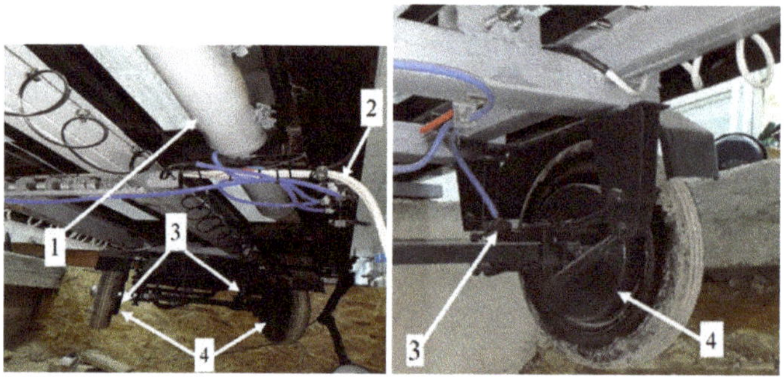

(**a**) (**b**)

Figure 15. Brake system of the mobile overpass: (**a**) pneumatic system components; (**b**) wheel braking mechanism; 1—air receivers; 2—distributor valve; 3—brake chambers; 4—brake drum.

The diaphragm brake chamber (Figure 15a,b, pos. 3), type 24, is designed to actuate the front and rear axle brakes.

Thus, the overpass undercarriage is an integrated mechanical–pneumatic–electrical system that ensures not only the movement of the structure, but also its reliable fixation, operational stability, and ease of operation. By combining cable and pneumatic actuators, as well as an autonomous electrical control system, the overpass is able to be moved quickly, positioned precisely, and brought into operation quickly without the need to connect to complex external infrastructure. The complex interaction of all components of the undercarriage guarantees high mobility, reliability, and safety when it is operated in urban areas, on construction sites or during road repair operations. This makes the mobile overpass a versatile engineering solution capable of adapting to different terrain conditions and traffic flow.

As a result, a mobile overpass test bench was developed, which recreates key structural elements and functional units at a scale of 1:4 (Figure 16).

Figure 16. Overpass test bench.

This bench is the basis for a series of experimental studies aimed at studying the behavior of the overpass under various loads, including moving masses, vibrations, and static

https://doi.org/10.3390/app16083879

effects. Experimental research carried out on these benches allows for registering deflections, displacements, vibrational characteristics, and stability of the structure.

Experimental studies were carried out on a physical model of the mobile overpass at a geometric scale of 1:4 relative to the full-scale structure. To ensure the correct transfer of results to the real object, the principles of similarity theory and structural scaling were applied.

During modeling, the following similarity conditions were satisfied:

- Geometric similarity: All linear dimensions of the model were reduced by a factor of 4;
- Material similarity: The model was made of the same material (steel) as the full-scale structure, ensuring preservation of the elastic modulus E;
- Force similarity: Loads were scaled in accordance with similarity laws for bending structures.

Considering the geometric scale $\lambda = \frac{1}{4}$ and identical material properties, load scaling was determined by the following relationship:

$$P_{\text{full-scale}} = P_{\text{model}} \cdot \lambda^2 \quad (9)$$

This relationship is explained by the fact that, in bending structures, the applied load is related to structural stiffness, which depends on the geometric characteristics of the cross-section, particularly the moment of inertia. When material properties remain unchanged, this leads to a quadratic relationship between load and the linear scale.

Thus, a load of 200 kg applied to the model corresponds to

$$P_{\text{full-scale}} = 200 \times 4^2 = 3200 \text{ kg} \approx 3.2 \text{ t} \quad (10)$$

Similarly, the load range of 50–200 kg applied to the model corresponds to 0.8–3.2 t at full scale.

Thus, the obtained values represent the initial operational loading conditions of the structure and allow assessment of its stiffness, deformability, and structural behavior under load. The linear load–deflection relationship confirms the correctness of the scaling approach and the applicability of the results to the full-scale structure.

2.4. *Experimental Research Plan and Procedure*

The main objective of the experimental studies on the developed test bench was to evaluate the performance of the overpass under the action of a moving load simulating the movement of a vehicle. This required determining the following:

- How loads are distributed on the upper platform of the overpass when a loading cart passes and what the deflection of the structure will be;
- Whether the structure of the overpass remains serviceable when the load is increased in steps up to 200 kg.

The preparation for the experiment involved several sequential steps that allowed for the creation of test conditions as close to realistic as possible. First, a loading cart in the form of a rigid rectangular frame on four wheels was made: the rear wheels were fixed and the front wheels were swivel, which provided maneuverability when moving. The cart was moved manually using a strong synthetic rope eight meters long, and bags of dry silica sand with stable bulk density were used as cargo. To ensure accuracy, the mass of each load unit (bag) was determined in advance using weighing scales. The use of bulk material made it possible to ensure a uniform distribution of load over the surface of the structure (Figure 17).

https://doi.org/10.3390/app16083879

For the accuracy of the tests, the mass of each bag was pre-checked on both electronic platform scales and mechanical scales with dial indicators, which eliminated possible discrepancies. Next, the measuring equipment was installed. A PSK-MG4 (SKB Stroypribor, Chelyabinsk, Russia) deflection meter was used to measure the deformations and deflections of the suspension beam under load. Structurally, the deflection meter consists of an electronic unit, a clamp, a string with a tensioning weight, a clamping screw, and a remote control panel (Figure 18).

(a) (b)

Figure 17. Loading cart with bags: (**a**) loaded condition; (**b**) unloaded condition.

Figure 18. PSK-MG4 deflection meter: 1—electronic unit; 2—clamp; 3—tensioning weight; 4—string; 5—clamping screw; 6—stand (user-supplied); 7—remote panel with a button.

Prior to testing, the instrument was fixed at the control point of the beam according to the developed installation scheme (Figure 19a). The readings were then checked at the zero position and, if necessary, calibrated (Figure 19b).

The assessment of total measurement uncertainty was carried out based on the analysis of the main sources of error, including deflection measurement error and load application error. According to the technical specifications of the PSK-MG4 deflection gauge, the absolute measurement error within the range up to 10 mm is ±0.05 mm. For measured deflection values of approximately 1–2 mm, this corresponds to a relative error of 2.5–5%.

https://doi.org/10.3390/app16083879

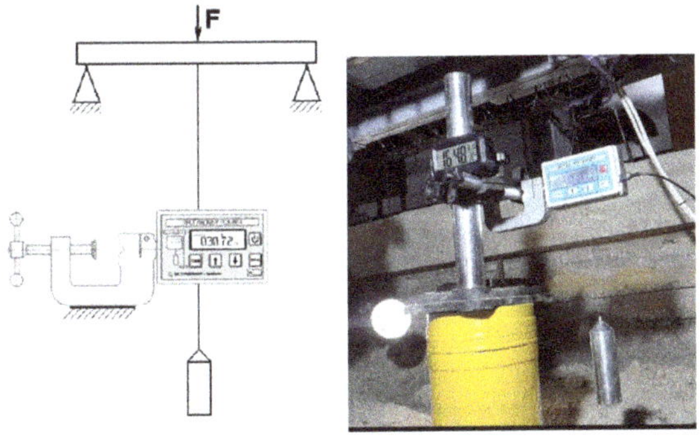

(a) (b)

Figure 19. (**a**) Deflection meter installation scheme. (**b**) Installed PSK-MG4 deflection meter for suspension beam deflection measurement.

The error in load application, associated with the use of pre-weighed bags filled with quartz sand, did not exceed ±1–2%. Taking into account the combined influence of these factors, the total relative error was estimated as the sum of the individual components and amounted to approximately 5–7%.

The obtained value corresponds to an acceptable level for engineering experimental studies and does not have a significant effect on the reliability of the identified relationships.

At the same time, the mobile overpass structure itself was prepared: it was transferred from the shipping position to the operating position, installed over the trench, aligned longitudinally and transversely, and supported, and all assemblies were fixed. Guardrails, signal lights, and entry ramps were mounted on the upper platform, which ensured compliance with safety requirements and completeness of simulation of real operating conditions (Figure 20).

Figure 20. Overpass test bench located over the trench.

Thus, the overpass, loading cart, and measuring instruments were brought into operating condition and prepared for experimental tests. The course of the experiment consisted of sequential loading of the mobile overpass structure using a specially prepared cart. Loads weighing 50, 100, 150, and 200 kg were alternately placed on the cart, after which it was manually moved along the bridge span using a synthetic rope. The movement was carried out smoothly, without jerks, in order to reproduce as accurately as possible the impact of a vehicle on the structure. To improve the reliability of the results and assess the repeatability of the experimental data, each load level was applied three times. Deflection measurements were performed for each repetition, followed by the determination of average values. Based on the test series, the mean deflection and standard deviation

https://doi.org/10.3390/app16083879

were calculated for each load level, characterizing the variability of the experimental data. The obtained results demonstrated a high degree of reproducibility: deviations between repeated measurements did not exceed acceptable limits, and the relative variation remained within the bounds of experimental error. Thus, the repeated tests confirm the reliability and reproducibility of the obtained data and allow their use for the analysis of the structural deformational behavior. While the cart was moving, the load was registered using the deflection meter. All measurement data were transferred to proprietary software provided with the PSK-MG4 device (SKB Stroypribor, Chelyabinsk, Russia) (Figure 21), where processing, plotting, and analysis of results were performed.

Experimental research was carried out sequentially with an increase in the mass of the load, starting from 50 kg. This approach made it possible to determine the dependence of the structure's deflection on the magnitude of the applied load and to evaluate its strength characteristics. Such a measurement method improved the precision and clarity of the obtained data while minimizing the influence of the human factor.

It should be noted that the conducted experimental study has several limitations associated with the applied methodology and modeling conditions.

First, the tests were performed on a physical model of the mobile overpass at a scale of 1:4. Despite the satisfaction of geometric and mechanical similarity, as well as the application of scaling theory, the transfer of results to the full-scale structure may involve additional factors not captured under laboratory conditions.

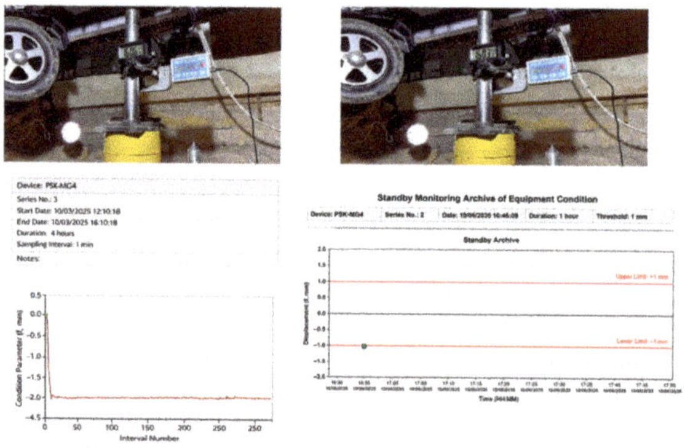

Figure 21. The process of measuring and processing the results of deflection of the suspension beam of a mobile overpass using the PSK-MG4 deflection meter (SKB Stroypribor, Chelyabinsk, Russia).

Second, the investigated load range (0.8–3.2 t in full-scale equivalent) corresponds to the initial operational level and does not cover the ultimate limit states of the structure. Therefore, the experimental results cannot be considered as confirmation of the load-carrying capacity under maximum design loads.

Third, the measurements were carried out under controlled laboratory conditions, which exclude the influence of external factors such as temperature-induced deformations, non-uniform support conditions, material wear, and real operating conditions in an urban environment.

Despite these limitations, the obtained results allow for a reliable assessment of the deformation behavior of the structure, confirm the validity of the adopted engineering solutions, and can serve as a basis for further analytical and experimental studies, including full-scale testing.

https://doi.org/10.3390/app16083879

3. Results

3.1. Analysis of the Experimental Research Result

The results of experimental research performed for the overpass bench are presented in Table 12.

Table 12. Experimental research results.

Load On the Overpass	Measured Points at the Overpass								
	1. Before Pulling Onto the Overpass	2. Pulling Onto the Overpass	3	4	5	6	7	8. Pulling Off the Overpass	9. After Pulling Off the Overpass
	Deflection Indicators (mm)								
50 kg	0	−0.17	−0.21	−0.29	−0.36	−0.38	−0.36	−0.35	0
100 kg	0	−0.32	−0.41	−0.54	−0.65	−0.67	−0.65	−0.60	0
150 kg	0	−0.44	−0.54	−0.76	−0.89	−0.93	−0.87	−0.82	0
200 kg	0	−0.57	−0.73	−0.95	−1.15	−1.19	−1.14	−1.06	0

Based on the results of the experimental research, a graph was drawn up showing the changes in the deflection of the overpass depending on the mass of the applied load and the point of its location along the overpass (Figure 22).

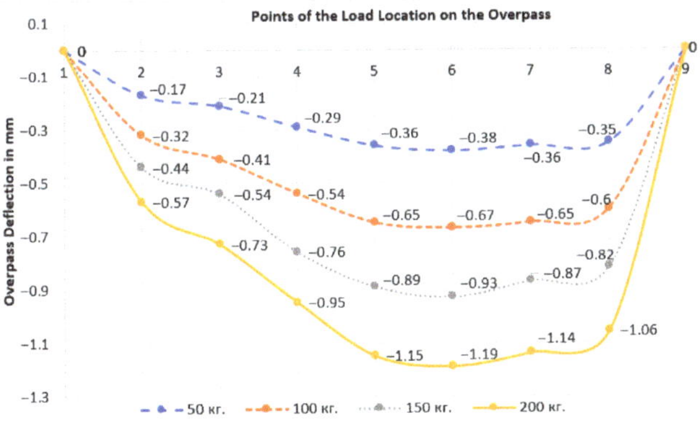

Figure 22. Change in the overpass deflection value depending on the load mass and points of its location along the overpass.

In the course of the experimental research, the dependence of the overpass deflection on the load magnitude was examined. The plotted curves for loads of 50, 100, 150, and 200 kg exhibit a parabolic shape, with maximum deflection values occurring at the midspan, while near the supports the deformations tend to zero. This trend is typical for beam-type structures subjected to bending, confirming the correctness of the experimental setup. As the load increased, a corresponding increase in the deflection magnitude was observed: for 50 kg the maximum deflection was approximately −0.36 mm; for 100 kg −0.65 mm; for 150 kg −0.95 mm; and for 200 kg −1.19 mm.

Thus, with a fourfold increase in the load, the maximum deformation increased by approximately 3.3 times, indicating a close to linear dependence between load and deflection within the studied range. Minor deviations from the strict proportionality can be explained by the elastic and plastic properties of the material and measurement features. At the same time, the shape of the curves remains symmetrical relative to the middle of the span, and the deflection values at symmetrical points coincide, which indicates uniform operation of the structure and the absence of distortions or asymmetries in the load

https://doi.org/10.3390/app16083879

distribution. The results obtained allow us to conclude that when exposed to loads of up to 200 kg, the overpass retains the elastic nature of deformations, and the increase in deflections occurs gradually, without sudden jumps, which confirms sufficient rigidity of the structure. The greatest deformations are concentrated in the central zone of the span, which should be taken into account when designing reinforcements and assessing the operational reliability of the structure.

Based on the results of the experimental research, a regression equation [79] was developed that describes the change in the overpass deflection depending on the mass of the applied load and the distance between points along the overpass x.

$$w(m,x) = 0.2073 - 0.003143\,m - 0.2789\,x + 0.000003\,m^2 + 0.05977\,x^2 - 0.000915\,(m \cdot x) \quad (11)$$

The deflection indicators were compared in the experiment and in the regression equation (Table 13).

Based on the compiled regression equation, a graph is presented showing a comparison of the actual (experimental) and estimated (according to the regression equation) deflection values (Figure 23).

According to the graph, there is a close convergence of the results from the experiment and the regression equation. In particular, the determination coefficient $R^2 = 0.97$ means that the regression equation explains almost 98% of the scatter of experimental data, which indicates the high accuracy of the selected equation. The correlation coefficient value $r = 0.992$ indicates a strong linear relationship between the estimated and actual deflection values. Moreover, the average approximation error is less than 6.6%, which indicates that the equation very accurately reproduces the experimental results.

Table 13. Comparison of the deflection indicators in the experiment and regression equation.

Load, kg	Distance, m	Actual Deflection (Experiment), mm	Estimated Deflection (Regression), mm	Error, mm
50	0.6	−0.170	−0.116	−0.054
50	1.2	−0.210	−0.246	0.036
50	1.8	−0.290	−0.333	0.043
50	2.4	−0.360	−0.377	0.017
50	3.0	−0.380	−0.378	−0.002
50	3.6	−0.360	−0.336	−0.024
100	0.6	−0.320	−0.278	−0.042
100	1.2	−0.410	−0.435	0.025
100	1.8	−0.540	−0.550	0.010
100	2.4	−0.650	−0.622	−0.028
100	3.0	−0.670	−0.650	−0.020
100	3.6	−0.650	−0.636	−0.014
150	0.6	−0.440	−0.425	−0.015
150	1.2	−0.540	−0.610	0.070
150	1.8	−0.760	−0.752	−0.008
150	2.4	−0.890	−0.851	−0.039
150	3.0	−0.930	−0.907	−0.023
150	3.6	−0.870	−0.920	0.050
200	0.6	−0.570	−0.557	−0.013
200	1.2	−0.730	−0.770	0.040
200	1.8	−0.950	−0.939	−0.011
200	2.4	−1.150	−1.066	−0.084
200	3.0	−1.190	−1.149	−0.041
200	3.6	−1.140	−1.190	0.050

https://doi.org/10.3390/app16083879

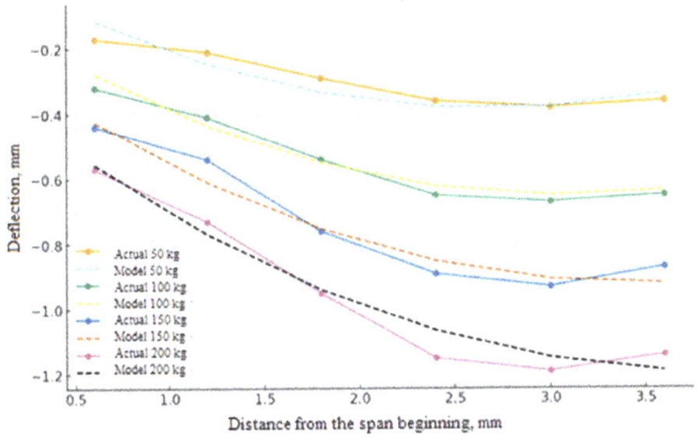

Figure 23. Comparison of actual (experimental) and estimated (according to the regression equation) deflection values.

A mathematical analysis of the regression equation [80] was carried out. The partial derivatives were obtained according to Equation (11).

By weight m:

$$\frac{\partial w}{\partial m} = -0.003143 + 2 \times 0.000003 \cdot m - 0.000915 \cdot x$$

By distance x:

$$\frac{\partial w}{\partial x} = -0.2789 + 2 \times 0.05977 \cdot x - 0.000915 \cdot m$$

Then, a system of two equations with two unknown values (m and x) was compiled:

$$\begin{cases} -0.003143 + 0.000006 \cdot m - 0.000915 \cdot x = 0, \\ -0.2789 + 0.11954 \cdot x - 0.000915 \cdot m = 0 \end{cases}$$

Solving this system gives

$$m = \frac{0.000273 \cdot x - 0.000737}{0.000002} = -17.25,$$

$$0.011082 - 0.006274 \cdot x - 0.000273 \times (-17.25) = 0, \quad x \approx 2.564 \text{ m}.$$

Thus, the critical point of the surface was found—the point where the function may have a minimum, maximum, or saddle point. The critical point at $m = -17.25$ means that at these values the weight and distance give the condition where the structure deflection ceases to change—it neither increases nor decreases.

However, the weight value $m = -17.25$ has no physical meaning, because the cart weight cannot be negative. Therefore, this extremum lies outside the acceptable physical domain. Within the actual ranges

$$m \in [50; 200] \text{ kg}, \qquad x \in [0.6; 3.6] \text{ m},$$

the function increases monotonically by weight and has a maximum by distance.

Hessian Determinant

The Hessian determinant was calculated to determine the curvature of the function surface near the critical point [81]. The matrix of second derivatives has the form:

$$H = \begin{bmatrix} 2 \times 10^{-6} & -0.000273 \\ -0.000273 & -0.006274 \end{bmatrix}$$

https://doi.org/10.3390/app16083879

$$\det(H) = (2 \times 10^{-6}) \times (-0.006274) - (-0.000273)^2 = -8.7077 \times 10^{-8} < 0 \quad (12)$$

Since the determinant of the Hessian matrix is negative, $\det(H) < 0$, the regression equation surface is characterized by a saddle point in the variable space (m, x). This means that the function is not strictly convex or concave but has directions in which it increases and directions in which it decreases.

Next, the Hessian trace was identified. The Hessian trace is the sum of the diagonal elements of the Hessian matrix, that is, the sum of the second derivatives with respect to each variable. It reflects the total "convexity" or "concavity" of the function in all directions. If the trace is positive, then on average the function at this point is curved upward; if it is negative, it means that the function is curved downward.

$$\mathrm{Tr}(H) = 2 \times 10^{-6} + (-0.006274) = -0.006272 \quad (13)$$

As a result of the mathematical analysis, it was established that the determinant of the Hessian matrix satisfies the condition $\det(H) < 0$. Therefore, the surface of the regression function has a saddle point. This means that the function increases in one direction and decreases in another. However, the saddle point found corresponds to negative load and distance values and, therefore, lies outside the physically feasible region.

Within the limits of real parameter values, the function demonstrates the expected physical behavior:

- The deflection increases monotonically with increasing load;
- Along the distance, the maximum deflection is observed in the middle part of the span, which corresponds to the classical picture of beam bending on supports.

Based on the above, a graph was developed that represents a three-dimensional surface of the change in the deflection $w(m, x)$. The surface was constructed using the regression equation with the overlay of contour lines and experimental data. The abscissa axis shows the load weight L in the range of 50 to 200 kg, the ordinate axis shows the distance along the span D (0.6–3.6 m), and the applicate axis shows the deflection of the structure in millimeters (Figure 24).

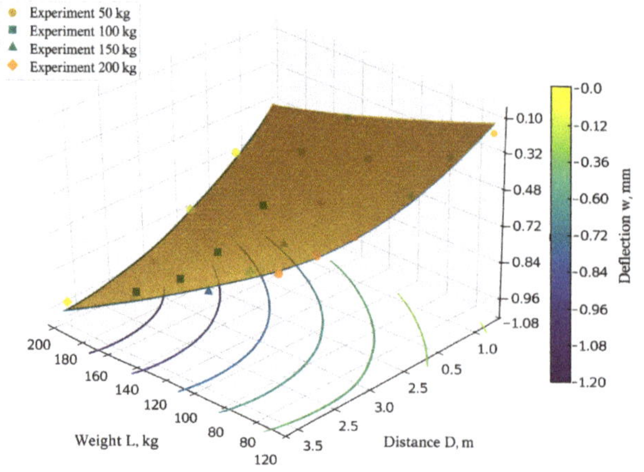

Figure 24. Change in deflection indicators depending on the load weight (m) and distance between points (x).

Analysis of the surface shows that the deflection naturally increases with increasing load mass, and a characteristic dependence is observed along the span coordinate: the

https://doi.org/10.3390/app16083879

maximum values occur in the middle of the span, while near the supports the deflection decreases. This effect fully corresponds to the classical picture of the operation of a beam on supports under the effect of a load.

Contour lines drawn on the projection of the "floor" make it possible to clearly trace the change in deflection depending on both parameters and confirm the natural increase in deformations with increasing weight. Experimental points, shown by various markers for the loads of 50, 100, 150, and 200 kg, are located in close proximity to the surface of the regression model, which indicates a high degree of agreement between the theoretical calculations and experimental data. Minor discrepancies are explained by the experimental error and the characteristics of the actual loading of the structure.

The presented graph demonstrates the correctness of the built regression dependence and its adequacy in describing the physical behavior of the structure: the deflection increases monotonically with increasing load and has a maximum in the central part of the span. Overlaying experimental points on the surface allows for a visual confirmation of the reliability of the results and for using the model to predict the behavior of the overpass under operational loads.

Thus, the analysis carried out confirms the adequacy of the developed regression equation. It correctly reflects the physical laws of the structure: the deflection increases with increasing load and has a parabolic distribution along the span. The presence of an extremum outside the permissible region does not affect the actual operation of the structure. The constructed model is stable and can be used to predict the behavior of the overpass under operational loads.

During the processing of experimental data obtained from measuring overpass deflections at various points of the span, a transition was made from local values to integral indices. For this purpose, the areas of geometric figures (triangles and trapezoids) bounded by the deflection line and coordinate axes were calculated. This method makes it possible to quantify not only the deflection at a specific point but also the total impact of the load on the span. The calculation was carried out using the trapezoidal area formula, where the values of deflections at adjacent points were taken as the bases, and the distance between them as the height.

Based on experimental measurements of deflections at control points of the span, an integral assessment of the effect of the load [82] was performed. For each combination of the cart weight (50, 100, 150, and 200 kg) and its position along the span length (0.6–3.6 m), the areas under the deflection curve were calculated using the trapezoidal method. The resulting areas reflect the integral effect of the load on the structure and make it possible to move from local deformations to a quantitative assessment of the overall structural performance.

The load force was determined as

$$P = m \cdot g, \tag{14}$$

and for a uniformly distributed load

$$q = \frac{P}{L}, \tag{15}$$

where $L = 3.0$ m is the section length.

The work of the distributed load was calculated using the following expression:

$$W_p = \frac{1}{2} q \int_0^L w(x)\, dx, \tag{16}$$

https://doi.org/10.3390/app16083879

and, for comparison, the work of a point load applied at the center of the span was calculated as

$$W_t = \frac{1}{2} P \cdot w_{\text{center}}. \tag{17}$$

Additionally, the equivalent rigidity at the central point was determined as the following ratio:

$$k = \frac{P}{w_{\text{center}}}. \tag{18}$$

The calculation results are presented in Table 14.

Table 14. Index calculation results.

Weight, kg	Area Under the Deflection Curve, $\sum \int w\,dx$ (m^2)	Load Force, $P = mg$ (N)	Equivalent Distributed Load, $q = P/L$ (N/m)	Distributed Load Work, W_p (J)	Deflection at the Center, w_{center} (m)	Work of Point Load, W_t (J)	Equivalent Rigidity, k_{center} (N/m)
50	0.000903	490.5	163.5	0.0738	0.00038	0.0932	1.29×10^6
100	0.001653	981.0	327.0	0.2703	0.00067	0.3286	1.46×10^6
150	0.002265	1471.5	490.5	0.5555	0.00093	0.6842	1.58×10^6
200	0.002925	1962.0	654.0	0.9565	0.00119	1.1674	1.65×10^6

Based on the obtained calculation results, graphs of the dependence of the load mass on the area under the deflection curve, load work, deflection at the center of the span, and equivalent rigidity were developed (Figure 25).

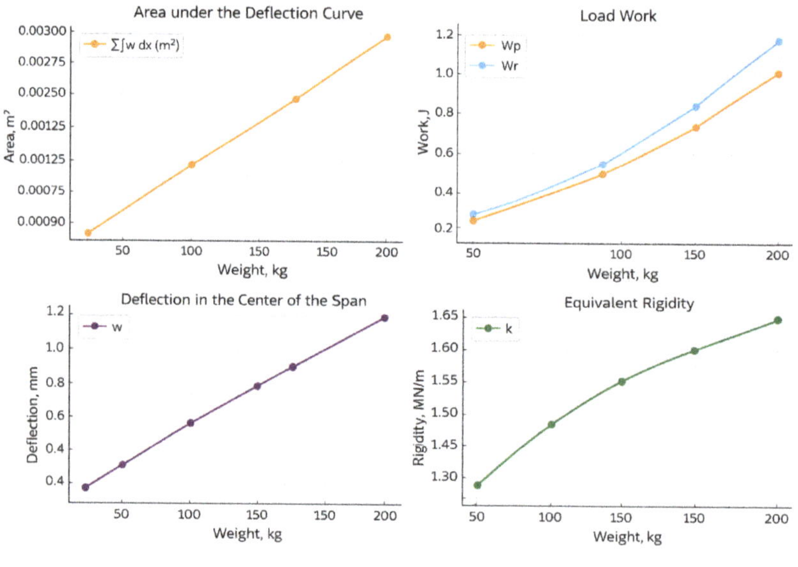

Figure 25. Graphs of the dependence of the load mass on the area under the deflection curve, load work, deflection at the center of the span, and equivalent rigidity.

According to the developed graphs, it follows that the total areas under the deflection curve increased from 0.0009 m^2 with a mass of 50 kg to 0.0029 m^2 with a mass of 200 kg. At the same time, the load work ranged from 0.07 to 0.96 J for the distributed diagram and from 0.09 to 1.17 J for the point diagram. The equivalent rigidity of the structure varied in the range of 1.3 to 1.65 MN/m, while demonstrating stability, and there was only a slight increase with increasing load, which can be explained by experimental errors and slight nonlinearity of the deformation.

https://doi.org/10.3390/app16083879

Thus, the integral assessment confirmed the pattern: with increasing weight, deflections and load work increase almost linearly. The work values are in the range of up to 1.2 J, which reflects the characteristic expenditure of energy of the structure for elastic deformation. The use of the method of areas under the deflection line made it possible to move from individual point measurements to a comprehensive analysis, which can be used to assess the durability and calculate the service life of bridge and overpass structures under operating conditions.

4. Discussion of the Research Results

The obtained research results are aimed at solving a scientific problem related to ensuring the continuity of traffic flow under roadwork conditions in dense urban environments. Unlike existing solutions focused on traffic flow redistribution or reducing road closure duration, the proposed approach is based on the physical preservation of road capacity directly within the work zone. This fundamentally changes the nature of the problem and makes it possible to eliminate the identified scientific and technical gap, which consists in the absence of scientifically substantiated undercarriage design solutions for mobile overpasses that ensure their efficient operation in dense urban environments, including structural transportation, positioning within the work zone, stability under loading, and interaction with the supporting base.

In contrast to existing studies, where bridge structures are predominantly considered as stationary infrastructure elements, the present work treats the mobile overpass as an integrated transport engineering system, in which the undercarriage acts as the functionally determining element ensuring its practical feasibility in urban conditions.

The scientific contribution of the study lies not only in the development of a specific structural solution but also in the formation of a methodological approach to the design of mobile overpass undercarriages. In particular, the use of morphological analysis is proposed for the systematic synthesis of design solutions, taking into account a combination of operational, technical, and economic factors. This approach enables a transition from the selection of individual components to the formation of an optimal configuration based on a set of criteria. In addition, a method of experimental verification based on scaled modeling is substantiated, ensuring the transfer of results to full-scale structures.

The obtained regression relationship describing the structural behavior under load represents a tool for quantitative assessment of overpass deformability and can be used for predicting its performance and optimizing undercarriage parameters at the design stage.

Thus, the results of the study demonstrate the possibility of transitioning from the concept of traffic flow adaptation to the concept of physical preservation of road capacity directly within the work zone. The proposed approach establishes a new direction in the field of transport infrastructure based on the development of mobile overpasses with a functionally integrated undercarriage system.

It should be noted that the conducted study has several limitations associated with the applied methodology and modeling conditions. The tests were performed on a physical model of the mobile overpass at a scale of 1:4, which, despite satisfying the conditions of geometric and mechanical similarity, does not fully account for all factors affecting the behavior of the full-scale structure. In addition, the investigated load range (0.8–3.2 t in full-scale equivalent) corresponds to the initial operational level and does not cover the ultimate limit states of the structure. The measurements were carried out under laboratory conditions without considering the influence of external operational factors such as temperature-induced deformations, support conditions, and material wear.

Despite these limitations, the obtained results allow for a reliable assessment of the structural behavior and confirm the validity of the adopted engineering solutions. They

https://doi.org/10.3390/app16083879

provide a foundation for further development of this research area, primarily in improving the undercarriage design of the mobile overpass as a key system element.

Promising research directions include the study of structure–foundation interaction, as well as the development of numerical models that account for complex operational scenarios.

Additionally, an important direction is the development of a regulatory and technical framework governing the design, operation, and application of mobile overpasses in accordance with the requirements of urban transport infrastructure.

5. Conclusions

The study addresses a relevant scientific and engineering problem aimed at ensuring the continuity of traffic flow during roadworks in dense urban environments through the development and substantiation of the undercarriage design of a mobile overpass. The analysis of existing organizational, technical, and engineering solutions has shown that they do not ensure the preservation of road network capacity, but rather adapt traffic flows to deteriorated operating conditions. This made it possible to justify the need for a fundamentally new approach based on the physical preservation of traffic directly within the work zone.

Within the framework of the study, a systematic approach to the design of the mobile overpass undercarriage has been proposed, based on morphological analysis and synthesis. This made it possible to define a space of feasible design solutions and to select a rational configuration considering a set of criteria, including load-carrying capacity, maneuverability, weight, manufacturability, and compatibility with transport vehicles. A structural scheme of the undercarriage has been developed, including a mechanical–pneumatic–electrical system that integrates transportation, installation, leveling, and fixation functions without the need for stationary supporting foundations.

Experimental studies have shown that the developed structure possesses sufficient stiffness and stability: the maximum deflection reached 1.19 mm under a load of 200 kg, and the load–deflection relationship is close to linear. The obtained regression model with a high level of reliability ($R^2 \approx 0.97$) confirms the adequacy of the adopted engineering solutions and allows its use for predicting the structural behavior under load.

At the same time, the applied model and the conducted studies have several limitations that should be considered when interpreting the results. First, the experimental tests were performed on a scaled model (1:4), which requires caution when extrapolating the results to a full-scale structure, especially considering possible nonlinear scaling effects. Second, the experimental results cannot be considered as confirmation of the load-carrying capacity of the overpass under maximum design loads. Third, the computational model includes simplifying assumptions related to uniform load distribution and idealized boundary conditions, which may differ from real operating conditions. In addition, the influence of operational factors such as material wear, temperature-induced deformations, and environmental effects was not considered within the scope of this study. The obtained results confirm that the developed mobile overpass represents an effective engineering solution that ensures the preservation of road network capacity within work zones, reduces traffic congestion, and improves traffic safety. The practical significance of the study lies in the possibility of implementing the proposed design in urban infrastructure, as well as in forming an engineering basis for further design and deployment of similar systems.

Future research directions are associated with further development and refinement of the obtained results. First, full-scale testing under real operating conditions is required to validate the model and verify its scalability. An important direction is the consideration of dynamic loads and the development of more advanced models accounting for the in-

https://doi.org/10.3390/app16083879

teraction between the structure and traffic flow. Another promising direction is the study of automation of positioning and control processes of the mobile overpass, as well as its integration into Intelligent Transportation Systems. In addition, the development of a regulatory and technical framework governing the application of mobile overpasses in urban environments is required, along with an assessment of their effectiveness at the level of the urban transport system.

Thus, the conducted study forms a scientifically grounded basis for the development and implementation of next-generation mobile overpasses aimed at improving the resilience and efficiency of urban transport infrastructure.

Author Contributions: Conceptualization, A.K. (Adil Kadyrov); Methodology, A.K. (Adil Kadyrov) and A.K. (Aliya Kukesheva); Software, A.Z.; Validation, A.K. (Aliya Kukesheva), A.G. and K.S.; Formal Analysis, A.K. (Aliya Kukesheva); Investigation, K.S., A.Z. and Z.Z.; Resources, A.Z.; Data Curation, K.S.; Writing—Original Draft Preparation, A.K. (Aliya Kukesheva); Writing—Review and Editing, A.K. (Adil Kadyrov) and A.G.; Visualization, A.K. (Aliya Kukesheva), A.Z. and N.Z.; Supervision, A.G.; Project Administration, A.K. (Adil Kadyrov); Funding Acquisition, A.G. All authors have read and agreed to the published version of the manuscript.

Funding: This research has been/was/is funded by the Science Committee of the Ministry of Science and Higher Education of the Republic of Kazakhstan (Grant No. AP23487832).

Institutional Review Board Statement: Not applicable.

Informed Consent Statement: Not applicable.

Data Availability Statement: The original contributions presented in this study are included in the article. Further inquiries can be directed to the corresponding authors.

Conflicts of Interest: The authors declare no conflicts of interest.

Abbreviations

ASTRA	Automated Safe Traffic Relief Arrangement Bridge (Switzerland)
CIS	Commonwealth of Independent States
DIN	Deutsches Institut für Normung (German Institute for Standardization)
HDGS	Hot-Dip Galvanized Steel
ITS	Intelligent Transportation Systems
KAMAZ	Kama Automobile Plant (Russia)
MGB	Medium Girder Bridge (UK, military)
MLC	Military Load Classification
PMP-3	Pontoon Bridge PMP-3 (Russia)
RDB	Rapidly Deployable Bridge (USA/Germany)
RF	Russian Federation
UK	United Kingdom
USA	United States of America
Volvo FMX	Volvo FMX Truck (Sweden)

References

1. Sabraliev, N.; Abzhapbarova, A.; Nugymanova, G.; Taran, I.; Zhanbirov, Z. Modern aspects of modeling of transport routes in Kazakhstan. *News Natl. Acad. Sci. Repub. Kazakhstan Ser. Geol. Tech. Sci.* **2019**, *2*, 62–68. [CrossRef]
2. Bräuninger, M.; Teuber, M.-O. The development of the passenger car population in Germany [Die Entwicklung des Pkw-Bestands in Deutschland]. *Wirtschaftsdienst* **2023**, *103*, 280–282. [CrossRef]
3. Gao, Y.; Zhu, J. Characteristics, impacts and trends of urban transportation. *Encyclopedia* **2022**, *2*, 1168–1182. [CrossRef]
4. Lu, J.; Li, B.; Li, H.; Al-Barakani, A. Expansion of city scale, traffic modes, traffic congestion, and air pollution. *Cities* **2021**, *108*, 102974. [CrossRef]

https://doi.org/10.3390/app16083879

5. Kumar, M.; Kumar, K.; Das, P. Study on road traffic congestion: A review. In *Recent Trends in Communication and Electronics*; Routledge: Oxfordshire, UK, 2021; pp. 230–240.
6. Mahasirikul, N.; Aksorn, P.; Kusonkhum, W. Driving speed and hazardous location in construction work zone: Case of highway 2 Hin Latnon–Sa At. *Geomate J.* **2021**, *20*, 143–151. [CrossRef]
7. Bagloee, S.A.; Sarvi, M.; Ceder, A. Transit priority lanes in congested road networks. *Public Transp.* **2017**, *9*, 571–599. [CrossRef]
8. Yue, W.; Li, C.; Mao, G. Urban traffic bottleneck identification based on congestion propagation. In *Proceedings of the 2018 IEEE International Conference on Communications (ICC)*; IEEE: New York, NY, USA, 2018; pp. 1–6. [CrossRef]
9. Boggio-Marzet, A.; Monzon, A.; Rodriguez-Alloza, A.M.; Wang, Y. Combined influence of traffic conditions, driving behavior, and type of road on fuel consumption: Real driving data from Madrid area. *Int. J. Sustain. Transp.* **2022**, *16*, 301–313. [CrossRef]
10. Al-Arkawazi, S.A.F. Measuring the influences and impacts of signalized intersection delay reduction on fuel consumption, operation cost and exhaust emissions. *Civ. Eng. J.* **2018**, *4*, 552–571. [CrossRef]
11. Marve, S.R.; Bhorkar, M.; Baitule, P. A survey on environmental impacts due to traffic congestion in peak hours. *Int. J. Sci. Technol. Eng.* **2016**, *2*, 151–154.
12. Yousif, S.; Nassrullah, Z.; Norgate, S.H. Narrow lanes and their effect on drivers' behaviour at motorway roadworks. *Transp. Res. Part F Traffic Psychol. Behav.* **2017**, *47*, 86–100. [CrossRef]
13. Tselentis, D.I.; Papadimitriou, E. Driver profile and driving pattern recognition for road safety assessment: Main challenges and future directions. *IEEE Open J. Intell. Transp. Syst.* **2023**, *4*, 83–100. [CrossRef]
14. Stueger, P.N.; Fehn, F.; Bogenberger, K. Minimizing the effects of urban mobility-on-demand pick-up and drop-off stops: A microscopic simulation approach. *Transp. Res. Rec.* **2023**, *2677*, 814–828. [CrossRef]
15. Jilani, U.; Asif, M.; Zia, M.Y.I.; Rashid, M.; Shams, S.; Otero, P. A systematic review on urban road traffic congestion. *Wirel. Pers. Commun.* **2025**, *140*, 81–109. [CrossRef]
16. Huang, Z.; Loo, B.P. Urban traffic congestion in twelve large metropolitan cities: A thematic analysis of local news contents, 2009–2018. *Int. J. Sustain. Transp.* **2023**, *17*, 592–614. [CrossRef]
17. Fu, Q.; Tian, Y.; Sun, J. Modeling and simulation of dynamic lane reversal using a cell transmission model. *J. Intell. Transp. Syst.* **2022**, *26*, 717–729. [CrossRef]
18. Kitano, A.; Yoshihiro, T. Controlling detour vehicle volume for stable congestion mitigation in route guidance systems. *Sens. Mater.* **2024**, *36*, 4483–4499. [CrossRef]
19. Stojanovic, N.; Grujic, I.; Boskovic, B. Application of intelligent transport systems in road traffic: A review. In *Proceedings of the International Conference "New Technologies, Development and Applications"*; Springer Nature: Cham, Switzerland, 2023; pp. 646–651. [CrossRef]
20. Pérez-Méndez, D.; Gershenson, C.; Lárraga, M.E.; Mateos, J.L. Modeling adaptive reversible lanes: A cellular automata approach. *PLoS ONE* **2021**, *16*, e0244326. [CrossRef]
21. Ampountolas, K.; Dos Santos, J.A.; Carlson, R.C. Motorway tidal flow lane control. *IEEE Trans. Intell. Transp. Syst.* **2019**, *21*, 1687–1696. [CrossRef]
22. Maksimchuk, O.V.; Borisova, N.I.; Panov, D.B.; Voronkova, G.V. Sustainable development of road transport infrastructure by including elements of reverse traffic in the "smart city" program. *IOP Conf. Ser. Earth Environ. Sci.* **2022**, *988*, 022072. [CrossRef]
23. Adanu, E.K.; Agyemang, W.; Lidbe, A.; Adarkwa, O.; Jones, S. An in-depth analysis of head-on crash severity and fatalities in Ghana. *Heliyon* **2023**, *9*, e18937. [CrossRef] [PubMed]
24. Babić, D.; Babić, D.; Fiolic, M.; Ferko, M. Road markings and signs in road safety. *Encyclopedia* **2022**, *2*, 1738–1752. [CrossRef]
25. He, Y.; Deng, X. Research on the application of reversible lanes in urban road traffic accidents emergency evacuation. *J. Wuhan Univ. Technol. Transportation Sci. Eng.* **2021**, *45*, 822–827. [CrossRef]
26. Liu, J.; Zheng, C.; Guo, L.; Liu, X.; Lu, Q. Evaluating routing asymmetry by passive flow measurements with spark. In *Proceedings of the 2018 IEEE 3rd International Conference on Cloud Computing and Big Data Analysis (ICCCBDA)*; IEEE: New York, NY, USA, 2018; pp. 414–419. [CrossRef]
27. Al-Bayati, A.J.; Ali, M.; Alhomaidat, F.; Bandara, N.; Chen, Y. The effects of temporary portable rumble strips on vehicle speeds in road work zones. *Safety* **2024**, *10*, 105. [CrossRef]
28. Valdés-Díaz, D.M.; López del Puerto, C.; Colucci-Ríos, B.; Figueroa-Medina, A.M.; Concepción-Carrasco, E.; Sierra-Betancur, L.; Taveras-Canela, Y. Driver compliance in work zones: Two-lane rural roads versus freeways. *J. Leg. Aff. Disput. Resolut. Eng. Constr.* **2021**, *13*, 04521036. [CrossRef]
29. Skovajsa, J.; Pribyl, O.; Pribyl, P.; Scerba, M.; Janota, A. Evaluation of a mobile highway management system at roadwork zones. *Int. J. Eng.* **2022**, *35*, 900–907. [CrossRef]
30. Acharya, S.; Mekker, M. *Evaluating the Impact of Detour Messaging on Actual Driver Detour Behavior*; No. UT-21.31; Utah Department of Transportation, Research Division: Salt Lake City, UT, USA, 2022.

https://doi.org/10.3390/app16083879

31. Morshedi, M.; Padhye, S.; Mwamba, I.; Kang, K.; Labi, S.; Hastak, M. Suitability assessment of detour routes for road construction projects: Framework and case studies. *J. Manag. Eng.* **2023**, *39*, 04022077. [CrossRef]
32. Kopelias, P.; Kepaptsoglou, K.; Sarantidis, V.; Theofilatos, A. Route Efficiency Index as an operational tool for detour decision making in an urban motorway. *Adv. Transp. Stud.* **2018**, *44*, 63–76.
33. Kido, R.; Nishimura, N.; Mitsutani, S.; Maruo, S.; Kimura, M. Measurement of pressure distribution on roadbed using soil bags during plate loading tests. In *Proceedings of the International Conference on Transportation Geotechnics*; Springer Nature: Singapore, 2024; pp. 69–78. [CrossRef]
34. Chu, D.; Cao, Y. Typical intelligent transportation applications. In *Intelligent Road Transport Systems: An Introduction to Key Technologies*; Springer Nature: Singapore, 2022; pp. 545–608. [CrossRef]
35. Schwarz, S.; Philosof, T.; Rupp, M. Signal processing challenges in cellular-assisted vehicular communications: Efforts and developments within 3GPP LTE and beyond. *IEEE Signal Process. Mag.* **2017**, *34*, 47–59. [CrossRef]
36. Formal model of a section of the road network with special congestion for intelligent transport systems. *Int. Multidiscip. Sci. Geoconference Sgem* **2021**, *21*, 403–410. [CrossRef]
37. Isradi, M.; Dwiatmoko, H.; Setiawan, M.I.; Supriyatno, D. Analysis of capacity, speed, and degree of saturation of intersections and roads. *J. Appl. Sci. Eng. Technol. Educ.* **2020**, *2*, 150–164. [CrossRef]
38. Kushchenko, L.; Kushchenko, S.; Novikov, A. A system for monitoring traffic parameters using intelligent transport systems. *AIP Conf. Proc.* 2022, *2503*, 080011. [CrossRef]
39. Pop, M.-D.; Pandey, J.; Ramasamy, V. Future Networks 2030: Challenges in intelligent transportation systems. In *Proceedings of ICRITO 2020—IEEE 8th International Conference on Reliability, Infocom Technologies and Optimization (Trends and Future Directions)*; IEEE: New York, NY, USA, 2020; pp. 898–902. [CrossRef]
40. Waqar, A.; Alshehri, A.H.; Alanazi, F.; Alotaibi, S.; Almujibah, H.R. Evaluation of challenges to the adoption of intelligent transportation systems for urban smart mobility. *Res. Transp. Bus. Manag.* **2023**, *51*, 101060. [CrossRef]
41. Hlinka, R.; Farbak, M.; Odrobinak, J. The use of up-to-date analyses for temporary bridge applications in the present. *Civ. Environ. Eng.* **2024**, *20*, 491–507. [CrossRef]
42. Yang, Y.Y.; Chang, C.M.; Kang, S.C.; Yeh, F.Y. Study of construction-oriented structural connectors for a temporary bridge. In *Proceedings of the International Symposium on Automation and Robotics in Construction (ISARC)*; IAARC Publications: Ann Arbor, MI, USA, 2019; Volume 36, pp. 1171–1175.
43. Osman, A.M.A. Design Optimization of Composite Deployable Bridge Systems Using Hybrid Meta-Heuristic Methods for Rapid Post-Disaster Mobility. Ph.D. Thesis, Concordia University: Montreal, QC, Canada, 22 August 2016.
44. Martins, H.M. Investigating the Behavior of Modular Truss and Lattice Bridges with Hybrid Design. Ph.D. Thesis, University of Notre Dame, Notre Dame, IN, USA, 26 August 2025.
45. Zumstein, M.; Chen, Q.; Adey, B.T.; Hall, D.M. A preliminary investigation of the potential benefits of using the ASTRA Bridge for short-span bridge deck refurbishment projects in Switzerland. *Struct. Infrastruct. Eng.* **2024**, *20*, 1629–1647. [CrossRef]
46. Meloni, M.; Ballegaard, E.; Zhang, Q.; Zhang, J.; Ma, R.; Cai, J. Shape and motion inverse design of an origami-based deployable structure for architectural applications. *J. Struct. Eng.* **2023**, *149*, 04023174. [CrossRef]
47. Zhang, Q.; Liu, Y.; Li, M.; Han, Y.; Cai, J.; Feng, J. Simulation of dynamics during deployment of foldable origami structures. *Int. J. Struct. Stab. Dyn.* **2020**, *20*, 2050058. [CrossRef]
48. Wang, J.; Liu, J.; Wang, Z.; Liu, T.; Liu, J.; Zhang, J. Cost-effective UHPC for accelerated bridge construction: Material properties, structural elements, and structural applications. *J. Bridge Eng.* **2021**, *26*, 04020117. [CrossRef]
49. Yin, Z.; Zhou, L.; Li, X.; Peng, J.; Yan, X.; Wu, Y. Research on key technologies for precast concrete bridge piers and bent caps. In *Bridge Safety, Maintenance, Management, Life-Cycle, Resilience and Sustainability*; CRC Press: Boca Raton, FL, USA, 2022; pp. 1528–1533.
50. Zamaliev, F.; Bikkinin, E.; Fekir, K. Numerical and experimental investigation of prestressed steel–concrete composite beams. *IOP Conf. Ser. Mater. Sci. Eng.* **2020**, *890*, 012048. [CrossRef]
51. Grau, L.W. Russian Deliberate River Crossings: Choreographing A Water Ballet. *Eng. Prof. Bull. Army Eng.* **2019**, *49*, 28–33. Available online: https://go.gale.com/ps/i.do?id=GALE%7CA655423285&sid=googleScholar&v=2.1&it=r&linkaccess=abs&issn=00461989&p=AONE&sw=w (accessed on 9 November 2025).
52. Ganyukov, A.A.; Sinelnikov, K.A.; Kabikenov, S.Z.; Karsakova, A.Z. Research and calculation of the deformed state of the roadway mobile overpass. *Mater. Mech. Eng. Technol.* **2020**, *3*, 88–95. [CrossRef]
53. Ganyukov, A.; Kadyrov, A.; Kukesheva, A.; Zhumabekov, A.; Sinelnikov, K.; Amanbayev, S.; Karsakova, A. Optimization of Mobile Overpass Support Placement Considering the Nonlinear Properties of the Soil Foundation. *Appl. Sci.* **2026**, *16*, 2075. [CrossRef]
54. Ritchey, T. General morphological analysis as a basic scientific modelling method. *Technol. Forecast. Soc. Change* **2018**, *126*, 81–91. [CrossRef]

https://doi.org/10.3390/app16083879

55. Luerkens, D.W.; Beddow, J.K.; Vetter, A.F. Theory of morphological analysis. In *Particle Characterization in Technology*; CRC Press: Boca Raton, FL, USA, 2018; pp. 3–14.
56. Vertakova, Y.; Klevtsova, M.; Babich, T. Identification of the new research areas and development of the existing ones by methods of morphological analysis and synthesis. *Econ. Ann. XXI* **2016**, *157*, 4–8. [CrossRef]
57. Pecheikina, M.A.; Rakov, D.L. Application of the Hierarchy Analysis Method for Morphological Analysis of Technical Systems. *J. Mach. Manuf. Reliab.* **2024**, *53*, 725–732. [CrossRef]
58. Aurier, L.; Hassan, M.; Jaworski, J.; Guizani, L. Review of accelerated bridge construction systems for bridge superstructures and their adaptability for cold weather. *CivilEng* **2023**, *4*, 83–104. [CrossRef]
59. Jang, H.L.; Han, C.H.; Kim, R.H.; Park, D.Y.; Kim, S.Y. Experimental and numerical analysis of a temporary bridge connector consisting of an upper bearing block–lower pin configuration. *Appl. Sci.* **2024**, *14*, 2012. [CrossRef]
60. Mohamadiazar, N.; Ebrahimian, A. A comprehensive decision support tool for accelerated bridge construction. *Infrastructures* **2025**, *10*, 265. [CrossRef]
61. Embaby, M.; El Naggar, M.H. Efficiency of modular bridge configurations: A study on the structural stability and capacity of single- and double-story modular bridges. *Buildings* **2025**, *15*, 1709. [CrossRef]
62. Oroho, S.; Kubicz, T. Accelerated bridge construction on Maui's Hana Highway. *Eng. Proc.* **2023**, *36*, 45. [CrossRef]
63. Prakhya, G.K.V. Temporary works in construction of bridges near third party assets. In *Structural Integrity and Failure*; IntechOpen: London, UK, 2021. [CrossRef]
64. Baek, S.H.; Lee, K.I.; Kim, S.J.; Lee, G.; Kim, S.M. Methodology for evaluating behavior of reinforced concrete slabs in temporary traffic bridge systems over uncured cement concrete pavements using small-scale experimental slabs. *Materials* **2026**, *19*, 1302. [CrossRef] [PubMed]
65. Kadyrov, A.S.; Ganyukov, A.A.; Amanbayev, S.S.; Bogdanova, A.A. Development of mobile communal overpasses applied during repair of urban utility networks. *Mater. Mech. Eng. Technol.* **2023**, *3*, 11–14.
66. Kadyrov, A.; Ganyukov, A.; Balabekova, K. Development of constructions of mobile road overpasses. In *Proceedings of the MATEC Web of Conferences*; EDP Sciences: Les Ulis, France, 2017; Volume 108, p. 16002. [CrossRef]
67. Kövesdi, B.; Kollár, D.; Dunai, L. Temporary structural health monitoring of historical Széchenyi Chain Bridge. *Buildings* **2024**, *14*, 535. [CrossRef]
68. Schwietering, C.; Feldges, M. Improving traffic flow at long-term roadworks. *Transp. Res. Procedia.* **2016**, *15*, 267–282. [CrossRef]
69. Hou, G.; Chen, S. Study of work zone traffic safety under adverse driving conditions with a microscopic traffic simulation approach. *Accid. Anal. Prev.* **2020**, *145*, 105698. [CrossRef] [PubMed]
70. Xu, Z.; Chen, D. Detection method for all types of traffic conflicts in work zones. *Sustainability* **2022**, *14*, 14159. [CrossRef]
71. Marzouk, M.; Elsayed, A. Framework for assessing bridges construction impact on work zone traffic using BrIM. *Sci. Rep.* **2024**, *14*, 83. [CrossRef]
72. Brusselaers, N.; Fredriksson, A.; Gundlegård, D.; Zernis, R. Decision support for improved construction traffic management and planning. *Sustain. Cities Soc.* **2024**, *104*, 105305. [CrossRef]
73. Vyas, T.; Varia, H.R. Impact of construction work zone on urban traffic environment. *Int. J. Smart Sens. Intell. Syst.* **2023**, *16*, 5. [CrossRef]
74. Gong, H.; Peng, Q.; Liu, L.; Yang, X. A decision-making system for traffic management during large-scale road network construction. *Expert Syst. Appl.* **2025**, *292*, 128527. [CrossRef]
75. Ramadan, O.E.; Sisiopiku, V.P. Evaluation of merge control strategies at interstate work zones under peak and off-peak traffic conditions. *J. Transp. Technol.* **2016**, *6*, 118–130. [CrossRef]
76. Technical Regulation of the Customs Union TR CU 018/2011 "On the Safety of Wheeled Vehicles". Available online: https://eec.eaeunion.org/comission/department/deptexreg/tr/bezopKolesnTrS.php (accessed on 1 January 2015).
77. Resolution of the Government of the Republic of Kazakhstan Dated 13 November 2014 No. 1196 (Abolished by Resolution of the Government of the Republic of Kazakhstan Dated 1 September 2023 No. 758) "On approval of the Traffic Rules, the Basic Provisions for the Admission of Vehicles to Operation, and the List of Operational and Special Services Requiring Vehicles to Be Equipped with Special Light and Sound Signals and Special Color Schemes". Available online: https://adilet.zan.kz/eng/docs/P1400001196 (accessed on 1 September 2023).
78. GOST 2349-75. Traction Coupling Devices of Automobiles and Road Trains. Available online: https://www.gostrf.com/normadata/1/4294830/4294830586.pdf (accessed on 1 January 2006).
79. Schroeder, L.D.; Sjoquist, D.L.; Stephan, P.E. *Understanding Regression Analysis: An Introductory Guide*; Sage Publications: Thousand Oaks, CA, USA, 2016; Volume 57.
80. Tyagi, K.; Rane, C.; Manry, M. Regression analysis. In *Artificial Intelligence and Machine Learning for EDGE Computing*; Academic Press: Cambridge, MA, USA, 2022; pp. 53–63, 943. [CrossRef]

https://doi.org/10.3390/app16083879

81. Gressman, P.T. Damping oscillatory integrals by the Hessian determinant via Schrödinger. *Math. Res. Lett.* **2016**, *23*, 405–430. [CrossRef]
82. Gholipour, G.; Zhang, C.; Mousavi, A.A. Effects of axial load on nonlinear response of RC columns subjected to lateral impact load: Ship–pier collision. *Eng. Fail. Anal.* **2018**, *91*, 397–418. [CrossRef]

https://doi.org/10.3390/app16083879

applied sciences

MDPI

Article

Optimization of Mobile Overpass Support Placement Considering the Nonlinear Properties of the Soil Foundation

Alexandr Ganyukov, Adil Kadyrov *, Aliya Kukesheva *, Aidar Zhumabekov, Kirill Sinelnikov, Sabit Amanbayev and Akbope Karsakova

Department of Transport Engineering and Logistics Systems, Abylkas Saginov Karaganda Technical University, 56 Nursultan Nazarbayev Avenue, Karaganda 100027, Kazakhstan; severchell@gmail.com (A.G.); aidar7478@gmail.com (A.Z.); coolzero7777@gmail.com (K.S.); director@kao-a.kz (S.A.); akbope321@gmail.com (A.K.)

* Correspondence: adil.suratovich@gmail.com (A.K.); alewa878@gmail.com (A.K.)

Abstract

This study addresses the problem of traffic congestion in large cities caused by long-term repairs of underground utility networks. An innovative mobile overpass is considered, which combines the functions of a vehicle and a temporary bridge, allowing passenger cars up to 3.5 t to pass directly over repair trenches without detours. The research focuses on optimizing the placement of overpass supports relative to the trench edge to reduce soil deformation and prevent trench wall instability. A numerical methodology is developed in ANSYS Workbench that integrates finite element analysis of the soil-support system with parametric optimization using the nonlinear Drucker–Prager elastoplastic model. The soil parameters are obtained from oedometer compression tests (KPr-1M) and direct shear tests (PSG-2M) on clayey soils and then used to calibrate the numerical model. The optimization results show that the optimal distance from the trench wall to the overpass support is $L_{\min} = 2.78$ m, which is 13.5% greater than the initial design value. This modification reduces the maximum horizontal displacement of the trench wall by more than a factor of two and ensures compliance with the displacement criteria. Comparison between experimental and numerical compression curves yields an average deviation of 37.55%, with errors below 5% at higher stress levels, confirming that the Drucker–Prager model is suitable for engineering optimization of mobile overpass support placement on similar soils. The proposed methodology can be applied to the design and verification of temporary bridge systems operating above utility trenches in urban environments.

Keywords: mobile overpass; temporary bridge; utility trench; soil–structure interaction; Drucker–Prager model; finite element method; parametric optimization; urban traffic management

1. Introduction

Modern megacities face the challenge of traffic congestion caused by increasing vehicle density, road accidents, infrastructure repairs [1], infrastructure deterioration, limited capacity, adverse weather conditions, and mass events [2]. This poses a threat to the sustainability of the transport system, resulting in significant social, environmental, and economic consequences [3]. Traffic jams increase travel time, raise stress levels, and reduce quality of life [4], and also cause economic losses averaging 1.5% of GDP across the EU [5], the USA [6], Japan [7], and ASEAN countries [8]. Modern solutions include intelligent transportation systems (ITS), big data analytics, and infrastructure optimization [9]. In addition to the general causes of congestion, a major contributing factor is roadwork related to the repair and modernization of underground utility networks, including those for heating, water, cable, and sewage [10]. In Kazakhstan, the seven-month heating season depends on combined heat and power plants (CHPs), whose thermal pipelines are laid beneath city roads. Excavation of repair trenches disrupts traffic flows, overloading streets and reducing throughput—especially where utility networks are suboptimally located [11,12]. Similar issues are reported in other countries, such as Bangalore (India), where utility replacement narrows roads and creates chronic gridlock [13].

One promising solution is the use of mobile overpasses [14,15]. A mobile overpass is a temporary bridge structure that combines the functions of a vehicle and a bridge. The structure is transported to the work site, deployed over a trench using concrete supports, and enables passenger vehicles to pass directly over excavations without detours; after repairs, it is folded back into transport mode.

A global review shows that such structures are almost absent in civil urban transport construction. Similar designs exist primarily in military engineering, such as tank bridge-layers and tracked folding bridges [16], as well as modular Bailey bridges and their derivatives, including Mabey and Acrow bridges [17,18]. Deployable bridge systems, such as the portable ERE Logistics beam bridge, can be hydraulically installed without heavy machinery [19]. Lederman et al. developed a roll-up arch bridge mounted on an armored vehicle for rapid deployment in military or rescue operations [20]. The mobile ASTRA bridge system, operated by the Swiss Federal Roads Office (FEDRO), allows highway repairs without traffic closure [21].

Analysis of existing analogues reveals that the proposed mobile overpass, which combines the functions of a bridge and a vehicle, offers unmatched mobility and rapid deployment—superior to prefabricated bridges that require heavy equipment or costly tank bridge-layers. Made of carbon steel S-245 (analogous to S245) with a reliability factor of 0.95 [22,23], it operates within a temperature range of −45 °C to +55 °C and has a service life exceeding 15 years with regular maintenance. Designed in LIRA-SAPR [24] in accordance with Kazakh and Eurocode standards [25], the structure is optimized for urban conditions.

Implementing mobile overpasses in real-world conditions requires studying their behavior under different operational modes. A key research objective is to analyze the interaction between overpass supports and the soil foundation in bridge mode [26,27], which is necessary to prevent trench wall collapse and ensure safe operation. This interaction must be modeled to optimize support placement, accounting for soil properties, stress–strain state, structural geometry, boundary conditions, and traffic loads. The underlying hypothesis is that such optimization is achievable.

To the authors' knowledge, no studies have addressed the optimization of mobile overpass support placement near utility trenches using experimentally calibrated nonlinear soil models and parametric finite-element optimization.

The novelty of this work lies in integrating experimentally calibrated Drucker–Prager soil parameters with a parametric FEM optimization framework in ANSYS to determine the optimal placement of mobile overpass supports near utility trenches—an approach not reported in previous studies. Therefore, this study develops a methodology for optimizing support placement to prevent trench wall instability and ensure reliable mobile overpass operation.

The developed optimization methodology accounts for the nonlinear behavior of the soil foundation [28,29] based on the Drucker–Prager model [30,31], is founded on parametric optimization techniques [32,33], and employs finite element numerical modeling using the ANSYS Academic Research, 2025 (R2) package. The methodology is further

https://doi.org/10.3390/app16042075

validated through experimental verification, which ensures the reliability and structural stability during operation.

The proposed methodology represents a systematic integration of experimentally calibrated soil parameters, parametric finite element optimization, and a practical engineering operating scenario of the mobile overpass. The suggested approach provides a well-founded assessment of the interaction within the "structure–soil" system and enables optimal support placement under varying geotechnical operating conditions.

2. Materials and Methods

2.1. Modeling the Interaction Between Overpass Supports and the Soil Foundation and Formulation of the Optimal Placement Problem

In the operational position, the overpass rests on two reinforced concrete supports located on both sides of the trench, forming the "overpass support–soil foundation" system. The supports act as entry ramps to the overpass (Figure 1, transmitting vertical loads from the structure's weight and vehicle traffic (Figure 2, positions 4 and 12). These loads cause soil compression (Figure 2, position 10) and horizontal deformations of the trench wall, which may lead to wall collapse (Figure 2, position 8) or soil displacement. A reduced distance between the supports and the trench edge amplifies these effects, whereas increasing this distance increases the structural span, thereby raising the weight and overall construction costs [34,35].

(**a**) (**b**)

Figure 1. Sketch 3D Model of the Mobile Overpass: (**a**) Transport Position (Trailer Coupled to the Tractor), (**b**) Operational Position (Bridge Mode).

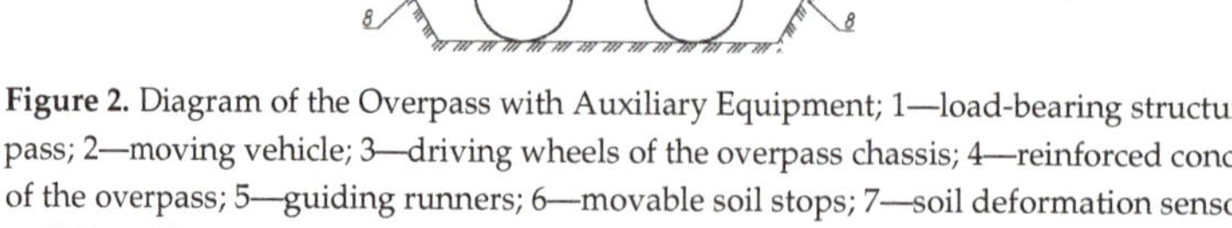

Figure 2. Diagram of the Overpass with Auxiliary Equipment; 1—load-bearing structure of the overpass; 2—moving vehicle; 3—driving wheels of the overpass chassis; 4—reinforced concrete supports of the overpass; 5—guiding runners; 6—movable soil stops; 7—soil deformation sensors; 8—trench wall; 9—utility networks; 10—soil foundation; 11—light and sound signal display; 12—vertical and horizontal load on the soil foundation.

The design (initial) distance between the supports and the trench wall edge is $L_p = 2.45$ m.

The condition of the soil mass is defined by its deformation and physico-mechanical parameters: deformation modulus E, cohesion C, Poisson's ratio ν, internal friction angle

https://doi.org/10.3390/app16042075

φ, and dilatancy angle ψ. The parameter ranges are relatively broad, since the overpass may operate on different types of soils, excluding structurally unstable (collapsible) soils that change their structure under relatively small external loads [36].

The distributed vertical load transmitted to the soil from the overpass support (reinforced-concrete ramp) is determined by the following expression:

$$q = (P_a + R_p)/l_a = 135.5kN/m, \tag{1}$$

where P_a = 33 kN is the load from the reinforced-concrete ramp; R_p = 265 kN is the load from the overpass structure, including the weight of the structure and the total load from moving vehicles; l_a = 2.2 m is the ramp length.

The horizontal load caused by vehicle braking or traction forces can be neglected, as it is insignificant given the low travel speed on the overpass (not exceeding 15 km/h).

The stiffness condition of the trench wall—considered as the displacement-limiting condition for u_{xi} of the vertical wall CD for different cases of overpass support placement—can be defined as [37,38]:

$$max(u_{x_i})|CD \leq [u_x], \tag{2}$$

where $\max(u_{x_i})$ | CD is the maximum horizontal soil displacement at the trench wall CD; $[u_x]$ is the allowable horizontal displacement of the trench wall boundary CD, which can be defined as $[u_x] = [h/\ (500 \div 1000)]$, where h is the height of the trench wall. Expression (2) specifies that the maximum horizontal soil displacements along the trench wall CD must not exceed the established limit values.

This condition serves as the basis for optimizing support placement. Accordingly, the optimization problem is defined as follows: to determine the optimal minimum distance L_{min} from the trench wall CD to the overpass support that ensures the minimal horizontal displacements u_{xi} of the trench wall CD among all possible configurations (2), taking into account the support load (1) acting on the soil, its physico-mechanical properties, the geometry of the computational domain, boundary conditions, and the stress–strain state (SSS) of the soil mass.

Thus, the optimization criterion is the minimum distance L_{min} of the support from the trench wall edge at which the stiffness condition (2) is satisfied, sufficient to prevent its collapse.

Conditionally, the objective function for optimizing the placement of supports can be represented as a multidimensional function P, depending on several parameters mentioned above:

$$P = P(q, L_{np}, E, C, \phi, \psi, GeometryoftheDomain,\ Stress - StrainState(SSS)oftheMass). \tag{3}$$

The stiffness condition of the wall (2) and the boundary conditions of the computational domain—EF, BA, and AF (Figure 3)—act as constraints of the objective function (3 Solving this objective function determines the optimum location of the support L_{min}, corresponding to the minimum of the function $L_{min} \rightarrow$ minP.

Depending on the given parameters and constraints, a single optimal value of the parameter L_{min} is obtained. When any parameter of function (3) changes, the optimal point L_{min} also shifts.

The accuracy of soil behavior prediction depends primarily on the deformation model used. Among known nonlinear criteria are the Coulomb–Mohr, Lade–Duncan, Matsuoka–Nakai, Drucker–Prager, Duncan–Chang, Schwer–Murray, and von Mises models, describing the SSS and failure behavior under plastic deformation [39–42]. The Coulomb–Mohr

https://doi.org/10.3390/app16042075

model, while widely used, does not account for nonlinear volumetric compression behavior, and its angular yield surface complicates numerical 3D analysis, often leading to predictions that are close to linear elastic behavior.

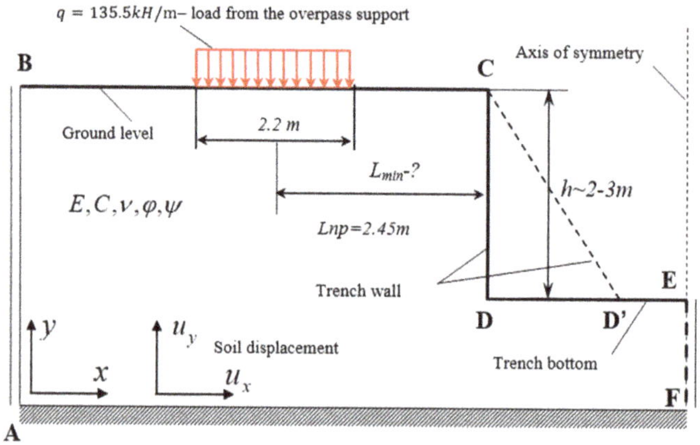

Figure 3. Computational Scheme of the Soil Foundation.: BC—ground level with the support positioned on it; CD—vertical trench wall; DE—trench bottom; EF—axis of symmetry of the domain and a sliding boundary condition; AB—sliding boundary of the computational domain; AF—fixed (rigid) boundary. The stress–strain state (SSS) of the modeled domain corresponds to plane strain conditions.

The Drucker–Prager model eliminates these drawbacks by employing a smoothed conical yield surface defined by C and φ, thereby improving numerical stability and providing a more accurate description of nonlinear SSS.

2.2. Parametric Optimization Using ANSYS Workbench

An analytical solution to the optimization problem for the objective function (3) is complicated due to its high complexity; however, numerical methods of parametric optimization implemented in ANSYS Workbench (WB) [43,44] provide an effective way to solve it through virtual experimentation and automated analysis [45,46]. This approach enables variation in key model parameters (such as dimensions or material properties) and the evaluation of their impact on target characteristics—such as strength or, in the present case, deformation [47].

In WB, the process begins by creating parametric geometry in DesignModeler or by importing a CAD model, during which the variables to be optimized are defined. The subsequent steps include generating a design of experiments, constructing response surfaces, and performing a numerical search for the optimum using built-in optimization algorithms [48,49]. The Drucker–Prager (DP) model is integrated into WB to describe the soil's nonlinear behavior, enabling determination of the soil's stress–strain state (SSS) under loads transmitted from the overpass supports using the finite element method (FEM). Based on the obtained SSS, WB's parametric optimization tools solve the optimal support placement problem, yielding more accurate positioning results.

Thus, the optimization procedure searches for the value of L_{min} that minimizes horizontal trench wall displacements while satisfying the constraints on soil properties, geometry, and boundary conditions [50] In brief, the workflow consists of: (i) defining geometry, material models, boundary conditions, and loads; (ii) selecting design and response parameters; (iii) planning and running the virtual experiment; (iv) constructing response surfaces; and (v) performing optimization and verifying the obtained solution [51].

https://doi.org/10.3390/app16042075

At the first stage, the stress–strain state (SSS) analysis was performed using the physico-mechanical characteristics of clay loam soils, which are predominant in the Karaganda region of Kazakhstan (Table 1).

Table 1. Physico-Mechanical Characteristics of the Soil.

Name	Accepted Values
Elastic modulus, E	23.544 MPa
Poisson's Ratio, ν	0.35
Cohesion, C	20.601 kPa
Internal Friction Angle, φ	18°
Dilatancy angle, ψ	2.5°

According to Figure 4, the design (computational) scheme of the soil and its finite-element approximation in ANSYS Workbench are shown in the figure.

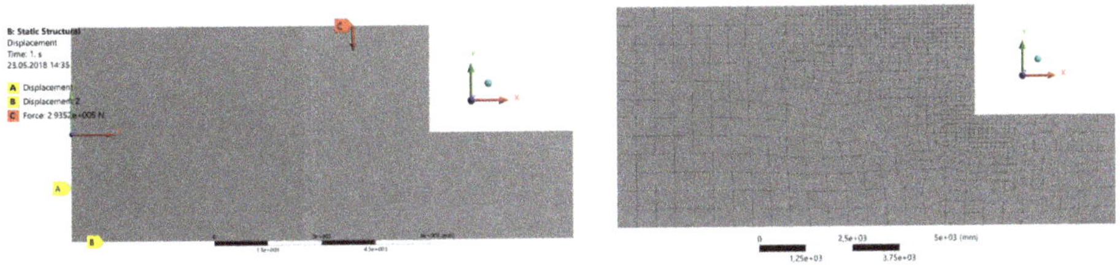

(**a**) (**b**)

Figure 4. Soil computational model in ANSYS Workbench: (**a**) Design (computational) scheme, (**b**) Finite element approximation.

For the approximation of the computational scheme of the soil foundation (Figure 4), a two-dimensional eight-node quadrilateral finite element PLANE82 [52] was used. The element is defined by eight nodes, each having two degrees of freedom: displacements along the global x and y coordinate axes. This element is suitable for nonlinear elasto-plastic problems, accounting for stiffness variation, large displacements, large strains, and orthotropy in 2D modeling.

The total number of finite elements was selected automatically based on an optimal mesh generation criterion. Mesh refinement was applied in the contact zone between the support and the soil mass, as well as near the trench wall, in order to improve the accuracy of stress and displacement determination. The model consists of 975 elements, 1023 nodes, and 2046 degrees of freedom. Loading applied to the computational model was defined according to expression (1). The geometry and boundary conditions were specified in accordance with Figure 3 and are described in Section 2.1.

2.3. *Justification of the Objective and Tasks of the Experiment*

The studies on optimizing the placement of overpass supports were based on a virtual numerical experiment in ANSYS Workbench (WB) using the Drucker–Prager model. The calculation methodology developed using this approach applies to various soil types but requires experimental verification of its reliability. Therefore, an experimental validation of the "overpass support–soil foundation" system was conducted to confirm the applicability of the adopted soil model.

The purpose of this experimental study is to verify the nonlinear Drucker–Prager soil deformation model implemented in ANSYS by comparing it with experimental data, thereby justifying its use in the engineering methodology for calculating and designing mobile overpasses.

https://doi.org/10.3390/app16042075

The research tasks include:

1. Describing the experimental setup and equipment;
2. Preparing for testing;
3. Conducting tests and processing results;
4. Performing mathematical modeling in ANSYS;
5. Carrying out a comparative analysis of field tests and numerical modeling results in ANSYS.

All experimental works were carried out using certified instruments (stabilometers): KPr-1M for compression testing [53] and PSG-2M for direct shear testing [54], both manufactured by *Pribor* (Uglich, Russia). The tests were performed in accordance with GOST 12248-2010, "Soils—laboratory methods for determining strength and deformability characteristics" [55]. Statistical data processing was performed in compliance with GOST 20522-2012 "Soils. Statistical processing methods for test results" [56].

The experiment consisted of two parts—physical and mathematical modeling—each including three stages:

Stage 1: Using the compression method on the KPr-1M device, soil deformations and settlements, along with parameters such as the modulus of deformation (E) and Poisson's ratio (ν), were determined.

Stage 2: Using the single-plane shear method on the PSG-2M device, strength parameters—cohesion (C) and internal friction angle (φ)—were recorded.

Tasks 1–3 were performed separately for each test type.

Stage 3: Soil compression was simulated in ANSYS using the Drucker–Prager model with the experimentally obtained parameters (E, ν, C, φ). A comparative analysis of settlement and deformation curves was then performed between the experimental and simulated results.

2.3.1. Experimental Studies of Soil Compression

Experimental investigations to determine soil deformations were carried out using the KPr-1M compression apparatus, provided by the certified laboratory of Karaganda GIIZiK LLP (Karaganda, Kazakhstan).

For testing, soil samples were taken either undisturbed from monoliths (for natural soils) or reconstituted in the laboratory (for fill or artificial soils Each specimen had a cylindrical shape with a diameter of 70–100 mm and a height equal to 1.5–2 times the diameter, in accordance with the standard. For this experiment, a clay soil sample with an undisturbed structure and natural moisture content was used.

The sample parameters are presented in Tables 2 and 3 and illustrated in Figure 5.

The load was gradually applied to the specimen (in stages).

The magnitude of each stage depended on the soil type and the test's purpose, typically ranging from 0.05 to 0.5 MPa.

Deformations were measured with high accuracy (typically up to 0.01 mm) using dial indicators and electronic sensors installed on the apparatus.

Table 2. Data on the Tested Soil Sample.

Soil Type	Soil Condition	Sample Height, mm	Sample Diameter, mm	Sample Cross-Sectional Area, mm^2	Sample Volume, mm^3
Clay	At Natural Moisture Content	h = 25	d = 87.43	S = 6000	V = 15,000

https://doi.org/10.3390/app16042075

Table 3. Physical Characteristics of the Soil.

Parameter	Value
Moisture, W,%	31.00
Soil Density, ρ, g/cm^3	1.92
Dry Soil Density, ρ_d d, g/cm^3	1.47
Soil Particle Density, ρ_s g/cm^3	2.77
Porosity, n, dimensionless units	0.47
Void Ratio, E	0.89
Degree of Saturation, Sr, dimensionless units	0.96
Liquid Limit, WL, %	77.00
Plastic Limit, WP, %	39.00
Plasticity Index, Ip, %	38.00
Coefficient Accounting for the Absence of Lateral Soil Expansion, β	0.4

Figure 5. Soil Sample under Compression in the KPr-1M Apparatus.

For each loading stage, the following parameters were calculated [55,56]:

- Absolute stabilized vertical deformation of the soil specimen (Δh, mm)—the arithmetic mean of the instrument readings after subtracting the correction for the deformation of the compression apparatus;
- Relative vertical deformation of the soil specimen.

$$\varepsilon_i = \Delta h_i / h, \tag{4}$$

where h—the height of the tested specimen;

- The soil porosity coefficient e_i is calculated using the formula:

$$e_i = e_0 - \varepsilon_i(1 + e_0), \tag{5}$$

where e_0 is the initial porosity coefficient in the absence of load.

Based on the calculated values, a graph of $\varepsilon = f(p)$ or $\Delta h = f(p)$ was plotted, known as the compression curve. A smooth average curve was drawn through the plotted points. The compressibility coefficient m_0, $MPa^{-1,}$ at each loading stage from p_i to p_{i+1} as calculated with an accuracy of 0.001 MPa^{-1} according to the formula:

$$m_0 = \frac{e_i - e_{i+1}}{p_{i+1} - p_i}, \tag{6}$$

where e_i and e_{i+1} are the void ratios corresponding to the pressures p_i and p_{i+1}.

https://doi.org/10.3390/app16042075

The oedometer modulus of deformation E_{oed} and the deformation modulus obtained from compression tests E_k within the given pressure interval Δp were calculated with an accuracy of 0.1 MPa according to formulas [55,56]:

$$E_{oed} = \Delta p / \Delta \varepsilon, E_k = E_{oed} \cdot \beta, \quad (7)$$

where $\Delta\varepsilon$ is the change in relative strain corresponding to Δp.

β is a coefficient accounting for the absence of lateral soil expansion in the compression device; in the absence of experimental data, it is allowed to take β = 0.4.

2.3.2. Experimental Study of the Direct Shear Test of Soil

Experimental investigations of soil strength characteristics were carried out using the PSG-2M direct shear apparatus, provided by the certified laboratory Karaganda GIIz & K LLP (Karaganda, Kazakhstan). The PSG-2M device is designed to determine soil shear strength parameters such as the internal friction angle φ and cohesion C in accordance with GOST 12248-2010 [55].

Soil specimens were placed in the ring mold and shaped as cylinders with a diameter of at least 70 mm and a height equal to one-third to one-half of the diameter. The maximum particle size (inclusions, aggregates) in the sample did not exceed one-fifth of its height.

Undisturbed natural clay soil specimens were prepared in a cylindrical form with the following geometric parameters: height h = 29 mm, diameter d = 71.4 mm, cross-sectional area S = 4000 mm^2, and volume V = 116,055.1 mm^3.

The soil's physical characteristics, including the liquidity index, served as the basis for selecting the normal pressure steps in accordance with the recommendations of GOST 12248-2010 [55]. The tested clay specimen is shown in Figure 6.

Figure 6. Tested soil specimen under shear.

The tests were conducted sequentially at three vertical (normal) stress levels, σ_n: 0.1 MPa, 0.3 MPa, and 0.5 MPa, which were selected for clay soils with $I_L \leq 0.5$ based on the tabulated data of the standard GOST 12248-2010 [55].

For each stress level, the specimen was placed in the shear box, where a vertical load was applied in a single stage through a system of weights and levers, ensuring a uniform pressure distribution over the sample surface. Immediately after applying σ_n, the horizontal shear mechanism was activated. The shearing load was applied statically in discrete steps, each not exceeding 10% of the current load value, with intervals of 10–15 s between steps.

The total shearing time was limited to 2 min, from the start of loading, to minimize the influence of consolidation processes.

During the test, key parameters were continuously recorded—horizontal displacement (l) (mm), shear stress τ (MPa), and elapsed time—which made it possible to track deformation dynamics in real-time.

The test was considered complete when one of two criteria was met: either a sudden failure (sliding of one half of the specimen relative to the other) or the attainment of 10% relative deformation, whichever occurred first.

The ultimate shear strength was defined as the maximum τ recorded during the test.

Based on the measured values of horizontal shear and normal loads, the shear and normal stresses τ and σ_n (MPa) were calculated according to the formulas provided in [55]:

$$\tau = 10\frac{Q}{A}, \sigma = 10\frac{P}{A}, \tag{8}$$

where Q and P are the horizontal shear and normal forces acting on the shear plane, respectively, kN; A—sample area, cm^2.

Determination of τ should be performed for at least three different values of P.

Based on the measured shear deformation values l corresponding to different stress levels, a dependence graph $l = f(\tau)$ was constructed.

The ultimate shear strength of the soil was taken as the maximum value of τ obtained from the shear diagram at the segment l_k, where the relative deformation does not exceed 10%.

From the obtained data, the dependence $\tau = f(\sigma)$ [55] was plotted.

The internal friction angle φ and cohesion C are determined as the parameters of the linear relationship between τ and σ.

$$\tau = \sigma \cdot tg\phi + c, \tag{9}$$

where τ and σ are determined according to expressions (8).

The internal friction angle φ and cohesion C were calculated using expressions (10), obtained by processing the experimental data points of the $\tau = f(\sigma)$ relationship with the least squares method, or determined graphically by drawing the best-fit straight line through the experimental points [55,56].

$$tg\phi = \frac{n\sum \tau_i\sigma_i - \sum \tau_i \sum \sigma_i}{n\sum (\sigma_i)^2 - (\sum \sigma_i)^2}, C = \frac{\sum \tau_i \sum \sigma_i^2 - \sum \sigma_i \sum \tau_i\sigma_i}{n\sum (\sigma_i)^2 - (\sum \sigma_i)^2}, \tag{10}$$

where τ_n are the experimental values of shear resistance determined at different normal stresses σ_n, and n is the number of tests.

2.3.3. Simulation of Soil Compression in ANSYS

Numerical modeling of clay soil compression in ANSYS WorkBench was carried out using the Drucker–Prager elastoplastic model, based on experimentally determined parameters: deformation modulus E, Poisson's ratio ν, internal friction angle φ, cohesion C, and dilatancy angle ψ.

The soil compresses without lateral expansion in the tested cylindrical mold, resulting in no radial deformations. Therefore, the computational scheme was implemented under plane strain conditions, reproducing the tests conducted with the KPr-1M apparatus, with the following specimen geometry: height h = 25 mm, diameter d = 87.43 mm, and loading stages P ranging from 0.05 to 0.5 MPa. These loads were converted into the equivalent linear load $q = P \cdot d$, N/m (corresponding to the midsection of the specimen along its diameter d, side BC) for all loading stages.

The computational model of the soil specimen simulating compression in ANSYS is shown in Figure 7.

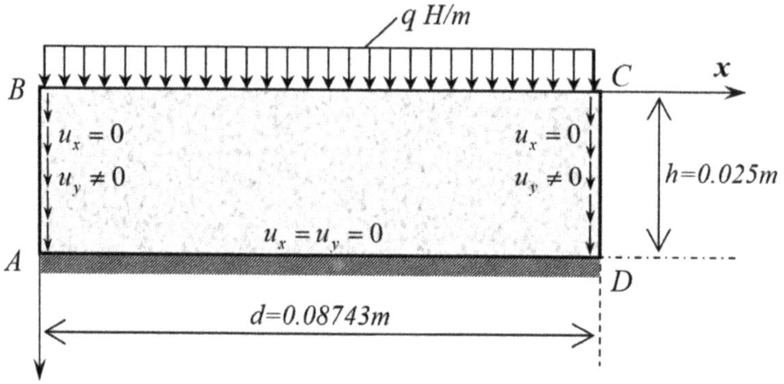

Figure 7. Computational model of the soil specimen in ANSYS. Boundary conditions: sides AB and CD—sliding; side AD—fixed (rigid constraint).

The soil model was defined as the Drucker–Prager elastoplastic model. To apply this model, it was necessary to specify the soil parameters obtained experimentally (see Section 3.3): internal friction angle $\varphi = 7.13°$, cohesion $C = 0.11$ MPa, deformation modulus E, Poisson's ratio ν, and dilatancy angle ψ.

According to GOST 12248-2010 [55], in the absence of experimental data, Poisson's ratio ν for clay soils was taken depending on the liquidity index I_l: 0,2–0,3 for $I_l < 0,5$.

The liquidity index I_l was determined from the expression based on the parameters of the tested soil as given in Table 2 [55,56]:

$$I_L = (W - W_p)/W_L - W_P = -0.21, \quad (11)$$

where $I_p = W_L - W_p$ is the plasticity index, equal to the difference between the liquid limit W_L and the plastic limit W_p.

As seen from the calculations, $I_L < 0$; therefore, ν (Poisson's ratio) was set to 0.3 as the upper limit, since the tested soil was non-moistened and had a lower density.

The deformation modulus E used in the numerical model was obtained from the oedometer modulus E_{oe}d, derived from compression test data and adjusted for the selected Poisson's ratio:

$$E = E_{oed}\nu = 40 \cdot 0.3 = 12\ MPa. \quad (12)$$

For clay soils, the dilatancy angle ψ was assumed to be $\psi = 2.5°$ in the calculations. An eight-node rectangular finite element (PLANE82) was selected to ensure higher computational accuracy [57].

The simulation was performed by applying load steps corresponding to the equivalent linear load q, as specified in Table 8.

3. Results

3.1. Parametric Optimization Solution

According to the parametric optimization algorithm implemented in ANSYS WorkBench, the first stage involved calculating the stress–strain state (SSS) of the soil mass under the effect of the bridge support (Figure 3). The results include stress and displacement fields, as well as zones of equivalent plastic deformation within the soil foundation. The subsequent optimization was performed in ANSYS DesignXplorer using the screening method

https://doi.org/10.3390/app16042075

to identify key influencing parameters, a central composite design (CCD) of experiments, the construction of response surfaces, and the formation of an optimization model aimed at minimizing the distance between the bridge support and the trench edge, subject to constraints on horizontal displacements, as defined by expressions (2) and (3). Figures 8–10 present the isofields of normal and shear stresses (σ_x, σ_y, τ_{xy}) along the corresponding axes.

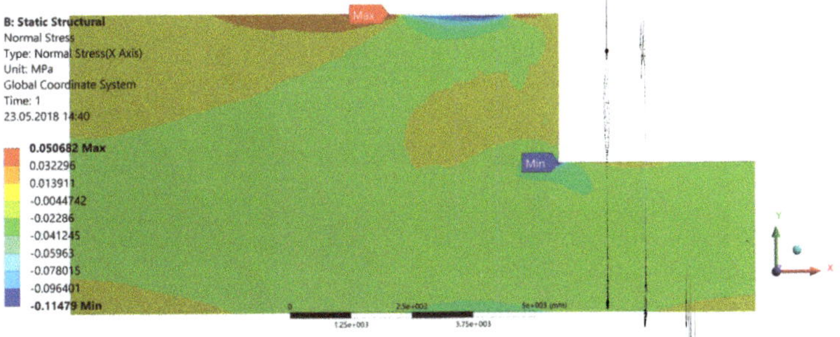

Figure 8. Isofield of normal stresses σ_x, MPa.

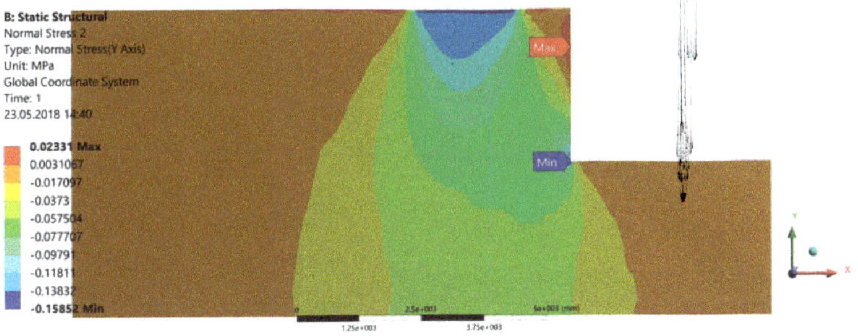

Figure 9. Isofield of normal stresses σ_y, MPa.

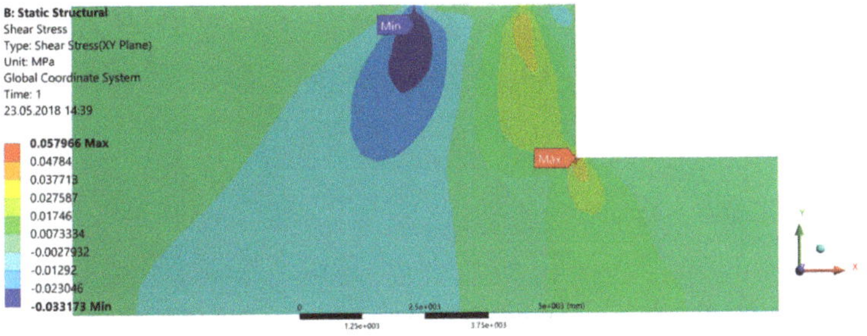

Figure 10. Isofield of shear stresses τ_{xy}, MPa.

Analysis of Figures 9 and 10 shows that the principal maximum stresses by absolute value are concentrated directly beneath the bridge support, which is especially characteristic for vertical compressive stresses σ_y.

Figures 11–13 illustrate the isofields of displacements (u_x, u_y) and the distribution zones of equivalent plastic strains (ε^p_{ed}) in the soil mass under the influence of the bridge support.

The analysis of Figure 11 demonstrates that the largest horizontal displacements of soil particles occur along the vertical trench wall, which may later cause its collapse during bridge operation. The most significant vertical displacements are observed in the area directly beneath the bridge support and gradually decrease with depth and distance from this zone.

https://doi.org/10.3390/app16042075

The plastic deformation zones of the soil, as expected, are located at shallow depths beneath the bridge support (Figure 13).

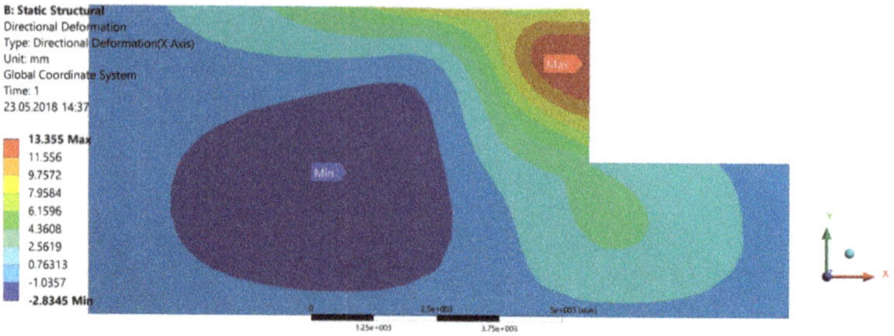

Figure 11. Isofield of horizontal displacements u_x, mm.

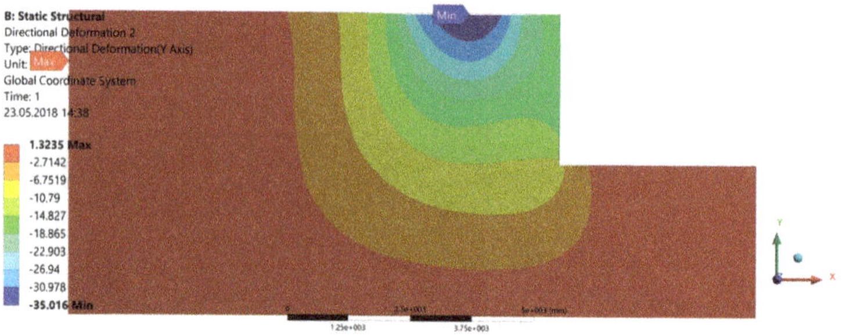

Figure 12. Isofield of vertical displacements u_y, mm.

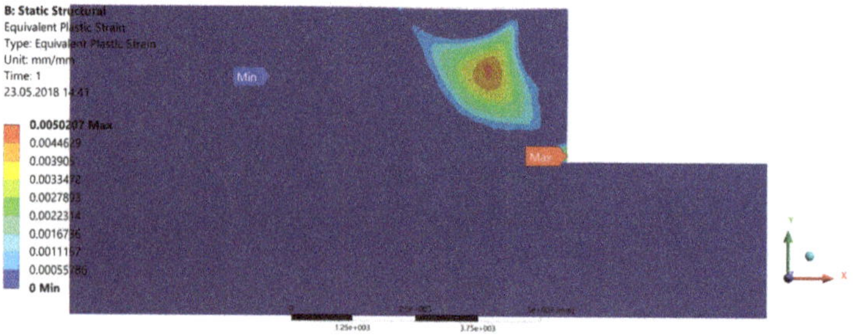

Figure 13. Zone of equivalent plastic deformations ε^p_{ed}.

After calculating the stress–strain state (SSS) of the soil mass, the second stage of the parametric optimization algorithm was performed, which involved determining the optimization parameters and planning the virtual experiment. This stage was conducted using ANSYS DesignXplorer (DX), a specialized module integrated into ANSYS WorkBench for performing multi-parameter optimization and parametric studies. As the optimization method, the built-in Screening Optimization Method [57,58] was used. This is one of the simplest methods, based on sampling and sorting, that supports multiple parameters, constraints, and input data types. The input optimization parameter, serving as the optimization criterion, was the minimum distance of the bridge support from the trench edge (L_{min} In DX). This parameter was designated as P1 in the experiment scheme tab "Outline of Schematic D2: Design of Experiments." Considering the change in the support position, the variable parameter was labeled P1-x. The default initial distance between the support and the trench edge was set to 1.0 m, with an upper boundary of 3.4 m. The output parameters acting as constraints included the maximum allowable horizontal displacement of the

https://doi.org/10.3390/app16042075

trench wall (P2), determined according to boundary condition (2), as well as the calculated stress–strain state (SSS) of the soil model. A central composite design (CCD) [59–61] was selected for two factors: P1 (L_{min}), the minimum distance of the bridge support from the trench edge, and P2 ($\max(u_{x_i})\,|\,CD$), the maximum deformation of the trench wall CD (see expression (2). The system automatically generated five design points (Table 4), which were used for calculations after determining the soil mass's stress–strain state at each iteration of the optimization process.

Table 4. Outline A2: Design points of the central composite design plan.

Name of Point	P1-x (m)	P2—Directional Deformation Maximum (mm)
1	2.2	6.3794
2	1	13.355
3	3.4	5.1637
4	1.6	7.7809
5	2.8	5.4981

Next, the third stage of the parametric optimization algorithm was carried out, which involved constructing the response line (Figure 4). In this case, for the two-factor design, the response surface represented a function of one variable, $P_2 = f(P_1)$ The response surface approximates the output parameter values across the entire analyzed parameter space, eliminating the need for complete calculations at every point. In ANSYS DesignXplorer (DX), several types of response surfaces (or lines) are available, including second-order polynomials [62], Kriging [63], nonparametric regression [64], neural networks, and sparse grid models [65]. The quality of the approximation was evaluated using the coefficient of determination (R^2), root-mean-square error (RMSE), and relative errors [66–68]. The corresponding results are presented in Table 5.

Table 5. Accuracy assessment of the distribution of design points on the response surface.

Accuracy Assessment	P2—Directional Deformation Maximum
Coefficient of Determination R^2 (Best Value = 1)	
Learning Points	1
Cross-Validation on Learning Points	1
Root Mean Square Error (Best Value = 0)	
Learning Points	2.6207×10^{-7}
Cross-Validation on Learning Points	8.5555×10^{-7}
Relative Maximum Absolute Error (Best Value = 0%)	
Learning Points	0
Cross-Validation on Learning Points	3.8653×10^{-5}
Relative Average Absolute Error (Best Value = 0%)	
Learning Points	0
Cross-Validation on Learning Points	2.3295×10^{-5}

The metrics in Table 5 indicate an almost perfect approximation ($R^2 = 1$, very low errors of the order 10^{-5}–10^{-7}). Therefore, a second-order polynomial response line, $P_2 = f(P_1)$, was adopted for further optimization. Based on these results, a response curve was constructed (Figure 14), illustrating the dependence of the horizontal displacement parameter of the trench wall (CD) on the position of the bridge support relative to the trench edge (CD) (Figure 3), considering the stress–strain state (SSS) of the soil mass modeled using the Drucker–Prager approach.

https://doi.org/10.3390/app16042075

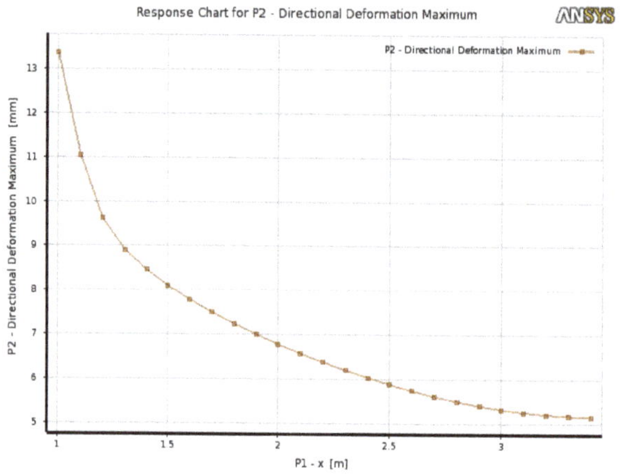

Figure 14. Response curve of horizontal displacement of the trench wall CD depending on the bridge support position.

The response curve is monotonically decreasing and nonlinear, and is approximated by a second-order polynomial with high accuracy, as confirmed by the metrics in Table 5. The coefficient of determination R^2 = 1 indicates perfect agreement between the model and the data, while the minimal errors ensure high prediction reliability. The decrease in P_2 from approximately 13 mm (at the minimum support position P_1 = 1 m from the edge) to about 5.1 mm (at the maximum P_1 = 3.4 m) reflects clear physical logic: increasing the distance between the bridge support and the trench edge enhances the trench wall's stability, reducing its horizontal displacement.

Next, the fourth stage of the parametric optimization algorithm was performed—the creation of the optimization model. This involved defining the optimization objective, constraints, and parameter admissible ranges (Table 6).

Table 6. Conditions of the studied optimization parameter.

Name	Parameter	Objective	Constraint	Lower Bound	Upper Bound
Minimize P1	P1-x	Minimize	No Constraint		
Minimize P2; 5.82 mm ≤ P2 ≤ 6.18 mm	P2—Directional Deformation Maximum	Minimize	Lower Bound ≤ Values ≤ Upper Bound	5.82	6.18

The problem defined the conditions for minimizing the horizontal displacement of the trench wall CD (Figure 3) within the range of 5.82–6.18 mm (Table 6), determined according to condition (2) based on the preliminary calculation of the stress–strain state (SSS) of the soil mass (Figure 11).

According to condition (2), $[u_x] = [h/(500 \div 1000)]$, where h is the height of the trench wall. In the present study, the maximum value of the trench depth was adopted as h = 3 m = 3000 mm (Figure 3). Therefore, $[u_x] = 3000/500 = 6$ mm. By selecting an admissible maximum deviation from this value of $\pm 3\%$, the resulting range is (5.82–6.18 mm).

At the fifth stage of the parametric optimization algorithm, the optimization model was launched using the screening method. As a result of the computation, three optimal variants (candidate points) of the support position were obtained (Table 7).

Table 7. Obtained optimal variants (candidate points).

Name	P1-x (m)		P2—Directional Deformation Maximum (mm)	
	Parameter Value	Variation from Reference	Parameter Value	Variation from Reference
Candidate Point 1	2.4196	0.00%	6	0.00%
Candidate Point 2	2.518	4.07%	5.8506	−2.49%
Candidate Point 3	2.6164	8.13%	5.7146	−4.76%

The final solution selected is Candidate Point 1, as it satisfies the displacement constraint while preserving the baseline value of distance P1, without increasing the span and the associated structural and economic costs. The remaining options provide only a minor reduction in displacements but require an increased support installation distance, which reduces their practical feasibility.

Based on the optimal position of the bridge support (Candidate Point 1), isofields of horizontal displacements of the soil mass were constructed, calculated with consideration of the stress–strain state (SSS) (Figure 15), as well as correlation dependencies between the support installation distance L and the maximum horizontal displacement of the trench wall CD (Figure 16).

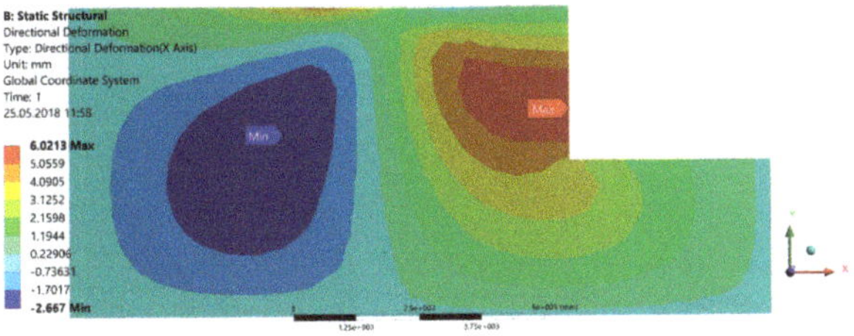

Figure 15. Candidate isofields of horizontal displacements ux, mm.

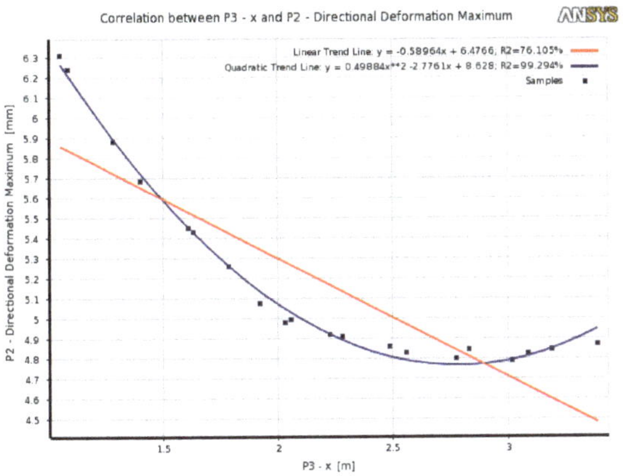

Figure 16. Dependence of the maximum horizontal displacement of the trench wall CD on the support position L.

By analyzing the soil displacement fields obtained for both the initial (design) support position (Figure 10) and the optimized configuration (Figure 14), it can be concluded that in both cases, the maximum displacements are concentrated in the zone of the vertical trench wall CD, which remains the most vulnerable area subject to potential collapse. However,

https://doi.org/10.3390/app16042075

after the optimization, the displacements of the trench wall CD, as well as those throughout most of the soil mass, decreased by more than half. This reduction significantly lowers the deformation intensity in the soil mass, thereby reducing the likelihood of trench wall failure during bridge operation.

Figure 15 presents the final results of the virtual optimization experiment—Samples (displayed as blue squares)—along with the quadratic (Quadratic Trend Line) and linear (Linear Trend Line) correlation curves. The determination coefficients (R^2) are shown as percentages, indicating the degree of agreement between the experimental Samples' output values and the corresponding correlation curves.

Each Sample value represents the maximum horizontal displacement of the vertical trench wall CD ($\max(u_{x_i}) \mid CD$), depending on the distance L from the bridge support to the trench edge CD, obtained from the stress–strain state (SSS) analysis of the soil using the nonlinear Drucker–Prager model. As shown by this relationship, it is possible to determine the minimum trench wall deformation and the corresponding optimal support position L_{min}.

The obtained correlation distributions of the Sample values define the law of variation of the maximum displacements of the vertical trench wall CD ($\max(u_{x_i}) \mid CD$) as a function of the bridge support placement L. The quadratic correlation dependence of the maximum displacements of the trench wall CD on the distance L from the support position is expressed as follows (Figure 16):

$$u_{xi,max} = 0.49884L_i^2 - 2,7761L_i + 8.628. \tag{13}$$

Linear correlation dependence (Figure 16):

$$u_{xi,max} - 0.58964L_i + 6.4766. \tag{14}$$

The determination coefficients for the quadratic and linear correlations are 99.3% and 76.12%, respectively. These values indicate a strong agreement between the output parameters of the virtual experiment and the fitted curve equations. The quadratic correlation demonstrates the highest accuracy of fit.

From Equation (13), the optimal minimum distance Lmin of the bridge support from the trench wall CD was determined, corresponding to the minimum value of the maximum horizontal displacement of the trench wall CD:

$$L_{min} = 2.78m, min(maxu_{xi}) = 4.77mm. \tag{15}$$

The design (initial) distance from the trench wall CD to the support was taken as L_{np} = 2.45 m. The difference between the design solution and the ANSYS DX optimization is 0.33 m, which is 13.5%. This distance L_{min} is a technological (constructability) factor and must be accounted for in the work execution plan. The calculation procedure for determining the optimal support location developed in this section is used in the overall engineering methodology for the mobile overpass when analyzing the "bridge support—soil foundation" system.

3.2. Results of Soil Compression Testing (Results from Section 2.3.1)

In the experimental study of soil compression using the KPr-1M oedometer, key deformation characteristics of undisturbed clay soil with natural moisture content were obtained. The tests were conducted in a stepwise manner, with applied pressures ranging from 0 to 0.5 MPa, allowing the determination of absolute and relative deformations, void ratios, compressibility coefficients, and deformation moduli.

https://doi.org/10.3390/app16042075

The collected data were processed in accordance with GOST 12248-2010 [55] and presented in both tabular and graphical form for comparison with the results of numerical modeling. Table 8 summarizes the parameters of the tested soil specimen at each loading step (0.1–0.5 MPa).

Table 8. Results of compression tests.

Load, MPa	Deformation, mm	Strain, mm	Compressive Deformation Modulus E_k, MPa	Void Ratio	Compressibility Coefficient, MPa^{-1}	Oedometer Modulus E_{oed}, MPa
0	0	0	0	0.89	0	0
0.050	0.23	0.009	2.222	0.873	0.340	5.556
0.100	0.40	0.016	2.941	0.860	0.257	7.353
0.200	0.50	0.020	10.000	0.853	0.076	25.000
0.300	0.57	0.023	14.286	0.847	0.053	35.714
0.400	0.60	0.024	28.571	0.845	0.026	71.429
0.500	0.65	0.026	22.222	0.841	0.034	55.556

From Table 8, using the iterative expressions (4)–(7), the following values were obtained for the final loading step of 0.5 MPa: compressibility coefficient m_0 = 0.047 MPa^{-1}; compressive deformation modulus E_k = 16 MPa; oedometer deformation modulus $E_{oe}d$ = 40 MPa.

Based on the values from Table 8, Figure 17 presents the experimental deformation curves for a single soil specimen. Since the specimen cannot fully return to its original state, a single representative point is taken.

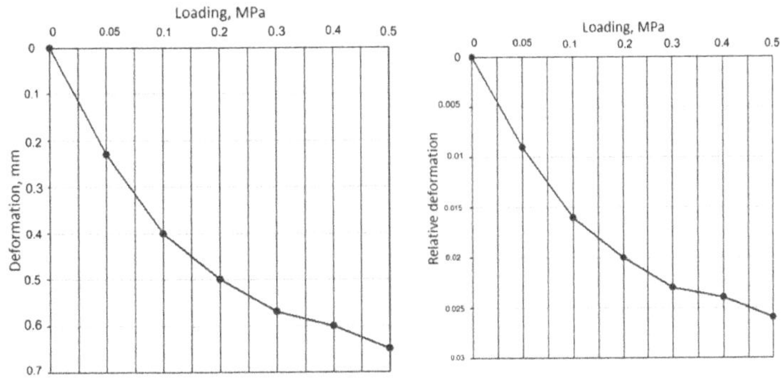

(**a**) (**b**)

Figure 17. Experimental deformation curves of the soil (**a**) deformation, (**b**) relative deformation.

The dependence of absolute deformation on load, Δh(P), shows a nonlinear increase in deformation with increasing load (Figure 17a). At low pressures (up to 0.1 MPa), deformation increases rapidly, indicating a high initial compressibility. Thereafter, the growth rate slows, reflecting soil densification and increased stiffness under load. The overall curve is concave downward, typical of plastic clays.

The dependence of relative deformation on load, ε(P), follows a similar pattern: the curve rises nonlinearly from 0 to 0.026 (Figure 17b). The steeper initial segment is followed by gradual flattening, indicating a reduction in deformation rate at higher loads. This behavior confirms the transition from elastic to plastic deformation, during which the soil densifies and resists further compression.

The obtained parameters ($E_{oe}d$ = 40 MPa, E_k = 16 MPa, m_0 = 0.047 MPa^{-1}, ν = 0.3, see Section 2.3.3) demonstrate pronounced nonlinear compressive behavior of the soil

https://doi.org/10.3390/app16042075

and justify their use in the ANSYS mathematical modeling and in the verification of the Drucker–Prager model, ensuring engineering calculation accuracy for the "bridge support–soil foundation" system.

3.3. Results of the Direct Shear Test of Soil (Results from Section 2.3.2)

Experimental studies of soil shear strength were conducted using the PSG-2M apparatus in the unconsolidated quick shear mode, allowing for the determination of key strength parameters of the clay soil, namely the internal friction angle (φ) and cohesion (C The tests were performed at three levels of everyday stress—0.1, 0.3, and 0.5 MPa—with continuous recording of shear deformation and shear stress over time.

The obtained data were processed using the least squares method in accordance with GOST 12248-2010 [55] and used to determine the shear strength parameters required for numerical modeling.

The results of the direct shear tests are presented in Tables 9–11.

Table 9. Results of the test at σ = 0.1 MPa.

Vertical Stress σ = 0.1 MPa		
Time from the Beginning of the Test, s	**Shear Deformation l, mm**	**Shear Stress τ, MPa**
122.0	0	0
122.0	0.14	0.005
141.6	0.26	0.01
202.1	0.29	0.015
211.6	0.32	0.02
273.5	0.34	0.025
290.2	0.37	0.03
354.8	0.4	0.035
374.1	0.44	0.04
438.1	0.47	0.045
483.4	0.55	0.05
618.3	0.68	0.055
656.4	0.78	0.06
726.8	0.86	0.065
788.5	0.95	0.07
863.5	1.03	0.075
919.2	1.08	0.08
995.8	1.18	0.085
1055.4	1.28	0.09
1163.6	1.4	0.095
1200.5	1.55	0.1
1294.5	1.78	0.105
1335.5	1.97	0.11
1459.7	2.1	0.115
1511.2	2.48	0.12
1615.5	7.1	0.125

In the last rows of Tables 9–11, the obtained values of shear stress τ correspond to the failure stage of the tested specimens—that is, the maximum shear load (ultimate shear strength. These results are summarized in Table 12.

Based on Table 8, the dependence τ = f(σ) was obtained and is shown in Figure 17.

According to expression (10) and the data from Table 12, the tangent of the internal friction angle was determined as tgφ = 0.125 and cohesion C = 0.11 MPa, from which φ = 7.13°. Thus, the required strength parameters of the tested specimen—internal friction angle and cohesion—were established.

https://doi.org/10.3390/app16042075

Table 10. Results of the test at σ = 0.3 MPa.

Vertical Stress σ = 0.3 MPa		
Time from the Beginning of the Test, s	**Shear Deformation l, mm**	**Shear Stress τ, MPa**
122.0	0	0
122.0	0.05	0.015
141.6	0.16	0.03
202.1	0.25	0.045
211.6	0.34	0.06
273.5	0.43	0.075
290.2	0.54	0.09
354.8	0.67	0.105
374.1	0.84	0.12
438.0	1.08	0.135
483.4	7.1	0.15

Table 11. Results of the test at σ = 0.5 MPa.

Vertical Stress σ = 0.5 MPa		
Time from the Beginning of the Test, s	**Shear Deformation l, mm**	**Shear Stress τ, MPa**
122.0	0	0
122.0	0.54	0.025
141.6	0.69	0.05
202.1	0.8	0.075
211.6	0.95	0.1
273.5	1.28	0.125
290.2	1.76	0.15
354.8	7.1	0.175

Table 12. Determined ultimate shear strengths.

Test Number	Vertical Stress, MPa	Shear Strength, MPa
1	$\sigma_1 = 0.1$	$\tau_1 = 0.13$
2	$\sigma_2 = 0.3$	$\tau_2 = 0.15$
3	$\sigma_3 = 0.5$	$\tau_3 = 0.18$

The obtained strength parameters (φ = 7.13° and C = 0.11 MPa) were derived from the ultimate shear strengths at normal stresses σ = 0.1, 0.3, and 0.5 MPa using the least squares method. The τ(σ) plot in Figure 18 exhibits a linear relationship, confirming the applicability of the Mohr–Coulomb model for this soil. The obtained values (φ and C) serve as input parameters for verifying the Drucker–Prager elastoplastic model in ANSYS, enabling more accurate modeling of the soil foundation behavior under shear loads and substantiating their use in engineering calculations for the "bridge support–soil foundation" system.

The consistency of the experimental results with regulatory standards confirms their suitability for numerical modeling and optimization of the mobile overpass design.

3.4. Results of Soil Compression Modeling in ANSYS (Results from Section 2.3.3)

The modeling results, which include vertical deformations (displacements) u_y and relative deformations ε_y along the loading boundary, were compared with the experimental data to assess the accuracy of the numerical model and its applicability for engineering analysis of the "bridge support–soil foundation" system.

https://doi.org/10.3390/app16042075

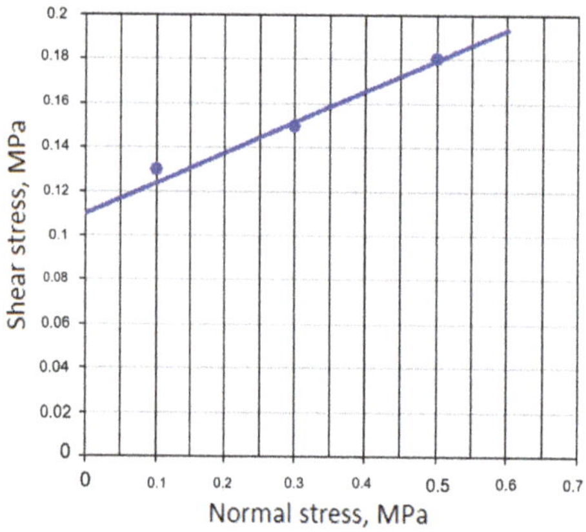

Figure 18. Experimental dependence of stresses $\tau = f(\sigma)$.

According to numerical simulation results, similar to the experimental studies (Section 3.2), the vertical displacements u_y and the relative deformations ε_y were obtained along the BC boundary (Figure 7). Along this line, the u_y deformations reach their maximum absolute values. In contrast, along the other sections parallel to BC, the u_y values decrease in magnitude.

The BC boundary was selected for validation, as deformation measurements along this line correspond directly to the experimental observations. The vertical relative deformation ε_y remains nearly constant throughout the specimen height, confirming a uniform stress distribution along the compression axis.

Figures 19 and 20 illustrate, respectively, the distribution of vertical displacements (in meters) and relative deformations at the final loading stage, corresponding to a load of q = 43.715 N/m. These results confirm the reliability of the Drucker–Prager soil model used in ANSYS for reproducing the experimental compression behavior of the clay soil.

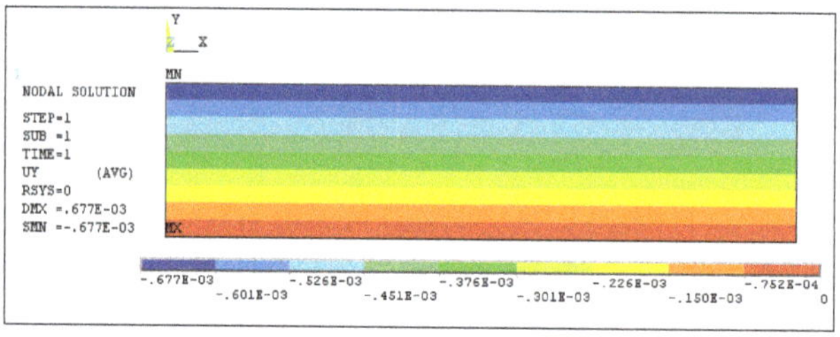

Figure 19. Vertical displacements u_y [m].

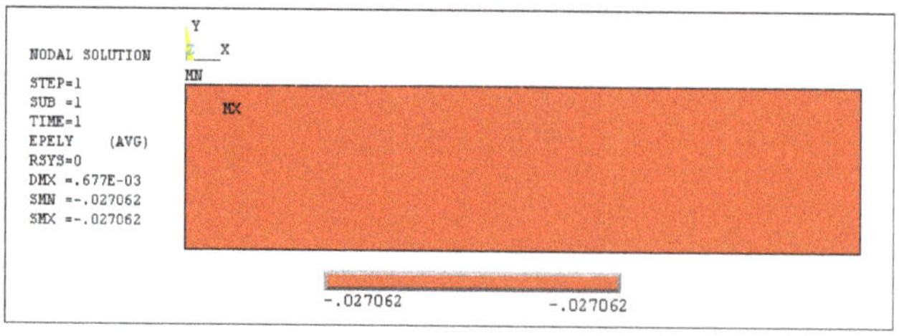

Figure 20. Relative deformations ε_y.

https://doi.org/10.3390/app16042075

3.5. Analysis of the Consistency Between Numerical Modeling in ANSYS and Experimental Test Results

A comparative analysis was carried out to evaluate the vertical deformations (u_y) and relative deformations (ε_y) obtained in the numerical simulation against the corresponding experimental results from compression tests. The comparison covered a load range of 0.05 to 0.5 MPa, allowing assessment of the numerical model's accuracy and validation of its applicability to the developed methodology for optimal bridge support placement using parametric optimization in ANSYS WorkBench (Sections 2.2 and 3.1).

The analysis of deviations across loading stages provides quantitative evidence of model accuracy and convergence between numerical and experimental results.

The comparative results with the experimental soil compression data (Section 3.2) are presented in Table 9 [68,69], while the graphical comparison is shown in Figure 20.

Analysis of Table 13 shows that at the initial loading stage, the deviation between experimental and numerical values reaches approximately 70%. As the load increases, this deviation decreases significantly, and at the final stages, it is just over 4%. At a load of 0.48 MPa, the experimental and numerical curves coincide, indicating complete agreement of the results (Figure 21).

Table 13. Comparison of ANSYS modeling results with experimental test data.

Loading q, N/m	Displacements, mm			Relative Deformations		
	Compression Test of the Specimen	ANSYS, Drucker–Prager Model	Deviation, %	Compression Test of the Specimen	ANSYS, Drucker–Prager Model	Deviation, %
4371.5	0.23	0.0677	70.6	0.009	0.002706	69.9
8743.0	0.40	0.135	66.2	0.016	0.005412	66.2
17,486	0.50	0.271	45.8	0.020	0.010825	45.9
26,229	0.57	0.406	28.8	0.023	0.016237	29.4
34,972	0.60	0.541	9.8	0.024	0.021649	9.8
43,715	0.65	0.677	4.2	0.026	0.027062	4.1

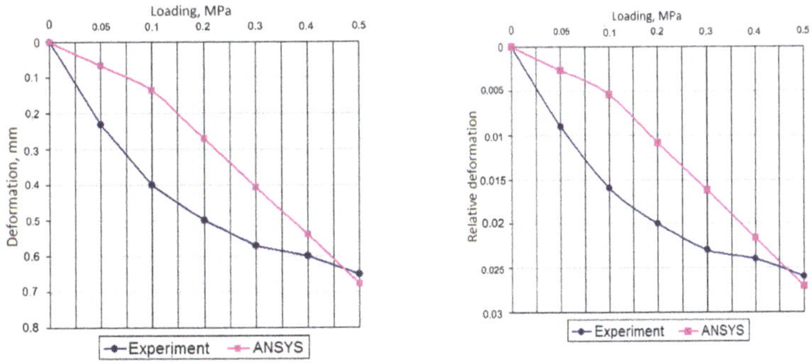

Figure 21. Comparison between experimental results and ANSYS modeling.

The significant deviation at the early stages is explained by the fact that the elastoplastic Drucker–Prager model behaves as an ideal elastic material under low loads. In contrast, the actual soil does not exhibit perfect elasticity. This is further confirmed by the deformation distribution patterns observed at the initial loading stages (Figure 21).

As the load increases, plastic potentials begin to act within the elastoplastic model, reflecting the nonlinear deformation behavior and leading to a closer convergence between calculated and experimental values. Such behavior is typical for any nonlinear soil deformability model based on an elastoplastic approach [70].

https://doi.org/10.3390/app16042075

The maximum absolute values of vertical stresses in the critical soil zone beneath the support are σ_y = 0.15 MPa (node 907, Figure 22). The average contact pressure is σ_{avr} = 0.138 MPa. The obtained stress level beneath the support does not fall within the low-stress region (≈0.05 MPa), where the maximum model error is observed, and corresponds to the range in which the discrepancy between experimental and numerical results does not exceed approximately 46% to 4% (Table 13, Figure 22).

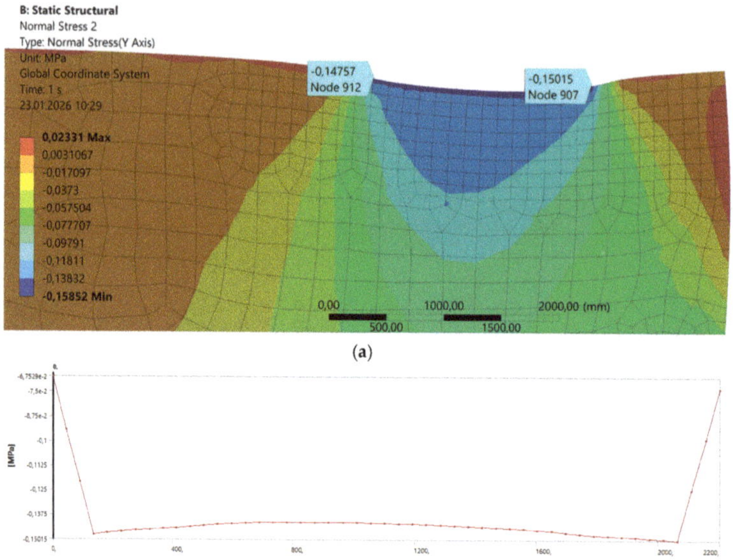

(a)

(b)

Figure 22. Distribution of vertical stresses in the soil beneath the support: (**a**) contour plot of vertical stresses σ_y obtained from finite element analysis; (**b**) diagram of vertical stress distribution σ_y along the contact zone between the support and the soil mass.

In addition, the finite element analysis results indicate that the maximum compressive stresses are localized directly beneath the support and are also outside the low-stress zone. That is, the operational stress level beneath the support does not correspond to the lowest loading stages. Local (peak) stresses beneath the edges of the support may exceed the average value (as evidenced by the stress contour plots beneath the support); therefore, the actual soil behavior partially occurs in the region of higher stresses. Consequently, the optimization results were obtained within the range of acceptable accuracy of numerical modeling.

The actual behavior of the structure under maximum external loading, as well as the stresses induced in the soil, generally correspond to the high-stress zone, in which the calculation error is considered acceptable (on the order of 46% to 4%). If this condition is not satisfied, it is necessary to introduce a safety correction factor k (≥1) into the optimization results to ensure additional reliability of the calculation. In this case, the minimum optimal support location is determined by the expression: $L'_{min} = k \cdot L_{min}$.

4. Discussion

In engineering practice, discrepancies between experimental and analytical results derived from classical linear-elastic soil mechanics can exceed 100%. To improve accuracy, advanced analysis methods and correction coefficients are introduced, resulting in better agreement with experimental data and sufficient precision for evaluating soil deformability.

https://doi.org/10.3390/app16042075

Deviations of 50–70% between analytical and experimental results are considered acceptable in soil mechanics, especially when safety coefficients for stability and strength are factored in [71].

In the present study, the average deviation between experimental data and ANSYS numerical modeling results was 37.55% (Table 9), which is significantly lower than that obtained with classical linear-theory methods. This level of accuracy is within the safety margins defined by Eurocode and other standards [72–74]. Hence, the nonlinear Drucker–Prager elastoplastic model shows good agreement with experimental results, consistent with other studies investigating similar models.

The developed calculation methodology is universal and intended for application not only to clay soils, but also to other soil types and a wide range of ground conditions. For each specific site of operation, the physical and mechanical properties of the soil are determined through core sampling followed by laboratory testing and subsequent calibration of the parameters. The experimentally obtained parameters (C, φ, E, ν, ψ) are then used as input data for numerical optimization for the given soil.

Thus, the optimal placement of supports is determined individually for each site, which ensures the correctness and reliability of the mobile overpass application under various geotechnical conditions.

Discrepancies between the experimental and numerical compression curves, especially in the low-stress range, are caused by several factors. First, the Drucker–Prager model has limitations in reproducing the nonlinear stiffness and structural sensitivity of clayey soils at small stress levels. Second, the boundary conditions and loading paths in laboratory tests differ from those adopted in the numerical model, which affects the initial portion of the deformation curve. Third, uncertainties in experimentally determined parameters are associated with sample heterogeneity, moisture variability, and parameter identification procedures.

Previous studies [75,76] report similar levels of agreement for elastoplastic soil models, with deviations up to 70% at low stress levels and improved accuracy at higher loads, emphasizing the importance of experimental calibration.

More advanced constitutive models, such as Hardening Soil and Hardening Soil Small [77,78], may further improve accuracy, particularly at low strain levels, but at the cost of increased model complexity and computational effort.

Recent studies employing advanced optimization and data-driven approaches [79–81] demonstrate improved accuracy; however, their implementation requires extensive experimental datasets and increased computational resources.

Despite its limitations at low stress levels, the Drucker–Prager model provides a reasonable balance between accuracy and computational efficiency for engineering applications involving soil–structure interaction.

Future work may refine the model through broader experimental validation and advanced calibration techniques.

In conclusion, the experimental verification of the Drucker–Prager elastoplastic model for the "bridge support–soil foundation" system confirms its scientific validity and engineering applicability. The model enables optimization of bridge support placement by considering key physical and mechanical soil properties, including deformability, strength, and stability. The achieved agreement with experimental data provides a reliable foundation for implementing this computational methodology in mobile bridge design.

The results obtained in this study are consistent with the findings of previous works in which the parameters of Drucker–Prager-type models are identified through numerical reproduction of laboratory tests followed by validation against experimental stress–strain curves [82,83]. Similar to [84], the parameter identification in the present work is effectively

https://doi.org/10.3390/app16042075

formulated as an optimization problem. Moreover, the sensitivity of approximation accuracy to the calibration procedure and parameter set, as reported in inverse analyses [85], confirms the necessity of explicit control over the sources of input parameters and their experimental justification.

The use of parametric optimization and response surface techniques is in line with modern approaches to inverse analysis and multi-criteria optimization of geotechnical systems [86]. While the consideration of parameter uncertainty and solution robustness, as discussed within Bayesian and surrogate-based frameworks [87–90], is only partially addressed in the present study, this limitation is mitigated by the fact that the operational stress levels lie outside the range of maximum model error, and the optimal solution exhibits low sensitivity to parameter variations. The application of screening and surrogate-based approaches is consistent with current trends aimed at reducing computational costs in finite element optimization [88].

The results also correlate with conclusions emphasizing the usefulness of monitoring data and parameter updating techniques to improve the reliability of predictive assessments [89], which is regarded as a promising direction for further development of the proposed methodologyx. Additional opportunities for extension are supported by studies on parameter inversion based on numerical data and machine learning methods [91], which may be employed for automated adaptation of the soil model during the operation of a mobile overpass under varying site conditions.

5. Conclusions

A scientifically grounded methodology has been developed to optimize mobile bridge support placement, preventing trench wall collapse and ensuring safe bridge operation, using parametric optimization in ANSYS Workbench. The study accounts for the nonlinear behavior of the soil foundation using the Drucker–Prager elastoplastic model, which provided a more accurate representation of the stress–strain state (SSS) of the soil, including its physical and mechanical parameters—deformation modulus, Poisson's ratio, internal friction angle, cohesion, and dilatancy angle. The optimization procedure aimed at minimizing horizontal displacements of trench walls, which constitute the primary failure criterion.

Numerical modeling of the interaction between the supports and the soil foundation was performed, accounting for both vertical and horizontal loads from the bridge structure and moving vehicles. It was established that supports located too close to trench walls significantly increase deformations, while excessive spacing enlarges the span and raises structural costs. The optimization determined an optimal minimum distance (L_{min} = 2.78 m), which is 13.5% greater than the original design value (L_{np} = 2.45 m). This adjustment reduced horizontal wall displacements by more than half, significantly decreasing the risk of wall failure and improving overall operational safety.

Correlation analysis revealed a pronounced quadratic relationship between the maximum trench wall displacements and the support distance, with a determination coefficient (R^2) of 99.3%. The screening optimization method in ANSYS DesignXplorer enabled selecting the optimal support configuration from numerous alternatives. These results provide a scientific and engineering basis for designing mobile bridge systems, offering an effective and reliable framework for optimizing their construction and ensuring safe operation. The proposed methodology can be adapted to different soil conditions, ensuring both structural reliability and versatility.

Experimental studies were conducted to verify the Drucker–Prager model and justify its use in the ANSYS WorkBench calculation methodology for optimal support placement. The tests included compression tests (KPr-1M apparatus) and direct shear tests (PSG-2M apparatus), as specified in GOST 12248-2010. The results obtained defined the deforma-

https://doi.org/10.3390/app16042075

tion parameters (deformation modulus and Poisson's ratio) and the strength parameters (cohesion and internal friction angle) of the clay soil.

A comparative analysis of the experimental and simulated data showed an average deviation of 37.55%, with a maximum of 70% at low loads and a minimum of 4% at higher loads. This confirms the high accuracy and applicability of the Drucker–Prager model for engineering calculations, surpassing classical linear models, which can yield discrepancies of up to 100%.

The obtained results have several limitations that should be considered when interpreting the conclusions. Experimental calibration and numerical verification were performed for clayey soils, which limits the direct transferability of the obtained parameters to other engineering–geological conditions. In addition, the Drucker–Prager model employed is characterized by increased uncertainty at low stress levels, while the numerical analyses are based on assumptions of material homogeneity, idealized boundary conditions, and a plane strain state, which may differ from real field conditions.

Therefore, prior to practical application of the methodology, it is necessary to conduct site-specific engineering–geological investigations, laboratory testing, and calibration of soil parameters for the particular location, followed by project-oriented optimization. Future research perspectives include extending the methodology to various soil types, applying more advanced constitutive models that account for small-strain behavior, and implementing sensitivity analysis and uncertainty quantification methods for determining the optimal distance L_{min}.

Author Contributions: Conceptualization, A.G. and A.K. (Adil Kadyrov); Methodology, A.G. and A.K. (Aliya Kukesheva); Software, K.S.; Validation, A.Z. and A.K. (Aliya Kukesheva); Formal analysis, A.K. (Adil Kadyrov), S.A. and A.Z.; Investigation, S.A., K.S., A.G. and A.K. (Akbope Karsakova); Data curation, A.K. (Aliya Kukesheva); Writing—original draft preparation, A.Z.; Writing—review and editing, A.G., A.K. (Adil Kadyrov), A.K. (Aliya Kukesheva), A.Z., K.S., S.A. and A.K. (Akbope Karsakova); Visualization, A.K. (Aliya Kukesheva); Supervision, A.G.; Project administration, A.Z. All authors have read and agreed to the published version of the manuscript.

Funding: This research has been/was/is funded by the Science Committee of the Ministry of Science and Higher Education of the Republic of Kazakhstan (Grant No. AP23487832).

Data Availability Statement: The original contributions presented in this study are included in the article. For further inquiries, please contact the corresponding authors.

Conflicts of Interest: The authors declare no conflicts of interest.

References

1. Saberi, M.; Hamedmoghadam, H.; Ashfaq, M.; Hosseini, S.A.; Gu, Z.; Shafiei, S.; González, M.C. A simple contagion process describes the spreading of traffic jams in urban networks. *Nat. Commun.* **2020**, *11*, 1616. [CrossRef]
2. Harris, N.; Shealy, T.; Klotz, L. Choice architecture as a way to encourage a whole systems design perspective for more sustainable infrastructure. *Sustainability* **2017**, *9*, 54. [CrossRef]
3. Falcocchio, J.C.; Levinson, H.S. The costs and other consequences of traffic congestion. In *Springer Tracts on Transportation and Traffic*; Springer: Cham, Switzerland, 2015; Volume 7, pp. 159–182. [CrossRef]
4. Serok, N.; Havlin, S.; Blumenfeld-Lieberthal, E. Identification, cost evaluation, and prioritization of urban traffic congestions and their origin. *Sci. Rep.* **2022**, *12*, 13026. [CrossRef] [PubMed]
5. European Commission. Reports on Urban Mobility and Congestion 2019–2023. Available online: https://transport.ec.europa.eu/transport-themes_en (accessed on 7 April 2025).
6. Texas A&M Transportation Institute. *Urban Mobility Report*; Texas A&M Transportation Institute: Austin, TX, USA, 2019. Available online: https://mobility.tamu.edu/umr/ (accessed on 9 April 2025).
7. Japan Institute for Labour Policy and Training. Reports on Urban Congestion and Economic Impacts 2010–2015. Available online: https://www.mlit.go.jp (accessed on 9 April 2025).

https://doi.org/10.3390/app16042075

8. Victoria, A.N.; Dela Cruz, O.G.; Muhi, M.M.; Bianes, J.R.A. A review on the analysis of evaluating traffic congestion. *Lect. Notes Civ. Eng.* **2024**, *466*, 1049–1058. [CrossRef]
9. Bellam, K.; Nagappan, K.; Sankaran, N.; Chowdhury, S.; Perakovic, D. *Urban Mobility and Challenges of Intelligent Transportation Systems*; IGI Global: Hershey, PA, USA, 2025; pp. 1–694. [CrossRef]
10. Hojjati, A.; Jefferson, I.; Metje, N.; Rogers, C.D. Sustainability assessment for urban underground utility infrastructure projects. *Proc. Inst. Civ. Eng.—Eng. Sustain.* **2017**, *171*, 68–80. [CrossRef]
11. Agilbayeva, M.; Kalinin, A. The development of the district heat supply system of the city of Karaganda. *E3S Web Conf.* **2019**, *135*, 01031. [CrossRef]
12. Silvestri, F.; Polimeni, A.; Belcore, O.M. Impact mitigation of tunnels maintenance works on highway performance. *Procedia Struct. Integr.* **2024**, *62*, 998–1005. [CrossRef]
13. Verma, A.; Anbazhagan, P.; Mayakuntla, S.K.; Bhat, F.A. Assessing the impact of underground utility works on road traffic and users: A study from an Indian city. In *International Conference on Transportation Infrastructure Projects: Conception to Execution*; Springer Nature: Singapore, 2022; pp. 243–259. [CrossRef]
14. Kadyrov, A.S.; Ganyukov, A.A.; Amanbayev, S.S.; Bogdanova, A.A. Development of mobile communal overpasses applied during repairing of urban communal networks. *Mater. Mech. Eng. Technol.* **2023**, *3*, 11–14. [CrossRef]
15. Ganyukov, A.A.; Sinelnikov, K.A.; Kabikenov, S.Z.; Karsakova, A.Z. Research and calculation of the deformed state of the roadway mobile overpass. *Mater. Mech. Eng. Technol.* **2024**, *3*, 88–95. [CrossRef]
16. ERE Systems Inc. *Corporate Profile and Technical Capabilities*; ERE Systems Inc.: Tokyo, Japan, 2021. Available online: https://erelogistics.com/wp-content/uploads/2021/11/ERE-Systems-Brochure-Final-2.pdf (accessed on 2 April 2025).
17. Prokop, J.; Odrobiňák, J.; Farbák, M.; Novotný, V. Load-carrying capacity of Bailey bridge in civil applications. *Appl. Sci.* **2022**, *12*, 3788. [CrossRef]
18. Russell, B.R.; Thrall, A.P. Portable and rapidly deployable bridges: Historical perspective and recent technology developments. *J. Bridge Eng.* **2013**, *18*, 1074–1085. [CrossRef]
19. ERE Logistics. *Flat Rack Portable Bridges*; ERE Logistics: Okotoks, AB, Canada, 2025. Available online: https://erelogistics.com/ (accessed on 2 April 2025).
20. Lederman, G.; You, Z.; Glišić, B. A novel deployable tied arch bridge. *Eng. Struct.* **2014**, *70*, 1–10. [CrossRef]
21. Federal Highway Administration. *ASTRA Bridge*; Federal Highway Administration: Washington, DC, USA, 2025. Available online: https://www.astra.admin.ch/astra/de/home/themen/nationalstrassen/baustellen/wissenswertes/astra-bridge.html (accessed on 5 March 2025).
22. Kang, W.H.; Hicks, S.J.; Uy, B.; Fussell, A. Design resistance evaluation for steel and steel–concrete composite members. *J. Constr. Steel Res.* **2018**, *147*, 523–548. [CrossRef]
23. Puklický, L. Time-dependent reliability analysis of stainless steel members and bridges. *AIP Conf. Proc.* **2023**, *2928*, 020013. [CrossRef]
24. Tatarchenko, G.O.; Chernih, O.A.; Sokolenko, V.M.; Tatarchenko, Z.S. Modern information technologies in system architecture—Urban planning—Building constructions. In *International Conference Building Innovations*; Springer International Publishing: Cham, Switzerland, 2019; pp. 483–490. [CrossRef]
25. Baláž, I.J.; Koleková, Y.P. New generation of Eurocodes. In *Production Management and Engineering Sciences: Proceedings of the International Conference on Engineering Science and Production Management (ESPM 2015)*; CRC Press: Boca Raton, FL, USA, 2016; pp. 345–350. [CrossRef]
26. Farouk, H.; Farouk, M. Effect of soil model on contact stress under strip footing. In *New Frontiers in Geotechnical Engineering*; ASCE: Reston, VA, USA, 2014; pp. 67–77. [CrossRef]
27. Ahmari, A.M.; Abdollahipour, A.; Kargar, A. Wellbore stability analysis using ground reaction curve and Mohr–Coulomb failure criterion. *J. Anal. Numer. Methods Min. Eng.* **2025**, *15*, 69–76. [CrossRef]
28. Fu, Y.; He, S.; Zhang, S.; Yang, Y. Parameter analysis on hardening soil model of soft soil for foundation pits based on shear rates in Shenzhen Bay, China. *Adv. Mater. Sci. Eng.* **2020**, *2020*, 7810918. [CrossRef]
29. Alsirawan, R.; Koch, E. Finite element modeling of rigid inclusion-supported embankment. *Pollack Period.* **2022**, *17*, 197–208. [CrossRef]
30. Zhang, T.; Zhou, X.P.; Qian, Q.H. Drucker–Prager plasticity model in the framework of OSB–PD theory with shear deformation. *Eng. Comput.* **2023**, *39*, 1395–1414. [CrossRef]
31. Liu, K.; Chen, S.L. Finite element implementation of strain-hardening Drucker–Prager plasticity model with application to tunnel excavation. *Undergr. Space* **2017**, *2*, 168–174. [CrossRef]
32. Weaver-Rosen, J.M.; Malak, R.J. An algorithm for multi-objective efficient parametric optimization. *J. Mech. Des.* **2023**, *145*, 031709. [CrossRef]
33. Galvan, E.; Malak, R.J.; Hartl, D.J.; Baur, J.W. Performance assessment of a multi-objective parametric optimization algorithm with application to a multi-physical engineering system. *Struct. Multidiscip. Optim.* **2018**, *58*, 489–509. [CrossRef]

https://doi.org/10.3390/app16042075

34. Chouw, N. Dynamics of bridge structures. In *Innovative Bridge Design Handbook*; Butterworth-Heinemann: Oxford, UK, 2022; pp. 141–168. [CrossRef]
35. Nowak, A.; Pipinato, A.; Stawska, S. Loads on bridges. In *Innovative Bridge Design Handbook*; Butterworth-Heinemann: Oxford, UK, 2022; pp. 21–45. [CrossRef]
36. Khitrov, E.G.; Andronov, A.V.; Martynov, B.G.; Spiridonov, S.V. Interrelations of various soil types mechanical properties. *J. Phys. Conf. Ser.* **2019**, *1177*, 012032. [CrossRef]
37. Cheng, K.; Xu, R.-Q.; Lin, C.-G.; Liang, R.-Z.; Shan, H.-F.; Gan, X.-L. Simplified method for predicting surrounding soil's horizontal displacement caused by arbitrary deflection of retaining wall. *J. Zhejiang Univ. (Eng. Sci.)* **2020**, *54*, 1012–1020. [CrossRef]
38. Wang, J.H.; Xu, Z.H.; Wang, W.D. Wall and ground movements due to deep excavations in Shanghai soft soils. *J. Geotech. Geoenvironmental Eng.* **2010**, *136*, 985–994. [CrossRef]
39. Lee, J.; Jin, H.; An, J.; Baek, Y.; Yoon, H. A study on the suitability of the Mohr–Coulomb model for numerical analysis of ground excavation. *J. Eng. Geol.* **2020**, *30*, 1–15. [CrossRef]
40. Liu, L.-Y.; Zhao, T.-B.; Wang, Q.-Y. Drucker–Prager models and its application in ANSYS and FLAC3D. *J. Northeast. Univ.* **2015**, *36*, 71–75. Available online: https://clk.li/iUAU (accessed on 12 July 2025).
41. Xu, X.; Dai, Z.H.; Chen, L.J.; Xu, X.T. Nonlinear elastic–perfectly plastic model of soil and its numerical implementation and application. *Soil. Mech. Found. Eng.* **2022**, *59*, 215–223. [CrossRef]
42. Liu, J.; Li, R.; Sun, P.; Wang, Z.; Luo, J.; Li, Z. Duncan–Chang nonlinear constitutive model based on joint strength theory of structural loess. *Chin. J. Geotech. Eng.* **2018**, *40*, 124–128. [CrossRef]
43. Weaver-Rosen, J.M.; Malak, R.J. Efficient parametric optimization for expensive single-objective problems. *J. Mech. Des.* **2021**, *143*, 031401. [CrossRef]
44. Tsai, Y.-K.; Malak, R.J. Surrogate-Assisted Constraint-Handling Technique for Parametric Multi-Objective Optimization. *Struct. Multidiscip. Optim.* **2024**, *67*, 38. [CrossRef]
45. Quliyev, S. Automated and automatic systems of management of a programs package for making optimal decisions. In *International Conference on Information Technologies and Their Applications*; Springer Nature: Cham, Switzerland, 2024; pp. 287–299. [CrossRef]
46. Belorit, M.; Hrček, S.; Bisták, M. Description of the methodology of the automated optimization process. In *Current Methods of Construction Design: Proceedings of the ICMD 2018*; Springer International Publishing: Cham, Switzerland, 2019; pp. 13–20. [CrossRef]
47. Fedorik, F.; Kala, J.; Haapala, A.; Malaska, M. Use of design optimization techniques in solving typical structural engineering related design optimization problems. *Struct. Eng. Mech.* **2015**, *55*, 1121–1137. [CrossRef]
48. Corriou, J.P. Numerical methods of optimization. In *Numerical Methods and Optimization: Theory and Practice for Engineers*; Springer International Publishing: Cham, Switzerland, 2021; pp. 505–574. [CrossRef]
49. Fajfar, I.; Puhan, J.; Bűrmen, Á. Evolving a Nelder–Mead algorithm for optimization with genetic programming. *Evol. Comput.* **2017**, *25*, 351–373. [CrossRef]
50. Ravichandran, K.; Masoudi, N.; Fadel, G.M.; Wiecek, M.M. Parametric optimization for structural design problems. In *International Design Engineering Technical Conferences and Computers and Information in Engineering Conference*; ASME: New York, NY, USA, 2019; Paper No. V02BT03A022. [CrossRef]
51. Khiste, S.G.; Kulkarni, P.; Katekar, S.D. Probabilistic design of composite leaf spring by using finite element method. *IOP Conf. Ser. Mater. Sci. Eng.* **2020**, *814*, 012019. [CrossRef]
52. Li, G. A four-node plane parametric element based on quadrilateral area coordinate and its application to coupled solid–deformation/fluid–flow simulation for porous geomaterials. *Int. J. Numer. Anal. Methods Geomech.* **2015**, *39*, 251–276. [CrossRef]
53. LLC Pribor. *Compression Device KPr-1M*; LLC Pribor: Moscow, Russia, 2017; Available online: https://clk.li/qhCa (accessed on 10 July 2025).
54. LLC Pribor. *Soil Shear Tester PSG-3M*; LLC Pribor: Moscow, Russia, 2017. Available online: https://clk.li/Escl (accessed on 10 July 2025).
55. *GOST 12248-2010*; Soils. Laboratory Methods for Determining the Strength and Strain Characteristics. Federal Agency for Technical Regulation and Metrology: Moscow, Russia, 2011. Available online: https://clk.li/CTNc (accessed on 12 July 2025).
56. *GOST 20522-2012*; Soils. Methods of Statistical Treatment of Test Results. Federal Agency for Technical Regulation and Metrology: Moscow, Russia, 2013. Available online: https://clk.li/DiDD (accessed on 12 July 2025).
57. Huang, Y.Q.; Yang, Y.F.; Wang, J.Z.; Liu, X.C.; Chen, H.B. Unsymmetric extensions of Wilson's incompatible four-node quadrilateral and eight-node hexahedral elements. *Int. J. Numer. Methods Eng.* **2022**, *123*, 101–127. [CrossRef]
58. Akhmatnurov, D.; Zamaliyev, N.; Mussin, R.; Demin, V.; Tolovkhan, B.; Ganyukov, N.; Skrzypkowski, K.; Korzeniowski, W.; Stasica, J.; Rak, Z. Geomechanical Modeling of the Northern Katpar Deposit (Kazakhstan): Assessing the Impact of Rock Mass Disturbance on Stability Safety Factor. *Mining* **2025**, *5*, 73. [CrossRef]

https://doi.org/10.3390/app16042075

59. Nichayarach, I.; Warisnoicharoen, W. A comparison of model fitting using different runs in Box–Behnken design for alginate film formulation. *Thai J. Pharm. Sci.* **2020**, *44*, 174–178. [CrossRef]
60. Lu, L.; Anderson-Cook, C.M.; Ahmed, T. Non-uniform space filling (NUSF) designs. *J. Qual. Technol.* **2021**, *53*, 309–330. [CrossRef]
61. Mee, R.; Eberhard, A. Response surface split-plot designs: A literature review. *Qual. Reliab. Eng. Int.* **2019**, *35*, 1645–1662. [CrossRef]
62. Kleijnen, J.P. Regression and Kriging metamodels with their experimental designs in simulation: A review. *Eur. J. Oper. Res.* **2017**, *256*, 1–16. [CrossRef]
63. Lucay, F.A.; Sales-Cruz, M.; Gálvez, E.D.; Cisternas, L.A. Modeling of the complex behavior through an improved response surface methodology. *Miner. Process. Extr. Metall. Rev.* **2021**, *42*, 285–311. [CrossRef]
64. Ghattas, B.; Manzon, D. Machine learning alternatives to response surface models. *Mathematics* **2023**, *11*, 3406. [CrossRef]
65. Cheng, C.L.; Garg, G. Coefficient of determination for multiple measurement error models. *J. Multivar. Anal.* **2015**, *126*, 137–152. [CrossRef]
66. Żądło, T. On asymmetry of prediction errors in small area estimation. *Stat. Transit.—New Ser.* **2017**, *18*, 413–432. [CrossRef]
67. Hodson, T.O. Root mean square error (RMSE) or mean absolute error (MAE): When to use them or not. *Geosci. Model. Dev.* **2022**, *15*, 5481–5487. [CrossRef]
68. Ganyukov, A. Research of operation of mobile overpass constructions. *Proc. Univ.* **2022**, *1*, 195–200. [CrossRef]
69. Wen, Y.; Yang, G.; Tang, L.; Zhong, Z.; Yao, J. Verification of non-coaxial constitutive model based on generalized potential theory. *J. Cent. South Univ. (Sci. Technol.)* **2017**, *48*, 1817–1823. [CrossRef]
70. Lei, Y.; Liu, H.; Lu, Z. Comparisons of two approaches for geotechnical model calibration with scarce data. *Adv. Civ. Eng.* **2021**, *2021*, 4245051. [CrossRef]
71. Tseshkovskaya, Y.; Tsoy, N.; Tseshkovskiy, V.; Ibrayev, M.; Zamaliyev, N.; Akhmatnurov, D.; Mussin, R.; Ganyukov, N.; Zakharov, A.; Skrzypkowski, K.; et al. Research Progress of Environmental Studies of a Mining Facility for Land Restoration (Using the Example of a Mining Enterprise in the Karaganda Region). *Resources* **2025**, *14*, 144. [CrossRef]
72. Song, E.X.; Kong, Y.F.; Yang, J. Discussion of safety factor definitions and computation methods for geotechnical structures. *Eng. Mech.* **2016**, *33*, 1–10.
73. Low, B.K. *Reliability-Based Design in Soil and Rock Engineering: Enhancing Partial Factor Design Approaches*; CRC Press: Boca Raton, FL, USA, 2021; pp. 1–399. [CrossRef]
74. Muizemněk, A.Y.; Boldyrev, G.G.; Aref'ev, D.V. Identification of soil model parameters. *Eng. Geol.* **2010**, *5*, 36–41. Available online: https://clk.li/Mfgt (accessed on 12 July 2025).
75. Boldyrev, G.G.; Idrisov, I.K.; Valeev, D.N. Determination of parameters for soil models. *Soil. Mech. Found. Eng.* **2006**, *43*, 101–108. [CrossRef]
76. Boldyrev, G.G.; Muyzemnels, A.J. The modeling of deformation process in soils with use of ANSYS and LS-DYNA programs. In *Proceedings of the International Conference on Case Histories in Geotechnical Engineering*; Missouri University of Science and Technology: Rolla, MO, USA, 2008; Paper No. 37. Available online: https://scholarsmine.mst.edu/cgi/viewcontent.cgi?article=3326&context=icchge (accessed on 12 July 2025).
77. Chen, Y.; Luo, M.M.; Xia, N.W.; He, P. Statistical analysis of existing test results of HSS model parameters for soft soils. *Chin. J. Geotech. Eng.* **2021**, *43*, 197–201. [CrossRef]
78. Wu, J.T.; Tung, C.Y.; Adams, M.T.; Nicks, J.E. Analysis of stress–deformation behavior of soil–geosynthetic composites in plane strain condition. *Transp. Infrastruct. Geotechnol.* **2018**, *5*, 210–230. [CrossRef]
79. Zhao, X.; Schorr, J.; Peña Olarte, A.A.; Cudmani, R. A step towards quantifying the uncertainty of the soil mechanical response through the use of genetic algorithms. In *Proceedings of the International Conference of the International Association for Computer Methods and Advances in Geomechanics*; Springer: Cham, Switzerland, 2022; pp. 263–271. [CrossRef]
80. Thapa, I.; Ghani, S. Applying optimized machine learning models for predicting unconfined compressive strength in fine-grained soil. *Transp. Infrastruct. Geotechnol.* **2024**, *11*, 2235–2269. [CrossRef]
81. Zhu, J.; Xu, R. Secondary development of modified Duncan–Chang model considering disturbance. *Chin. J. Geotech. Eng.* **2015**, *37*, 84–88. [CrossRef]
82. Hasanpouri Notash, N.; Dabiri, R.; Hajialilue Bonab, M.; Khodadadi, L.; Behrouz Sarand, F. A Finite Element Modeling of Drained Triaxial Test on Loose Sand Using Different Constitutive Models. *AUT J. Civ. Eng.* **2022**, *6*, 339–358. [CrossRef]
83. Okimura, Y.; Imamura, R.; Shimo, K.; Hanai, T.; Kato, Y.; Hashimoto, K.; Faiz, M.K.; Okane, T.; Miyashita, T.; Yoshida, M. Experimental Construction and Validation of Revised Drucker–Prager Model Using Finite Element Method for Moisture Condensation Zone in Bentonite-Bonded Silica Sand. *J. Mater. Eng. Perform.* **2024**, *33*, 9145–9162. [CrossRef]
84. Davarpanah, S.; Allili, M.; Mousavi Ajarostaghi, S.S. Numerical Optimization of Drucker–Prager–Cap Model Parameters in Powder Compaction Employing Particle Swarm Algorithms. *Fluids* **2024**, *9*, 262. [CrossRef]
85. Buljak, V.; Baivier-Romero, S.; Kallel, A. Calibration of Drucker–Prager Cap Constitutive Model for Ceramic Powder Compaction through Inverse Analysis. *Materials* **2021**, *14*, 4044. [CrossRef] [PubMed]

https://doi.org/10.3390/app16042075

86. Zhao, C.; Chen, L.; Ni, P.; Xia, W.; Wang, B. A Modified Back Analysis Method for Deep Excavation with Multi-Objective Optimization Procedure. *J. Rock Mech. Geotech. Eng.* **2024**, *16*, 1373–1387. [CrossRef]
87. Zhang, T.; Abdallah, A.; Cuisinier, O.; Masrouri, F. Polynomial Chaos Kriging-Assisted Reliability-Based Design Optimization for a Deep Mixing Wall. *Comput. Geotech.* **2025**, *179*, 107044. [CrossRef]
88. Ding, J.; Zhou, J.; Cai, W. An Efficient Variable Selection-Based Kriging Model Method for the Reliability Analysis of Slopes with Spatially Variable Soils. *Reliab. Eng. Syst. Saf.* **2023**, *235*, 109234. [CrossRef]
89. Chen, R.; Zhou, X.; Yu, M.; Wu, J.; Tao, Y.; Xue, F.; Zhang, Y. Bayesian Updating for Ground Surface Settlements of Shield Tunneling. *Front. Earth Sci.* **2023**, *11*, 1321883. [CrossRef]
90. Zhang, W.; El Naggar, M.H.; Ni, P.; Zhao, M.; Du, X. Bayesian Updating of Geotechnical Parameters with Polynomial Chaos Kriging Model and Gibbs Sampling. *Comput. Geotech.* **2025**, *180*, 107087. [CrossRef]
91. Liu, D.; Shen, X.; Pan, D. Soil Parameter Inversion Considering the Influence of Temperature Effects. *Appl. Sci.* **2025**, *15*, 12511. [CrossRef]

https://doi.org/10.3390/app16042075

applied sciences

MDPI

Article

Adaptive MEC–RBF Neural Network-Based Deflection Prediction for Prestressed Concrete Continuous Rigid Frame Bridges During Construction

Chunyu Zhou [1], Qingfei Gao [1,*], Qijun He [1], Liangbo Sun [2] and Dewei Tian [2]

[1] School of Transportation Science and Engineering, Harbin Institute of Technology, Harbin 150090, China; zhoucy060228@126.com (C.Z.); 24s032005@stu.hit.edu.cn (Q.H.)

[2] The First Company of China Eighth Engineering Division Ltd., Jinan 250014, China; sunlb1012@outlook.com (L.S.); 18644071076@163.com (D.T.)

* Correspondence: gaoqingfei@hit.edu.cn

Abstract: A deflection prediction approach based on an adaptive MEC–RBF neural network was developed in this study. By dynamically optimizing the centres, widths, and weights of the RBF network, the proposed method substantially increases the prediction accuracy, and it achieves an R^2 of 0.9789 and an *RMSE* of 1.4978 on the training dataset. It effectively resolves the stability challenges that are associated with nonlinear construction conditions in traditional models. An orthogonal experimental design analysis revealed that the girder block length and the cantilever-to-span length ratio (d/L) were the most influential factors that affected deflection, whereas the effects of uniformly distributed loads and temperature were negligible, thereby providing a sound basis for parameter simplification. The application of the model to the Hannan Yangtze River Bridge yielded a maximum discrepancy of only 5.56 mm (17.7% error rate) between the predicted and measured values, thus demonstrating its practical engineering reliability. By innovatively integrating intelligent optimization techniques with neural networks, this approach overcomes the limitations in terms of real-time responsiveness and long-term stability of conventional methods and offers an efficient and reliable technical tool for the control of large-scale bridge construction.

Keywords: deflection prediction model; adaptive MEC–RBF neural network; prestressed concrete continuous rigid frame bridge; construction stage

1. Introduction

In the global infrastructure construction sector, bridge engineering plays a crucial role as both a vital link that connects regional economies and an important indicator of a nation's civil engineering capabilities [1,2]. With sustained global economic development and accelerated urbanization, increasingly sophisticated demands have been placed on bridge span lengths, design, and construction technologies [3,4]. In particular, under complex geological and hydrological conditions, the demand for long-span bridges continues to increase annually. Owing to their excellent structural performance and adaptability, prestressed concrete continuous rigid frame bridges have become a mainstream choice in modern bridge engineering [5]. In China, bridge construction has achieved globally recognized successes and has resulted in numerous internationally influential bridge projects [6,7]. These successes are attributable to the nation's strong emphasis on and continuous investment in infrastructure development, as well as ongoing innovations in bridge engineering technology. Under the guidance of the "Strong Transportation Nation" strategy, China's bridge construction is advancing towards intelligent, green, and sustainable development [8].

In recent years, onsite dynamic identification and numerical model updating have been increasingly employed for monitoring the structural health of bridges. These approaches enable the continuous calibration of analytical models using measured modal frequencies and mode shapes, improving the accuracy of deformation and stress prediction. Such integration of field monitoring and simulation represents an essential link for reliable construction control. Relevant advances can be found in recent studies [9,10]. However, controlling deflection during the cantilever construction stage of prestressed concrete continuous rigid frame bridges remains a long-standing challenge in civil engineering. While cantilever construction technology effectively reduces formwork and falsework requirements, thereby reducing construction costs, structural deflections during construction are influenced by multiple factors and are difficult to predict accurately [11]. Insufficient control of the deflection of a bridge may compromise both the durability and safety of the bridge during service [12].

In bridge construction control and deformation adjustment, Zhou [13] studied prestressing camber regulation for long-span prestressed concrete continuous rigid frame bridges during cantilever construction by employing finite element methods and creep testing. The study confirmed that the maximum cantilever end deflection occurs near the mid-span and side-span closure regions under a concrete self-weight of 1.10γ and revealed that the proposed creep correction model improved the accuracy of predicting deformation and stress. Zhou [14] applied Bayesian updating to control the construction of a continuous rigid frame bridge with high piers and combined finite element models with field monitoring data to correct internal forces at control sections, which significantly reduced stress errors. Wan [15] developed a GM(1,1)-based grey-prediction scheme, calibrated with a finite-element model, to forecast the main-girder vertical deflection during staged tendon tensioning and cantilever erection of a rigid-frame–continuous girder bridge. Comparisons with field measurements revealed a maximum absolute error of 1.73 mm, underscoring the method's value for precamber setting and displacement control in construction monitoring.

Although these studies have made progress in terms of prestressing camber control, internal force correction, and prediction methods, the adaptability of these methods to complex construction conditions is limited. Considering the coupling effects of multiple factors under extreme environments remains insufficient in existing methods, which makes addressing comprehensive effects under special conditions such as large temperature differentials and complex geological settings difficult. Moreover, existing methods lack sufficient dynamic adjustment mechanisms for addressing scenarios that involve multiple dynamically varying parameters. Under strongly nonlinear or abrupt changes in construction conditions, the prediction stability and control capability are inadequate; thus, these methods fail to provide systematic solutions for complex and dynamic construction scenarios.

With respect to finite element model optimization, Dong et al. [16] updated continuous girder bridge models via a cubic response surface method in combination with environmental vibration testing and achieved frequency errors below 3% and a mode assurance criterion (MAC) value above 85%. Wang [17] compared response surface models of various orders and reported that while higher-order polynomials increase computational costs, accuracy improvements are limited, and they proposed a robust estimation method for suppressing error propagation. However, existing model optimization research faces clear trade-offs between computational efficiency and accuracy. Although high-order models can improve accuracy to some extent, the accompanying increase in the number of parameters leads to significantly higher computational costs, thus conflicting with economic requirements. Moreover, existing approaches are highly dependent on engineering conditions and sample quality, which limits their adaptability for large-scale complex bridges or insufficient sample sizes and makes maintaining stable optimization performance across diverse engineering scenarios challenging.

To address these challenges, the integration of advanced numerical analysis techniques with artificial intelligence algorithms for high-precision deflection prediction during bridge construction has become a focal research topic in civil engineering [18,19]. Zhu [20] developed a safety prediction framework for prestressed steel structures that is based on BP neural networks and digital twins and validated the good agreement between predicted and measured values through field verification. Li [21] proposed a hybrid GM–BP model that reduced the average relative error in camera prediction from 8.69% to 3.01%. Wang [22] presented a deflection prediction framework for wave-shaped steel web prestressed concrete rigid frame bridges that is based on an MEC–BP neural network and demonstrated superior performance over conventional BP models in controlling the deflection of the Liangqugou Bridge through MEC optimization.

While intelligent prediction technologies have achieved certain improvements in terms of accuracy, challenges remain regarding real-time capability and long-term stability [23]. Some models lack sufficient real-time mechanisms for coupling virtual structures and dynamic field environments, which results in inadequate speeds in response to abrupt changes in construction conditions [24]. Additionally, under long-term construction conditions, the ability to control cumulative errors is weak, and maintaining stability in long-term predictions is difficult, which results in the failure to fully meet the dual engineering demands for immediate dynamic response and reliable long-term forecasting [25].

This study focuses on deflection prediction during the cantilever construction stage of prestressed concrete continuous rigid frame bridges and aims to develop a prediction model that balances accuracy and efficiency. An MEC–RBF neural network that integrates adaptive MEC with an RBF is proposed for deflection prediction. To increase the prediction accuracy and generalization ability, the adaptive MEC algorithm is employed to globally optimize the centres, widths, and connection weights of the RBF network. Data samples are generated by using a three-dimensional finite element model to simulate the cantilever construction process, with the model accuracy validated against field tests. Latin hypercube sampling is applied to address sample sparsity in high-dimensional spaces. A sensitivity analysis that involves an orthogonal experimental design is conducted to identify the parameters that exert the greatest influence on the construction deflection. Model performance is assessed in terms of R^2, *RMSE*, and *MAE*. Finally, a case study on the Hannan Yangtze River Bridge is conducted to verify the effectiveness of the proposed method. The results of this study provide effective technical support for intelligent prediction and construction process control during long-span bridge construction stages. The remainder of this paper is organized as follows. Section 2 describes the proposed adaptive MEC–RBF framework and algorithmic design. Section 3 presents the case study setup, dataset preparation, model validation, and discussion of the results, while Section 4 concludes the study and outlines future research directions.

2. Deflection Prediction Method Based on an Adaptive MEC–RBF Neural Network

The main innovation of this study lies in the integration of adaptive mind evolutionary computation (MEC) with a radial basis function (RBF) neural network to enable real-time parameter evolution and self-tuning capability. Unlike conventional MEC–BP hybrids, the proposed approach dynamically adjusts the search variance and subgroup structure during optimization, resulting in greater convergence stability and robustness under nonstationary construction conditions.

2.1. RBF Neural Network

The radial basis function (RBF) neural network is a three-layer feedforward model that comprises input, hidden, and output layers. It addresses high-dimensional nonlinear approximation problems through mapping via nonlinear basis functions [26]. The core concept involves leveraging localized response characteristics to partition the input space into multiple subregions, with each hidden layer neuron responsible for function approximation within its respective local region [27]. A typical choice for the radial basis function is the Gaussian function. We let n, m, and l denote the numbers of neurons in the input, hidden, and output layers, respectively; the predicted output is then expressed by Equation (1).

$$o_k = \sum_{j=1}^{m} w_{kj} h_j = \sum_{j=1}^{m} w_{kj} \exp\left(- \frac{\left\| x - \kappa_j \right\|^2}{2\varsigma_j^{\ 2}} \right) \tag{1}$$

where x denotes the input vector, κ_j represents the centre of the radial basis function for the jth hidden layer neuron, ς_j represents the width of the radial basis function for the jth hidden layer neuron, and w_{kj} represents the weight that connects the jth hidden layer neuron to the kth output layer neuron. The cost function of the network output is given by Equation (2).

$$E = \frac{1}{2S} \sum_{i=1}^{S} (T_i - O_i)^2 \tag{2}$$

where S denotes the number of samples, T_i represents the target output from the experiments, and O_i denotes the predicted output.

As indicated in Equation (1), the model's accuracy depends on the hidden neuron count m, the basis function width, and the weights between the hidden and output layers. The values of the basis functions are influenced by their centres and widths, whereas the output is controlled by the connection weights. Improper selection of these parameters can directly impair network performance and lead to reduced convergence speed or even divergence [28]. To address this issue, in this study, an adaptive mind evolutionary computation (MEC) algorithm is employed to determine the parameters of the RBF neural network, and a deflection prediction method for long-span continuous girder bridge construction that is based on the adaptive MEC–RBF neural network is proposed.

2.2. Adaptive Mind Evolutionary Computation Algorithm

Mind evolutionary computation (MEC) is a swarm intelligence optimization framework that was inspired by cognitive science and involves the construction of optimization models by simulating the "convergence–divergence" mechanisms of human thought processes [29]. The core idea involves guiding individuals towards local optima within subgroups through convergence operations, while diversification operations enable individuals to break through the current search space and explore new regions. Additionally, by combining mechanisms of competition and cooperation between subgroups, MEC forms an optimization paradigm that coordinates global exploration and local exploitation [30]. However, the adaptability of traditional MEC is limited when it is applied to high-dimensional nonconvex optimization problems and dynamic environments because of fixed parameter settings and static population structures [31]. Adaptive MEC introduces dynamic parameter adjustment and structural adaptation mechanisms [32], thereby enabling real-time tuning of convergence search scales and diversification probabilities. Moreover, by integrating "dynamic group replacement" and "maturity evaluation" strategies, it achieves global propagation of superior solutions and intelligent avoidance of ineffective searches [33]. The specific process used to construct the adaptive MEC algorithm is detailed in the following sections.

(1) Population and Subpopulations

A population $P = [p_1, \ldots, p_n]$ is formed by randomly generating n samples. The score for each sample is calculated via Equation (3). The population is divided into multiple subpopulations, which are classified into w winning subpopulations P_{supi} and r temporary subpopulations P_{temi} on the basis of their scores, as shown in Equation (4).

$$S = 1/(1+E) \tag{3}$$

$$\begin{cases} P_{\text{supi}} = \{p_1, \ldots, p_k\} & (i = 1, \ldots, w) \\ P_{\text{temi}} = \{p_1, \ldots, p_h\} & (i = 1, \ldots, r) \\ n = w \cdot k + r \cdot h \end{cases} \tag{4}$$

(2) Bulletin Boards

Bulletin boards are used to record essential information about the population and are classified into global bulletin boards and local bulletin boards that correspond to each subpopulation. The global bulletin board stores information from local bulletin boards, differentiation history, iteration details, and the average Euclidean distance among individuals. Each local bulletin board records individual parameters and scores, the position of the subpopulation winner, maturity status, historical records of the winner, individual distribution statistics, and the search scale.

(3) Adaptive Convergence Operation

The scores for all the individuals within each subpopulation are calculated, and the optimal sample p_{win} is continuously sought within each subpopulation. Centring on p_{win}, new samples p_{new} are generated according to a normal distribution:

$$p_{new} \sim N\left(p_{win}, \sigma_t^2\right) \tag{5}$$

where σ_t represents the adaptive variance range.

$$\sigma_t = \sigma_0 \cdot (1 - \beta t/T) \tag{6}$$

where σ_0 denotes the initial variance range, t represents the current iteration count, and T indicates the maximum number of iterations. The scores for all individuals within the subpopulation are repeatedly calculated, and the operation terminates once no individuals with higher scores can be found. At this point, the corresponding subpopulation is considered "mature".

(4) Adaptive Diversification Operation

The process of continuously searching for the optimal subpopulations within the population is referred to as diversification. During this process, multiple subpopulations with high scores become new winning subpopulations, and the remaining subpopulations are discarded on the basis of a probability $P_{\text{diss}}(t)$. After the diversification operation concludes, new temporary subpopulations are randomly generated within the search space to maintain the total number of subpopulations.

$$P_{\text{diss}}(t) = P_0 \cdot e^{-\gamma t/T} \tag{7}$$

where γ denotes the decay rate and P_0 represents the initial diversification probability.

2.3. Framework for the Construction of a Deflection Prediction Method Based on an Adaptive MEC–RBF Neural Network

The framework of the adaptive MEC–RBF neural network-based deflection prediction method is shown in Figure 1. It consists of dataset generation, adaptive MEC–RBF model training, and model application for deflection prediction. The workflow is presented in detail below.

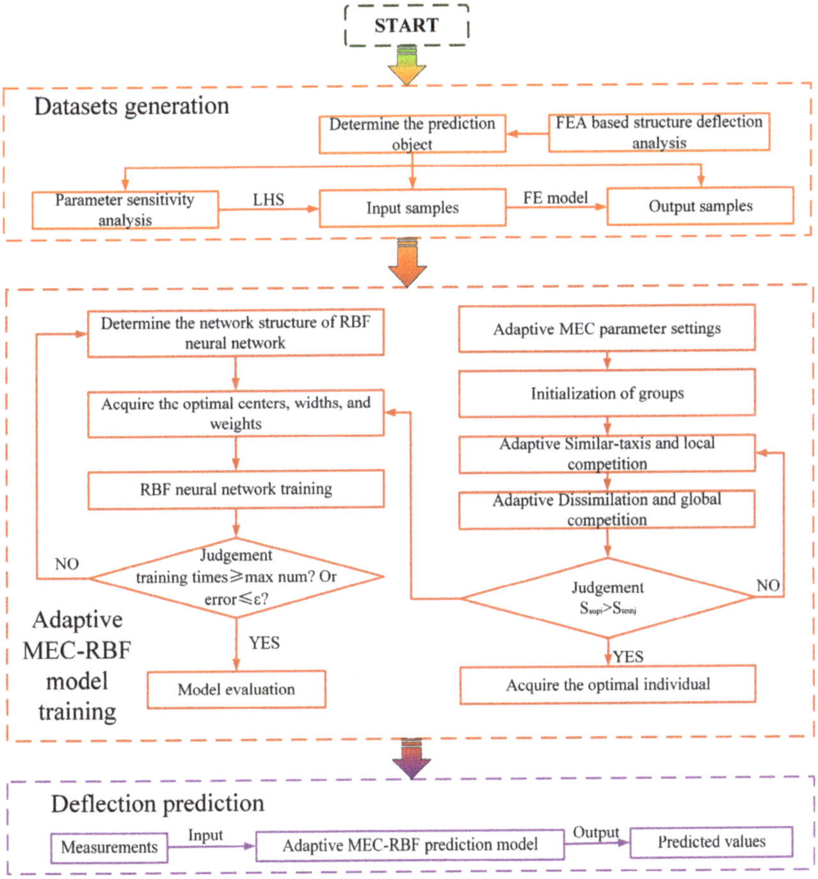

Figure 1. Framework of the adaptive MEC–RBF neural network for deflection prediction.

(1) Dataset Generation

First, finite element analysis (FEA) is used to identify the primary deflection targets that exhibit significant deformation during the construction stages of the structure. Next, sensitivity analysis is employed to extract the key input parameters that affect these deflection targets. To address local sparsity in high-dimensional random sampling, input samples are generated from the identified parameters via the Latin hypercube sampling (LHS) technique [34]. The method involves mapping n random variables into an n-dimensional space, partitioning each dimension into m intervals, and randomly selecting one point per interval, thereby yielding m nonoverlapping sample sets. This approach improves the uniformity and representativeness of the samples. The corresponding output responses for these input samples are then computed via finite element analysis, thereby forming a dataset to support subsequent model building and predictive analysis. To ensure that the generated samples adequately represented the probabilistic properties of the variables, each parameter is sampled according to its mean and standard deviation derived from engineering material tests and historical monitoring data. The resulting sample distributions are compared with the target distributions using the Kolmogorov–Smirnov test, showing deviations less than 5%, which confirms satisfactory probabilistic consistency.

(2) Setting of the Neural Network Training Parameters

To ensure prediction accuracy, the training parameters of the adaptive MEC–RBF neural network, such as the number of neurons per layer, group size, and other learning-related factors, are determined.

(3) Optimization of Centres, Widths, and Weights

Via the adaptive MEC algorithm, the centres κ_j, widths of the radial basis functions ς_j, and connection weights between the hidden and output layers of the RBF neural network w_{kj} are optimized. Detailed descriptions are provided in Section 3.2. The optimized widths ς_j and weights w_{kj} are then used as initial parameters for the RBF network.

(4) Model Development and Validation

The training samples are fed into the adaptive MEC–RBF neural network, where the objective function E (Equation (2)) is evaluated. The network is iteratively updated until E satisfies the convergence criterion or until the maximum number of iterations is reached. Model performance is then validated on test samples via predefined metrics; further details are presented in Section 3.4.

(5) Model Application

The adaptive MEC–RBF model is applied to predict structural deflections. The predicted results are compared with actual response values obtained from field test data to verify the model's accuracy.

2.4. Evaluation Criteria

To quantitatively assess the prediction performance of the adaptive MEC–RBF model, multiple statistical metrics, which are shown in Equations (5)–(8), are selected to evaluate the quality of the predicted data. Specifically, these metrics include the coefficient of determination (R^2), root mean square error ($RMSE$), and mean absolute error (MAE).

$$R^2 = \frac{\sum\limits_{i=1}^{N} (t_i - o_i)^2}{\sum\limits_{i=1}^{N} \left(o_i - o_{avg}\right)^2} \tag{8}$$

$$RMSE = \sqrt{\frac{1}{N}\sum_{i=1}^{N}(t_i - o_i)^2} \tag{9}$$

$$MAE = \frac{1}{N}\sum_{i=1}^{N}|t_i - o_i| \tag{10}$$

where o_i represents the predicted value of the i-th sample, t_i represents the actual measured value, o_{avg} represents the mean of the predicted outputs, and N represents the total number of samples.

3. Case Study

3.1. Bridge Overview

The Hannam Yangtze River Bridge, which is located in Hubei Province, China, is a prestressed concrete continuous rigid-frame bridge with a span configuration of 95 m + 160 m + 95 m. The main girder adopts a variable cross-section concrete box girder design, with C55 concrete used as the material. The girder height reaches 3.6 m at the central span and up to 10 m near the piers. The construction method was the cantilever casting method, with hanging scaffolds that feature a diamond-shaped structure. The maximum number of cantilever segments is 20, with standard segment lengths of 3 m, 3.5 m, and 4 m, as illustrated in Figure 2b.

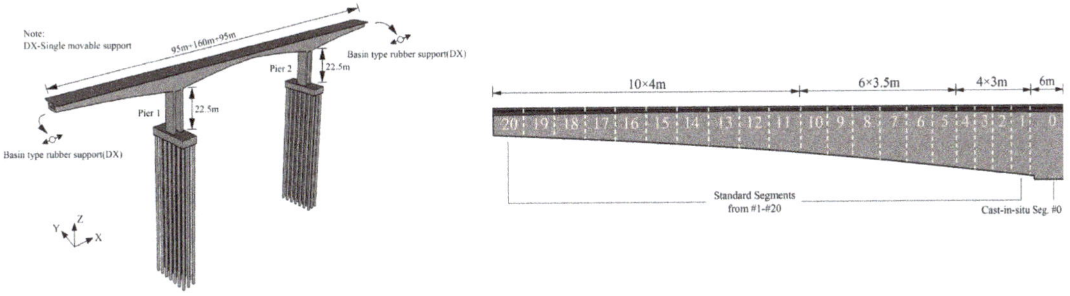

(a) Overall view (b) Segment division

Figure 2. Hannan Yangtze River Bridge.

3.2. Finite Element Model

The 3D finite element model in ANSYS 2024 R1 that was used to perform linear elastic analysis of the deformation of the main girder at each construction stage is shown in Figure 3. In the finite element analysis, the ANSYS Parametric Design Language (APDL) [35] was employed, which accounts for parameter uncertainties. The C55 concrete main girder—including the web, top flange, and bottom flange of the variable-section box girder—was modelled by using SOLID65 solid elements, whereas the prestressing tendons were numerically simulated with LINK180 elements. A multipoint constraint (MPC) method was adopted, with the degree of freedom compatibility between different elements coordinated via MPC184 elements to ensure force transmission between the prestressing tendons and the concrete. The element birth and death technique [36] was employed to simulate the construction sequence of the hanging-scaffold cantilever casting method.

Figure 3. Finite element model.

The model accounts for dead loads, including the self-weights of structural members, the self-weight of the hanging scaffold (modelled as concentrated loads), temporary loads (modelled as uniformly distributed loads), the initial tension of the prestressing tendon, and the temperature loads that affect the structure. Actual boundary conditions were modelled by fixing all degrees of freedom at the pier bases [37]. Notably, a linear elastic material model was adopted in the finite element analysis. This assumption is reasonable because the deflection of cantilever segments was evaluated within short construction intervals (approximately 24–48 h between successive segment castings), during which time-dependent effects such as creep and shrinkage have only a minor influence. According to field monitoring, the ambient temperature variation during the actual construction period was within approximately ±5 °C, indicating limited thermal gradients and negligible thermal curvature. Nevertheless, to examine a broader range of possible conditions, the orthogonal experimental design (Section 3.3) considered temperatures ranging from 2.5 °C to 27.5 °C. The close agreement between the predicted and measured deflections verifies that these simplifications provide sufficient accuracy for the present analysis. Field deflections were measured using a Leica TS60 high-precision total station equipped with reflective prisms installed at the tip of each cantilever segment. Measurements were repeated three times for each segment during the concrete curing stage and averaged to reduce random errors. The measurement accuracy was within ±0.5 mm, which ensured reliable validation of the numerical predictions presented in Figure 4. The computations were carried out on a system that was equipped with an Intel Xeon E5-2680 2.4 GHz processor and 64 GB of memory. A single finite element model run consumed approximately 0.5 h of CPU time, thus indicating a high computational cost. Finite element model parameters were iteratively adjusted to reflect actual construction conditions while balancing computational efficiency and accuracy. Hence, the adaptive MEC–RBF method offers practical guidance for bridge construction management.

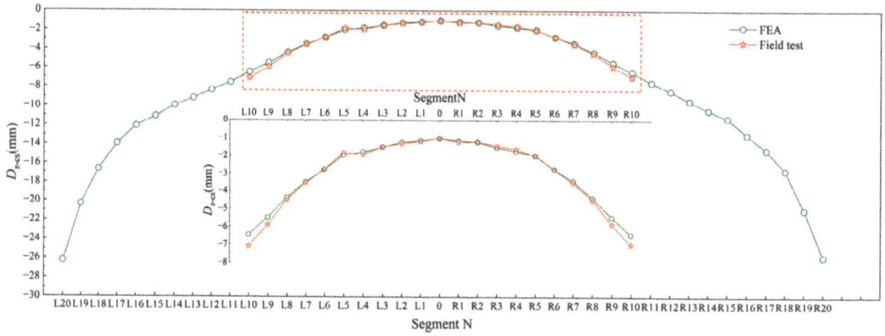

Figure 4. Comparison of segmental deflections during the concrete casting stage.

Model accuracy was assessed through a comparison of field test and finite element results for segmental deflections during construction ($D_{Z\text{-}CS}$), as shown in Figure 4. The analysis results are in good agreement with the measured deflections of the completed 10 segments. A maximum error of 0.8 mm was observed between the finite element results and field data. However, as the cantilever length L increased, the finite element analysis effects became more pronounced in the unconstructed segments, segments #11 to #20. Specifically, the $D_{Z\text{-}CS}$ value for segment #20 differs from that of segment #9 by approximately 19 mm, which indicates the necessity of deflection prediction for unfinished segments [38].

3.3. Parameter Selection and Sensitivity Analysis

A sensitivity analysis was performed to assess the importance of the input parameters, with the aims of reducing the computational load and increasing the efficiency of the adaptive MEC–RBF model [39]. In this study, eight parameters with significant variations during the construction of the prestressed concrete continuous rigid frame bridge were considered [40]. These parameters included the Young's modulus values of the concrete and steel, E_c and E_s, respectively; the densities of the concrete and steel, γ_c and γ_s, respectively;

the tensile stress of the prestressed tendon, σ; the structural temperature, T; the uniformly distributed load, P; and the ratio of the beam segment length to the cantilever length, d/L.

The range (R_j), which was used as the sensitivity index for evaluating the orthogonal experimental results, is defined in Table 1 as follows:

$$R_j = \max\{K_{1j}, K_{2j}, \ldots\} - \min\{K_{1j}, K_{2j}, \ldots\} \tag{11}$$

$$K_{ij} = \frac{1}{n}\sum_{k=1}^{n} Y_k \tag{12}$$

where K_{ij} represents the average value of parameter j at level i, n represents the number of trials, and Y_k represents the ith value of $D_{Z\text{-}CS}$.

Table 1. Orthogonal experimental design and calculation results.

No.	1 E_c(GPa)	2 E_s(GPa)	3 γ_c(kN/m^3)	4 γ_s(kN/m^3)	5 σ(GPa)	6 *P*(kN/m)	7 *d*/*L*	8 *T*(C°)	D_{Z-CS} (mm)
1	1 (25.075)	1 (167.25)	1 (22.125)	1 (67.625)	1 (1.2255)	1 (3.75)	1 (0.063)	1 (2.5)	−15.08
2	1	2 (189.9)	2 (23.85)	2 (73.25)	2 (1.3302)	2 (10.5)	2 (0.060)	2 (10)	−19.57
3	1	3 (205)	3 (25)	3 (77)	3 (1.40)	3 (15)	3 (0.056)	3 (15)	−23.45
4	1	4 (220.1)	4 (26.15)	4 (80.75)	4 (1.4698)	4 (19.5)	4 (0.053)	4 (20)	−29.57
5	1	5 (242.75)	5 (27.875)	5 (86.375)	5 (1.5745)	5 (26.25)	5 (0.051)	5 (27.5)	−31.55
6	2 (29.47)	1	2	3	4	5	1	2	−4.06
7	2	2	3	4	5	1	2	3	−6.28
8	2	3	4	5	1	2	3	4	−48.87
9	2	4	5	1	2	3	4	5	−56.85
10	2	5	1	2	3	4	5	1	−38.60
11	3 (32.4)	1	3	5	2	4	4	1	−43.92
12	3	2	4	1	3	5	5	2	−50.58
13	3	3	5	2	4	1	1	3	−23.80
14	3	4	1	3	5	2	2	4	−10.34
15	3	5	2	4	1	3	3	5	−35.86
16	4 (35.33)	1	4	2	5	3	5	3	−35.12
17	4	2	5	3	1	4	1	4	−28.00
18	4	3	1	4	2	5	2	5	−19.88
19	4	4	2	5	3	1	3	1	−23.78
20	4	5	3	1	4	2	4	2	−28.50
21	5 (39.725)	1	5	4	3	2	4	3	−38.57
22	5	2	1	5	4	3	5	4	−22.32
23	5	3	2	1	5	4	1	5	−6.61
24	5	4	3	2	1	5	2	1	−28.49
25	5	5	4	3	2	1	3	2	−32.39
26	1	1	1	4	5	4	3	2	−17.83
27	1	2	2	5	1	5	4	3	−55.68
28	1	3	3	1	2	1	5	4	−60.21
29	1	4	4	2	3	2	1	5	−25.18
30	1	5	5	3	4	3	2	1	−33.01
31	2	1	2	1	3	3	2	4	−22.85
32	2	2	3	2	4	4	3	5	−28.91
33	2	3	4	3	5	5	4	1	−35.53
34	2	4	5	4	1	1	5	2	−72.53
35	2	5	1	5	2	2	1	3	−17.35
36	3	1	3	3	1	2	5	5	−50.93
37	3	2	4	4	2	3	1	1	−22.79
38	3	3	5	5	3	4	2	2	−31.22
39	3	4	1	1	4	5	3	3	−18.86
40	3	5	2	2	5	1	4	4	−20.02

Table 1. *Cont.*

No.	1 E_c(GPa)	2 E_s(GPa)	3 γ_c(kN/m^3)	4 γ_s(kN/m^3)	5 σ(GPa)	6 P(kN/m)	7 d/L	8 T(C°)	D_{Z-CS} (mm)
41	4	1	4	5	4	1	2	5	−20.28
42	4	2	5	1	5	2	3	1	−26.04
43	4	3	1	2	1	3	4	2	−31.25
44	4	4	2	3	2	4	5	3	−42.37
45	4	5	3	4	3	5	1	4	−17.00
46	5	1	5	2	2	5	3	4	−37.19
47	5	2	1	3	3	1	4	5	−18.28
48	5	3	2	4	4	2	5	1	−26.76
49	5	4	3	5	5	3	1	2	−9.34
50	5	5	4	1	1	4	2	3	−30.52
K_1	−311.13	−285.81	−209.79	−316.10	−397.21	−292.64	−169.21	−293.98	
K_2	−331.83	−278.44	−257.55	−288.11	−352.50	−292.11	−222.44	−297.27	
K_3	−308.32	−307.58	−297.02	−278.35	−289.50	−292.83	−293.18	−291.99	
K_4	−272.22	−317.32	−330.83	−287.07	−236.07	−297.56	−358.16	−296.38	
K_5	−250.46	−284.80	−378.76	−304.32	−198.67	−298.81	−430.96	−294.32	
R_j	8.14	3.89	16.90	3.78	19.85	0.67	26.18	0.53	

The calculation results are presented in Figure 5. According to these results, d/L is the parameter that has the greatest influence on the response of $D_{Z\text{-}CS}$, followed by σ, γ_c, γ_s, E_c and E_s, which also have significant effects. Parameters P and T were insignificant, which is attributed mainly to the relatively small temperature gradients observed during the construction period. Although the ambient temperature levels explored in the orthogonal design ranged from 2.5 °C to 27.5 °C, representing an extended envelope for analysis, the corresponding thermal curvature remained minor compared with the effects of self-weight, prestressing, and construction sequencing. These findings are consistent with the justification provided in Section 3.2 and indicate that temperature influences play a secondary role in short-term construction deflection. The model, however, can be readily retrained with augmented temperature data should a project experience larger gradients or seasonal variations. Thus, the neural network inputs were narrowed to six parameters: d/L, σ, γ_c, γ_s, E_c, and E_s. Notably, the range-based index (R_j) adopted in this study reflects only the first-order (main) effects of each input variable and does not explicitly capture potential interactions among variables. Nevertheless, the method provides an intuitive and computationally efficient means to identify the dominant parameters influencing bridge deflection. Future work will extend the analysis using variance-based techniques such as the Sobol or Morris sensitivity methods to quantify higher-order interaction effects.

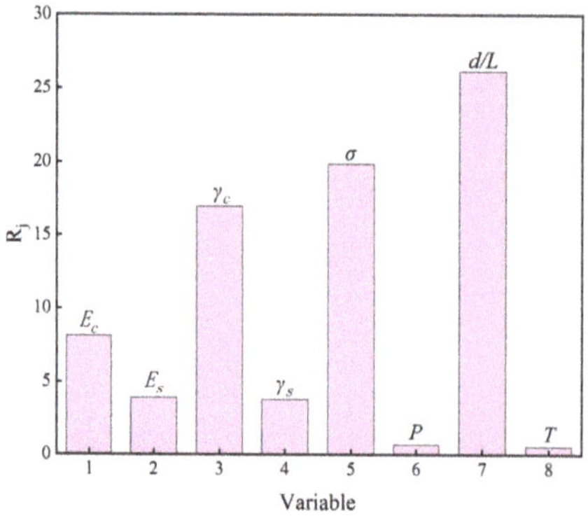

Figure 5. Histogram of R_j.

3.4. Adaptive MEC–RBF Neural Network Prediction Model

3.4.1. Model Construction

Developed in Python 3.12.7, the adaptive MEC–RBF prediction model was trained according to the procedure described in Section 3.3. Multiple training parameters, as listed in Table 2, were employed in the experimental setup to develop the adaptive MEC–RBF model.

Table 2. Parameter settings of the adaptive MEC–RBF model.

Algorithm	Parameter	Definition	Levels
Adaptive MEC	M	The population size	100, 200,. . ., 500
	M_{sup}	The superior group size	6
	M_{tem}	The temporary group size	6
	I_{max}	The maximum number of iterations	300
RBF	N_1	The number of neurons in the input layer	6
	N_2	The number of neurons in the hidden layer	4,. . ., 18
	N_3	The number of neurons in the output layer	1
	ep	The number of iterations	600
	G	The training accuracy	1×10^{-5}

The *RMSE* values of the adaptive MEC–RBF models with various architectures are shown in Figure 6. The *RMSE* decreased as the number of hidden neurons increased and reached a minimum at 16 neurons with a population size of M = 500. Thus, the 6–16–1 architecture with M = 500 was chosen as the optimal model, as illustrated in Figure 7.

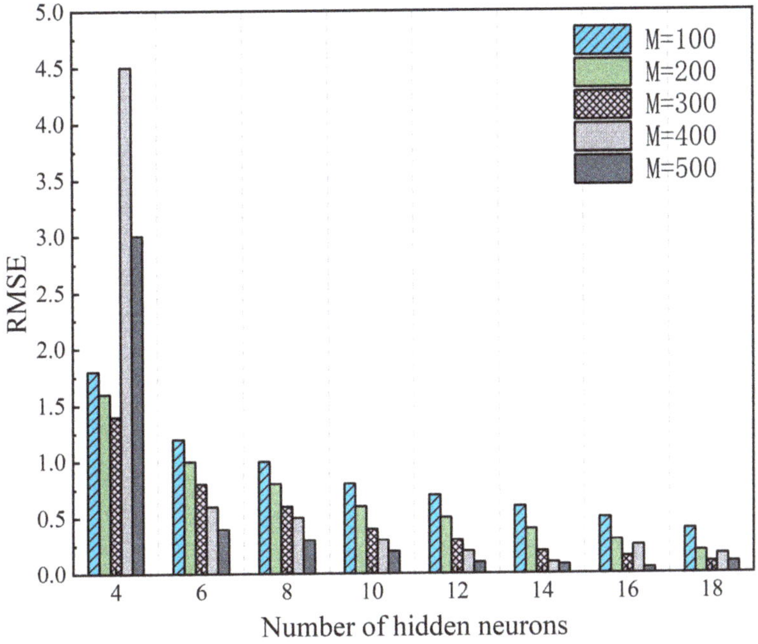

Figure 6. Performance of the adaptive MEC–RBF model at various N_2 values.

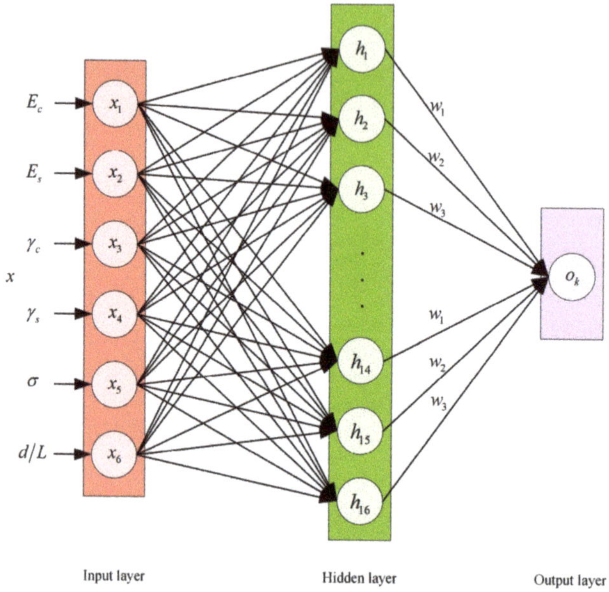

Figure 7. Structure of the three-layer RBF neural network.

Considering the probabilistic properties of the variables listed in Table 3, 300 input samples were randomly generated via Latin hypercube sampling (LHS) techniques for analysis. Among these 300 input samples, five levels of d/L (0.063, 0.060, 0.056, 0.053, and 0.051) were randomly distributed. The corresponding output samples were derived from the validated finite element model calculations.

Table 3. Properties of random variables.

Variable	Mean μ	Standard Deviation σ	Distribution Type	Remarks
E_c	32.4	2.93	Normal distribution	Young's modulus (concrete)
E_s	205	15.1	Normal distribution	Young's modulus (steel)
γ_c	25	1.15	Normal distribution	Density (concrete)
γ_s	77	3.75	Normal distribution	Density (steel)
σ	1.40	0.0698	Normal distribution	Prestressing tendon tensile stress
T	15	4.5	Normal distribution	Structural temperature
P	700	32.5	Normal distribution	Uniformly distributed load

The statistical properties of the input samples (excluding d/L) are shown in Figure 8. Among them, 250 samples were used for training, and 50 were used for testing the adaptive MEC–RBF model. To further verify the robustness and generalizability of the adaptive MEC–RBF model, a fivefold cross-validation procedure was performed. The entire dataset of 300 samples was randomly partitioned into five equal subsets, where four subsets were used for training and one was used for testing in each run. The process was repeated five times so that every subset was used once as the testing set. The mean coefficient of determination (R^2) across the five folds was 0.972, and the corresponding mean *RMSE* was 2.18 mm, indicating that the model maintained stable performance on unseen data without significant overfitting. The average training time of the adaptive MEC–RBF model was approximately 138 s, which represented a substantial reduction compared with that of the finite element analysis, which required approximately 0.5 h per run. Once trained, the

MEC–RBF model generated deflection predictions in less than 1 s, confirming its capability for near-real-time application in construction control.

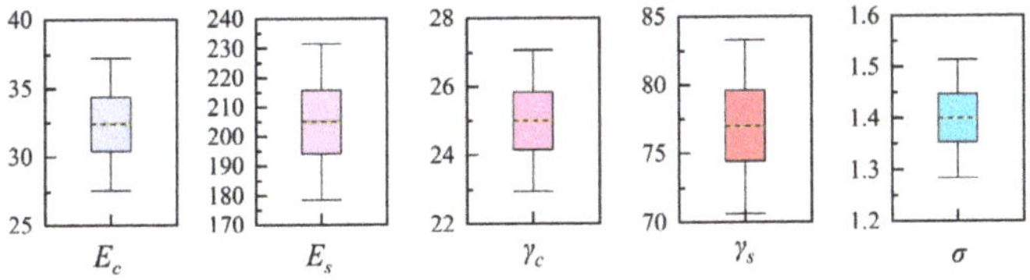

Figure 8. Statistical characteristics of the input samples.

3.4.2. Model Validation

The performance of the adaptive MEC–RBF model was compared with that of the conventional RBF model during the training and testing phases, as shown in Figure 9a and 9b, respectively. The residual distributions illustrated in Figure 10a,b reveal that most residuals lie within ±0.5 mm and are randomly scattered around zero, confirming the model's stability and the absence of systematic bias. These results demonstrate that, in both phases, the DZ–CS values predicted by the adaptive MEC–RBF model exhibit much closer agreement with those obtained from the finite element simulations. Moreover, based on the comparison between the adaptive MEC–RBF and FEM-derived DZ–CS results, the average prediction error of the adaptive MEC–RBF model was less than 5%, significantly outperforming the conventional RBF model. This clearly confirms the superiority of the adaptive MEC–RBF model over the traditional RBF approach.

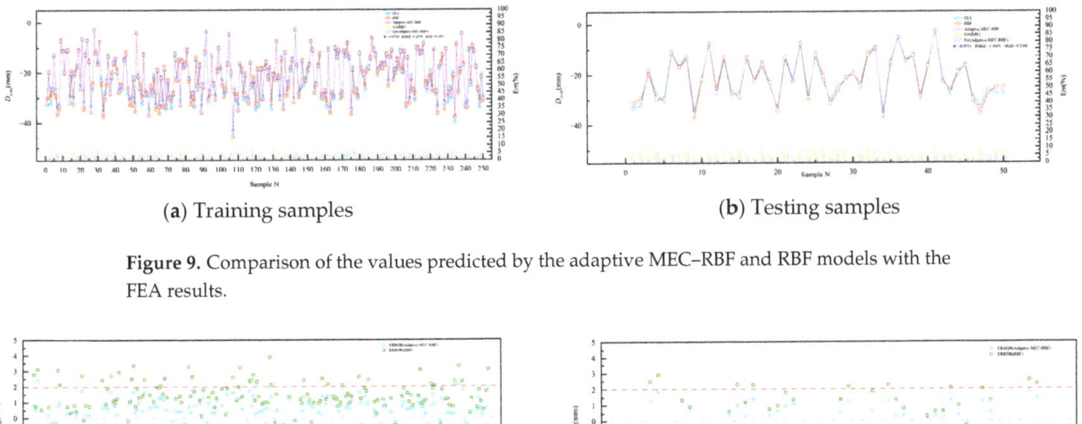

(**a**) Training samples (**b**) Testing samples

Figure 9. Comparison of the values predicted by the adaptive MEC–RBF and RBF models with the FEA results.

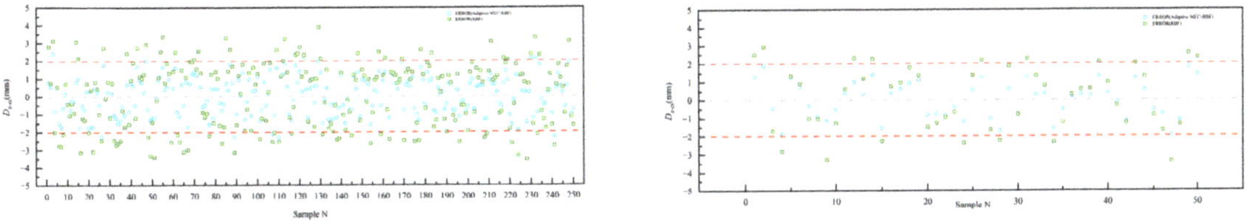

(**a**) Training samples (**b**) Testing samples

Figure 10. Comparison of residuals between the adaptive MEC–RBF and conventional RBF models with respect to the FEA results.

Furthermore, the evaluation metrics [41] for both the adaptive MEC–RBF model and the conventional RBF model are presented in Table 4. On the training samples, the adaptive MEC–RBF model achieved a coefficient of determination R^2 = 0.9789, an *RMSE* of 1.4978, and an *MAE* of 0.1411. On the testing samples, R^2 = 0.9751, the *RMSE* was 1.9491, and the *MAE* was 0.2341. In terms of all the metrics, the adaptive MEC–RBF model outperformed the conventional RBF model. These results confirm that the adaptive MEC–RBF model offers superior efficiency and accuracy, which makes it well suited for deflection prediction. To further evaluate the significance of the performance improvement achieved by the adaptive MEC–RBF model, a paired t test was conducted on the prediction errors obtained from the adaptive MEC–RBF and conventional RBF models. The resulting p value (<0.05) indicates that the difference in performance between the two models is statistically significant. This confirms that the observed improvement is not due to random variation but reflects the enhanced learning capability of the adaptive mechanism.

Table 4. Statistical metrics of the $D_{Z\text{-}CS}$ prediction model.

Model	Training Samples			Testing Samples		
	R^2	*RMSE*	*MAE*	R^2	*RMSE*	*MAE*
Adaptive MEC–RBF	0.9789	1.4978	0.1411	0.9751	1.9491	0.2341
RBF	0.9711	2.7869	0.4211	0.9679	3.1807	0.5678

3.4.3. Analysis and Discussion of the Results

The application of the trained adaptive MEC–RBF model to the Hannam Yangtze River Bridge yielded accurate deflection predictions. As illustrated in Figure 11a,b, the predicted $D_{Z\text{-}CS}$ values obtained by the adaptive MEC–RBF model on the basis of field-measured input parameters are in near-perfect agreement with the finite element analysis results across most cantilever segments. Most residuals lie within ±0.5 mm and are randomly scattered around zero in Figure 12a,b. The maximum discrepancy between the adaptive MEC–RBF model predictions and the measured field deflections of the Hannan Yangtze River Bridge was 5.56 mm, corresponding to a 17.7% relative error. These findings confirm that the proposed model provides satisfactory accuracy for engineering applications. Given the complexity of the construction environment and the effectiveness of the neural network, this accuracy is acceptable for practical deflection control.

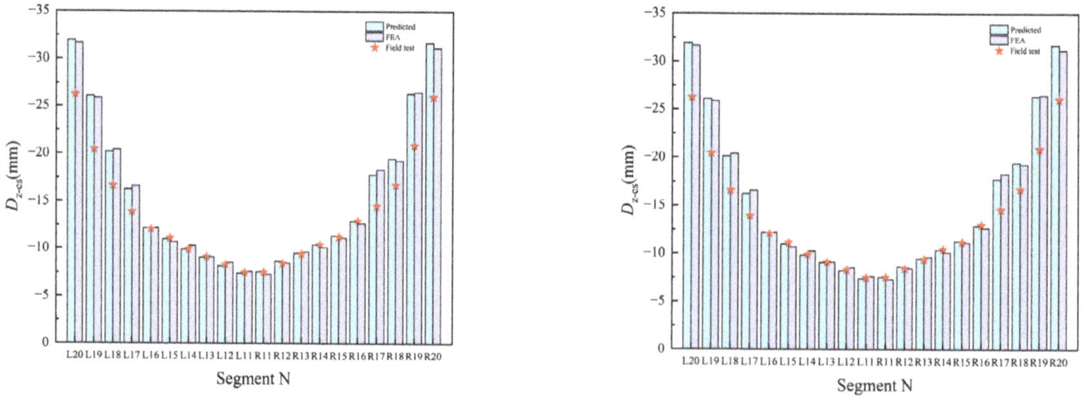

(**a**) Pier 1 (**b**) Pier 2

Figure 11. Comparison of cantilever beam deflection development between pier 1 and pier 2.

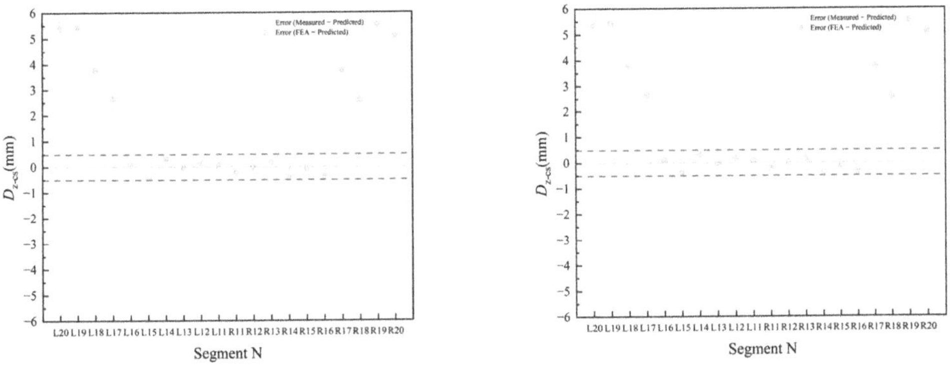

(**a**) Pier 1 (**b**) Pier 2

Figure 12. Comparison of measured–predicted and FEA–predicted deflection residuals between pier 1 and pier 2.

4. Conclusions

To increase control efficiency and prediction accuracy when construction schedules are complex and time-constrained and large-scale finite element analyses of intricate bridge structures are required, this study proposes a deflection prediction framework that integrates sensitivity analysis with an adaptive MEC–RBF neural network. The main contributions of this research are summarized as follows:

- The sensitivity analysis revealed that the parameter d/L had the greatest influence on segment deflection, followed by the parameters σ, γ_c, γ_s, E_c, and E_s, whereas the other parameters exhibited relatively minor effects.
- A comparison with finite element analysis revealed that the adaptive MEC–RBF model achieved higher accuracy and computational efficiency than the conventional RBF model did, providing more precise deflection predictions during construction.
- Application of the adaptive MEC–RBF model to the Hannan Yangtze River Bridge demonstrated that its predictions were in good agreement with field measurements, confirming the accuracy and practical applicability of the proposed framework. These findings suggest that the approach can effectively support deflection control for prestressed concrete continuous rigid-frame bridges.

The adaptive MEC–RBF framework was validated using data from the Hannan Yangtze River Bridge. It can be retrained and applied to other continuous rigid-frame or prestressed concrete bridges by updating the input parameters and training datasets. Although this flexibility allows the proposed model to serve as a general framework for deflection prediction during construction, several limitations remain and will be addressed in future studies. Verification of data reliability is required during the dataset generation phase, so the finite element model may require adjustments and lead to additional computational costs. Moreover, given the complexity of bridge construction, the probabilistic characteristics of variables may vary under different practical conditions. Future work will focus on extending the adaptive MEC–RBF model to stress prediction in prestressed concrete continuous rigid-frame bridges and comparing its performance with that of other hybrid optimization models, such as particle swarm optimization and grey wolf optimizer–radial basis function neural networks. In addition, the proposed model was developed and validated within a temperature range covered by both field observations (±5 °C) and an orthogonal experimental design (2.5–27.5 °C). For bridge projects subjected to wider temperature variations or pronounced through-depth thermal gradients, the same framework can be extended by incorporating additional temperature data to preserve predictive accuracy.

Author Contributions: Conceptualization, Q.G.; methodology, Q.G. and L.S.; software, Q.H.; validation, C.Z. and L.S.; formal analysis, Q.H. and D.T.; investigation, C.Z. and D.T.; resources, C.Z. and L.S.; data curation, Q.H. and D.T.; writing—original draft preparation, Q.G.; writing—review and editing, C.Z., Q.G., Q.H., L.S. and D.T.; visualization, Q.H.; supervision, Q.G. and D.T.; project administration, C.Z.; funding acquisition, Q.G. All authors have read and agreed to the published version of the manuscript.

Funding: This research received no external funding.

Institutional Review Board Statement: Not applicable.

Informed Consent Statement: Not applicable.

Data Availability Statement: The data, finite-element models, and codes that support the findings of this study are available from the corresponding author upon reasonable request.

Conflicts of Interest: Authors Liangbo Sun and Dewei Tian were employed by the company The First Company of China Eighth Engineering Division Ltd. The remaining authors declare that the research was conducted in the absence of any commercial or financial relationships that could be construed as a potential conflict of interest.

Abbreviations

The following abbreviations are used in this manuscript:

MEC	Mind evolutionary computation
RBF	Radial basis function
R^2	Coefficient of determination
RMSE	Root mean square error
MAE	Mean absolute error

References

1. Petroski, H. Engineering: Infrastructure. *Am. Sci.* **2009**, *97*, 370–374. [CrossRef]
2. Perez, A.I.; Ardaman, A.K. New infrastructure: Civil engineer's role. *J. Urban Plan. Dev.* **1988**, *114*, 61–72. [CrossRef]
3. Ge, Y.; Xiang, H. Concept and requirements of sustainable development in bridge engineering. *Front. Archit. Civ. Eng. China* **2011**, *5*, 432–450. [CrossRef]
4. Paunova, D.; Tzonevska, D. Strategic development of the national road network. *IOP Conf. Ser. Mater. Sci. Eng.* **2023**, *1297*, 012016. [CrossRef]
5. Peng, Y.; Zhang, Z. Development of a novel type of open-web continuous reinforced-concrete rigid-frame bridge. *J. Bridge Eng.* **2020**, *25*, 05020005. [CrossRef]
6. Zhang, S.; Yu, B.; Li, R. Research on the intelligent construction and development of bridge engineering safety management. *E3S Web Conf.* **2024**, *565*, 03013. [CrossRef]
7. Yang, J.; Xiang, F.; Li, R. Intelligent bridge management via big data knowledge engineering. *Autom. Constr.* **2022**, *135*, 104118. [CrossRef]
8. Zhang, J.; Liu, L.; Yang, H. An overview of intelligent construction and maintenance technology for highway subgrade engineering. *Intell. Transp. Infrastruct.* **2023**, *2*, liad019. [CrossRef]
9. Carbonari, S.; Nicoletti, V.; Martini, R.; Gara, F. Dynamics of bridges during proof load tests and determination of mass-normalized mode shapes from OMA. *Eng. Struct.* **2024**, *310*, 118111. [CrossRef]
10. Ferrara, M.; Bertagnoli, G.; Giordano, L. Model updating of different bridge types using ambient vibration and OMA identification. *Procedia Struct. Integr.* **2024**, *62*, 773–780. [CrossRef]
11. Li, X.; Wang, M.; Liang, Y. Research into design and construction key technologies of prestressed concrete continuous rigid frame bridge. In Proceedings of the 2011 IEEE International Conference on Remote Sensing, Environment and Transportation Engineering, Nanjing, China, 24–26 June 2011; pp. 617–621.

12. Li, H.; Huang, Y.; Guo, E. Construction stage seismic vulnerability evaluation of a continuous girder bridge with the cast-in-place cantilever construction method. *Adv. Civ. Eng.* **2021**, *2021*, 9915947. [CrossRef]
13. Zhou, Y. Pre-camber control of large-span prestressed concrete continuous rigid-frame bridges under cantilever construction. *Arch. Civ. Eng.* **2025**, *71*, 469–487. [CrossRef]
14. Zhou, X.; Deng, T.; Chen, L. Application of Bayesian update method in the construction control of continuous rigid frame bridge girders with high piers and large spans. *Buildings* **2023**, *13*, 1556. [CrossRef]
15. Wan, L.; Qiao, J.; Sun, Q.; Liu, J. Construction control of rigid frame-continuous girder bridge based on grey theory. *J. Eng. Appl. Sci.* **2024**, *71*, 171. [CrossRef]
16. Dong, F.; Shi, Z.; Zhong, R. FE model updating of continuous beam bridge based on response surface method. *Buildings* **2024**, *14*, 960. [CrossRef]
17. Wang, X.G.; Ma, M.; Zhao, B. Response surface model optimization algorithm for structural assessment based on vibration frequency. *Shock Vib.* **2025**, *2025*, 5446251. [CrossRef]
18. Kovačević, M.; Ivanišević, N.; Stević, D. Decision-support system for estimating resource consumption in bridge construction based on machine learning. *Axioms* **2022**, *12*, 19. [CrossRef]
19. Pan, Y.; Zhang, L. Integrating BIM and AI for smart construction management: Current status and future directions. *Arch. Comput. Methods Eng.* **2023**, *30*, 1081–1110. [CrossRef]
20. Zhu, H.; Wang, Y. Intelligent prediction of prestressed steel structure construction safety based on BP neural network. *Appl. Sci.* **2022**, *12*, 1442. [CrossRef]
21. Li, Q.; Xie, J. Bridge alignment prediction based on combination of grey model and BP neural network. *Appl. Sci.* **2024**, *14*, 7955. [CrossRef]
22. Wang, X.; Miao, C.; Wang, X. Prediction analysis of deflection in the construction of composite box-girder bridge with corrugated steel webs based on MEC-BP neural networks. *Structures* **2021**, *32*, 691–700. [CrossRef]
23. Yang, Y.; Xia, X.; Lo, D. Predictive models in software engineering: Challenges and opportunities. *ACM Trans. Softw. Eng. Methodol.* **2022**, *31*, 1–72. [CrossRef]
24. Liao, S.; Li, S.; Liu, J. A zeroing neural dynamics-based acceleration optimization approach for optimizers in deep neural networks. *Neural Netw.* **2022**, *150*, 440–461. [CrossRef] [PubMed]
25. Mao, Y.; Ma, Z.; Liu, X. A long-term prediction method of computer parameter degradation based on curriculum learning and transfer learning. *Mathematics* **2023**, *11*, 3098. [CrossRef]
26. Zhang, Z. Pattern classification based on radial basis function neural network. In Proceedings of the 2020 IEEE 5th International Conference on Smart Grid and Electrical Automation (ICSGEA), Zhangjiajie, China, 13–14 June 2020; pp. 213–216.
27. Dai, Y.; Wu, Q.; Zhang, Y. Generalized sparse radial basis function networks for multi-classification problems. *Appl. Soft Comput.* **2024**, *154*, 111361. [CrossRef]
28. Chen, S.; Bai, W. Performance analysis of typical linear augmented observers for a class of MIMO systems with nonlinear uncertainty. *ISA Trans.* **2022**, *128*, 316–327. [CrossRef]
29. He, X.J.; Zeng, J.C.; Jie, J. Artificial neural network weights optimization design based on MEC algorithm. In Proceedings of the 2004 IEEE International Conference on Machine Learning and Cybernetics, Shanghai, China, 26–29 August 2004; Volume 6, pp. 3361–3364.
30. Jie, J.; Zeng, J.; Zhang, S.L. Mind evolutionary computation for the constrained optimization. In Proceedings of the IEEE 2002 International Conference on Machine Learning and Cybernetics, Beijing, China, 4–5 November 2002; Volume 3, pp. 1582–1585.
31. Sun, C.; Wang, J. Performance of MEC in high-dimensional space. In Proceedings of the 2002 IEEE International Conference on Control and Automation (ICCA), Xiamen, China, 19 June 2002; p. 231.
32. Qin, J.; Huang, C.; Luo, Y. Adaptive multi-swarm in dynamic environments. *Swarm Evol. Comput.* **2021**, *63*, 100870. [CrossRef]
33. Wang, H.; Yang, S. Memetic algorithms for dynamic optimization problems. In *Evolutionary Computation for Dynamic Optimization Problems*; Springer: New York, NY, USA, 2013; pp. 137–170.
34. Helton, J.C.; Davis, F.J. Latin hypercube sampling and the propagation of uncertainty in analyses of complex systems. *Reliab. Eng. Syst. Saf.* **2003**, *81*, 23–69. [CrossRef]
35. Zheng, B. Parametric modeling and finite element analysis of the camshaft based on ANSYS APDL. In Proceedings of the 2018 IEEE 4th Information Technology and Mechatronics Engineering Conference (ITOEC), Chongqing, China, 14–16 December 2018; pp. 1805–1808.
36. Nie, J.G.; Zhu, L. Beam-truss model of steel–concrete composite box-girder bridges. *J. Bridge Eng.* **2014**, *19*, 04014023. [CrossRef]
37. Zhang, W.; Chang, J.; Tian, G. FEM-based shape-finding and force-assessment of suspension bridges via completed loop adjustment. *J. Bridge Eng.* **2022**, *27*, 04021098. [CrossRef]
38. Li, J.; Li, D.; Li, F. Finite element analysis of displacement and vibration characteristics of cantilevered plate. *J. Phys. Conf. Ser.* **2020**, *1578*, 012241. [CrossRef]
39. Zhao, T.; Zheng, Y.; Wu, Z. Improving computational efficiency of machine-learning modeling of nonlinear processes using sensitivity analysis and active learning. *Digit. Chem. Eng.* **2022**, *3*, 100027. [CrossRef]

40. Hisam, M.W.; Dar, A.A.; Elrasheed, M.O. The versatility of the Taguchi method: Optimizing experiments across diverse disciplines. *J. Stat. Theory Appl.* **2024**, *23*, 365–389. [CrossRef]
41. Tong, Y.L. *The Multivariate Normal Distribution*; Springer: New York, NY, USA, 2012.

applied sciences

MDPI

Article

Structural Safety Assessment of an Existing Steel Bridge According to the New Polish National Guidelines Based on prEN 1990-2

Mateusz Rajchel and Tomasz Siwowski *

The Faculty of Civil and Environmental Engineering and Architecture, Rzeszow University of Technology, Al. Powstancow Warszawy 12, 35-959 Rzeszow, Poland; mrajchel@prz.edu.pl
* Correspondence: siwowski@prz.edu.pl; Tel.: +48-17-865-15-96

Abstract: The paper examines a thin-walled steel road bridge that has been in service for over fifty years. Due to damage observed during detailed inspections, a comprehensive investigation was necessary to evaluate the structural safety of the superstructure. For this purpose, the new national guidelines for assessing the safety of existing road bridges were used for the first time. These guidelines are based on the new Eurocode prEN 1990-2, which provides the foundation for assessing existing structures. To enable reliable and rational decisions regarding repair or strengthening, a finite element analysis was performed considering the condition survey, NDT, and material testing. The analysis showed that the resistance of some superstructure elements was exceeded by over 600%, and about 180 elements are inadequately safe to carry the actual minimum traffic loads according to the safety standards mandated by the new national guidelines. A comparison between the analysis results and the condition survey identified the same elements where local plastic deformations were observed. Based on the experimental and numerical results within this new assessment framework, the final decision was made to close the bridge for service and replace the existing steel structure.

Keywords: load-bearing capacity; thin-walled steel shell; finite element analysis; national guidelines; maintenance decision

1. Introduction

Despite substantial funds from the European Union spent over the past twenty years on Polish road infrastructure, a significant portion of Poland's bridge stock remains relatively old. It requires regular maintenance to guarantee adequate structural safety and serviceability. Structural safety is a key assessment parameter for existing bridges, guiding decisions on their continued use. Bridges built after World War II, during the reconstruction of war-damaged transport infrastructure, are now especially approaching the end of their service life. Efficient and reliable safety assessments are crucial for maintaining and managing the bridge stock. The condition of bridges directly affects traffic safety and transportation costs, which tend to decline over time due to physical damage such as corrosion, fatigue, overloading, or accidental impacts. Meanwhile, increasing traffic volumes necessitate upgrading the load-bearing capacity of existing bridges. To prevent the high costs of repairs, strengthening, or replacements, accurate assessments of bridge safety are becoming increasingly important worldwide [1–6]. Road authorities can make informed maintenance decisions and implement effective measures based on structural safety evaluations. Hence, there is a need for a systematic, cost-effective, quick, and precise method for assessing bridge safety to support effective management.

A key part of bridge management is assessing the safety of bridge structures. In simple terms, a bridge is considered safe if its load-bearing capacity exceeds the traffic load it carries. It can be regarded as safe when there is a sufficiently low chance that the traffic load will exceed its capacity. Extensive research has been conducted on methods to evaluate the structural safety of bridges and their associated uncertainties, ranging from basic approaches such as performance observation and design standards to more advanced techniques, including the analytical hierarchy process (AHP), fuzzy assessment, and neural networks. Recent assessment methods were briefly described, with their characteristics summarised and compared, by Lu et al. [7] and Ferrara et al. [8]. Comparing various bridge safety assessment methods shows both similarities and differences. For instance, the design standard method has a solid theoretical foundation and has been widely used in Poland until now [8–12]. However, notable differences exist between design standards and assessment, especially in terms of information availability and updating, structural condition, reliability requirements, verification procedures, and decision options. These differences, briefly outlined in the contribution [13], can often lead to a high level of conservatism when applying design standards for assessment. Typically, bridge safety assessments are based on calculation methods, where inspection findings on local damages, supported by non-destructive testing (NDT), are used to develop a realistic numerical model for finite element analysis (FEA). In a standard assessment process, condition surveys, material testing, and NDT provide valuable data for calibrating models, which in turn leads to a comprehensive structural analysis [14–17]. A combined, rational approach to these steps enables the detection, characterisation, and quantification of useful reserves in load-carrying capacity or the identification of the weakest structural points. Nonetheless, a calculation-based analysis begins with a deterministic calculation. A more advanced method becomes necessary if the structural safety standards are not met. Probabilistic methods are among such options [18–21]. Using a probabilistic approach involves an analysis tailored to the specific bridge, incorporating information on load and material property variations into the calculations. This method is well-suited for evaluating the structural safety of existing bridges.

The field-loading test method is the most direct and reliable approach for assessing bridge structural safety [22–26]. Strain data obtained from static load testing of the bridge, together with modal information identified from the measured acceleration responses under operational traffic, are used simultaneously to update the numerical bridge model's material properties and dynamic characteristics. Based on this revised model, the load-bearing capacity is calculated to evaluate the bridge's performance under the current design traffic loading specified in various national guidelines or standards. However, the field-loading test method's cost is relatively high, and the technique is not very widespread. Additionally, it may disrupt regular traffic, limiting its frequent application on highway bridges. It may also cause severe structural damage, particularly for existing bridges that lack essential original data. Another approach is to implement the structural health monitoring (SHM) system [27–30] on a bridge. Through monitoring and evaluating the structure's condition, early warning signals are issued in cases of extreme climate conditions, severe traffic, or operational issues. This provides the basis and guidance for maintenance and management decisions. However, due to its high cost, SHM is mainly used for critical bridges and is rarely applied to small and medium-span bridges. To lower the costs of SHM in such cases, Jamali et al. [31] suggest a multi-level assessment procedure for evaluating the load-bearing capacity of existing bridges, combining SHM techniques, advanced nonlinear analysis, and probabilistic approaches to address safety effectively. Recently, Ko et al. [32] proposed a framework for assessing bridge structural safety using AI technology. This method allows evaluation of bridges with limited information, such as missing design documents, and does not require field tests. The framework relies on computational design strength and includes procedures for calculating calibration factors to reflect current conditions. Meanwhile, Liu & Zhang [33] have developed a deep learning-based method to model bridge condition ratings using selected data from the National Bridge Inventory (NBI) database. This research aims to develop a data-driven approach that predicts the future conditions of highway bridge components based on archival inspection data, trained using an online-accessible NBI dataset.

Over the past twenty years, extensive research efforts have been undertaken worldwide to develop new assessment methods for existing bridges. The primary outcomes of this research are recent standards and manuals for existing bridges approved for use in the USA [34], Canada [35], China [36], Italy [37], Switzerland [38], and the UK [39]. There are also two international standards [40,41]. For a review and discussion of most of these standards, guidelines, and their frameworks, see [42,43]. These various national standards and guidelines particularly include key elements such as: (a) the possibility of directly applying probability-based methods, (b) the inclusion of structural system redundancy and robustness criteria, (c) the use of site-specific live loads and dynamic amplification factors, (d) the integration of field measurements and diagnostic test data, and (e) the consideration of proof load testing. Wisniewski et al. [42] review how these key elements are addressed in existing assessment standards and suggest possible improvements. Allaix & Bigaj-van Vliet [43] recommend amending Eurocode design rules for assessment situations, including the use of measured strengths and imperfections in calculations. Actions and their combination formats are also examined, with recommendations for adjustments to these aspects in bridge assessments. Ultimately, an overall evaluation of the scope of work needed to develop an assessment suite based on Eurocodes is presented. However, the new Eurocode prEN 1990-2 [44] for assessment only establishes principles for specific reliability aspects. The consistent application of these general assessment rules in practice requires further assumptions or even case-specific developments. Therefore, practical applications are essential for identifying relevant issues in the future development of assessment methods and standards. This paper describes a new national method for the structural safety assessment of existing bridges that combines condition surveys, material testing, NDT, FE analysis, and Eurocodes' calculation-based verification of ULS/SLS. The new method is the national implementation of the provisions of the new Eurocode prEN 1990-2 [44], to be published on 30 January 2026, which extends the basis of structural design for new structures, as specified in standard EN 1990 [45], to existing structures. The assessment framework is presented within the context of a case study involving a steel road bridge that has been in service for half a century. In the 1960s, a unique and modern thin-walled steel shell superstructure for highway bridges was designed and implemented for several road bridges in Poland. The fully welded superstructure consisted of a thin-walled external shell in a basket curve shape, stiffened with longitudinal trapezoidal ribs, and internal longitudinal and transverse trusses that stiffened the shell and ensured its overall stability. However, the superstructure has several structural and technological weaknesses that, in a short time, have led to severe structural damage, as observed in these bridges just a few years after service (e.g., weld cracks, local buckling, etc.). The national framework for the structural safety assessment, as applied to the bridge in question, is presented in the article. The research work confirmed, on the one hand, the need for quick maintenance decisions regarding other bridges of this type, and on the other hand, the usefulness of the new framework for assessing the structural safety of existing road bridges.

2. The National Framework for Structural Safety Assessment of Existing Road Bridges

A common issue in bridge management in Poland is determining the structural safety of existing, deteriorated bridges and their actual load-bearing capacity. It is crucial to make timely decisions on whether to continue, limit, or stop traffic operations, as this is vital for effective maintenance and management of bridges. However, Poland has lacked a consistent methodology for assessing the structural safety of road bridges. Consequently, developing a simple, fast, cost-effective, and reliable method to evaluate the safety of existing bridges, especially those that are damaged or deteriorated, has become increasingly important and urgent.

The origins of the Polish system for assessing the technical condition of road bridges date back to 1985, when the Act on Public Roads was enacted. This Act outlines the responsibilities of the road authority, including conducting periodic inspections of the technical state of road bridges. Following Poland's political transformation, a new "Construction Law Act" came into force in 1994, which mandated and clarified the road authority's obligation to carry out regular inspections, primarily to assess the technical condition of bridges. However, neither Act provided detailed rules for how these inspections should be performed. It was not until the 2004 regulation by the Minister of Infrastructure that the fundamental requirements regarding the scope and criteria for assessing the technical condition of key components and the entire bridge were defined.

Based on the regulation of the Minister of Infrastructure, the Polish national road authority, the General Directorate for National Roads and Motorways (GDDKiA), commissioned Rzeszów University of Technology to develop an instruction for conducting road bridge inspections, which came into force in 2005 [46]. The instruction introduced a comprehensive inspection system for road bridges managed by GDDKiA, including current inspections, basic inspections, extended inspections, detailed inspections, and expert assessments. It also outlined, among other things, the frequency, purpose, methods, and documentation procedures for these inspections. To standardise the process of assessing road bridges by different bridge inspectors, in 2008, GDDKiA implemented rules for using a scale to rate the technical condition and suitability for continued use of road bridges, which were also developed by the Rzeszów University of Technology. The document provides guidelines and examples that, considering the inspector's knowledge and experience, should be followed when assessing individual elements of road bridges during inspections (basic, extended, and detailed). The authors of both primary documents, which contain the Polish system for evaluating the technical condition of road bridges (instructions and rules), revised them in 2011 and 2020, respectively.

The system for assessing the technical condition of road bridges, which remains in use in Poland today, is based on a six-point scale aligned with the criteria in Table 1 [46]. Elements of a road bridge evaluated under these criteria include, in particular: main girders, deck slab, cables and their anchorages, expansion joints, road and pavement surfaces, drainage system, kerbs, railings, protective barriers, cornices, abutments and pillars, bearings, embankments and slopes of approach roads, riverbed and underbridge space, and environmental protection devices. The exception is the insulation of the deck slab, which is assessed on a three-point scale as specified in Table 2 [46]. Besides assessing the technical condition of individual bridge elements as detailed in Tables 1 and 2, two additional evaluations are performed: the average bridge rate and the main bridge rate, as described in Table 3 [46]. This assessment system is supplemented by annual bridge inspection training organised by the Rzeszów University of Technology on behalf of the road authority. Bridge inspections and evaluations of the technical condition of road bridges in Poland can only be carried out by qualified bridge inspectors who hold an appropriate training certificate and a construction licence.

Table 1. Scale and criteria for evaluating the technical condition of road bridges [46].

Rate	Name	Description of the Technical Condition of the Element
5	appropriate	no damage or contamination was found during inspection
4	satisfactory	shows impurities or the first signs of damage that deteriorate the aesthetic appearance
3	anxious	shows damage, the failure to repair of which will shorten the period of safe operation
2	insufficient	shows damage that reduces the service life, but can be repaired
1	failure	shows irreversible damage, disqualifying the further service life of the bridge
0	emergency	has been destroyed or ceased to exist

Table 2. Scale and criteria for the evaluation of the deck insulation of road bridges [46].

Rate	Name	Description of the Technical Condition of the Insulation
5	appropriate	no symptoms indicating insulation leakage
2	insufficient	there are a few minor stains; local repair can stop the process of deterioration of the element
0	emergency	there are extensive leaks causing a reduction in the service life of the element

Table 3. Criteria for the assessment of the average and main bridge rating of road bridges.

Bridge Rating	Description of How the Rating Is Determined
average	arithmetic mean rating of all graded items
main	the lowest of the two: • the arithmetic mean of all assessed elements; • deck slab assessment; • evaluation of the main girders; • the arithmetic mean of the abutments and piers, that is, half of the sum of the lowest abutment rating and the lowest pillar rating (in the case of a single-span structure, this will be the lowest abutment rating);

As in other countries, the Polish system for evaluating the technical condition of road bridges serves as the basis for planning the funds required for bridge maintenance (repairs, rehabilitation, strengthening, and reconstruction), as well as for planning investments related to the replacement or construction of new bridges. This system is also frequently used to assess the load-carrying capacity and the safe operational lifespan of the bridge. The construction authority in Poland typically makes decisions on the further use of the bridge (and, in extreme cases, on decommissioning the bridge) solely based on the average or primary bridge rating (a single score).

However, the current Polish system for evaluating the technical condition of road bridges, as presented in Refs [9,46], is increasingly criticised for being imprecise. The final assessment does not account for all necessary criteria or their relative weights when evaluating the condition of individual elements or the entire bridge. It is suggested that the weights assigned to criteria related to load-bearing capacity or traffic safety should be higher, while those associated with durability or aesthetics should be lower. Additionally, a single overall score should not be the only criterion for determining whether a bridge can continue to be used or under what conditions. Instead, each subsequent inspection should be based on the assessments and conclusions from the previous one. The rating should be considered only as an auxiliary tool. The primary evaluation of a bridge's technical condition should be a professional opinion issued by an experienced bridge inspector holding a valid certificate and a building licence. This is especially important for large bridges, often with unusual structures.

To meet the bridge administration's expectations, the Ministry of Infrastructure commissioned the Rzeszów University of Technology to develop a national method for assessing the structural safety of existing road bridges. This method aims to complement the current rating system and reflect the current state-of-the-art and latest trends in assessing existing bridges. The technique is used to evaluate the safety of future operations of existing bridges, considering actual service live loads and the condition survey, and to assess the structural adequacy and fitness for the intended use of existing bridges, particularly in light of increased traffic volumes. An assessment of structural safety involves a series of activities carried out to ensure that relevant limit states are not exceeded, thereby demonstrating, with sufficient reliability, the safety of an existing bridge. Such an assessment should verify that the structure maintains adequate reliability without necessarily calculating the reliability index; this can, for instance, be achieved through verification using partial factors or qualitative methods. The assessment method for an existing bridge at relevant limit states can be: a quantitative assessment based on calculations, a qualitative evaluation based on past performance, or a combination of both approaches.

The method, which forms the basis for the new national guidelines [47], is based on general recommendations and principles outlined in Eurocodes and is intended to be used in conjunction with other Eurocodes. Existing structures designed and constructed according to withdrawn standards may have material properties, detailing arrangements, and execution tolerances that do not comply with, or are not covered by, current standards. The Eurocodes primarily provide rules for designing new structures; however, the principles of EN 1990 [45] can also be applied to existing structures, with additional provisions. The new standard prEN 1990-2 [44] offers further provisions that enable the structural assessment of existing structures, including updates to basic data and structural models. The method covers general principles regarding actions for assessment, complementing standard EN 1991-2 [48], and the rules for determining the ultimate load-bearing capacity of existing elements are based on the provisions of design standards, i.e., EN 1992-2 [49], EN 1993-2 [50], and EN 1994-2 [51], applied according to the type of material from which the existing bridge is built. The method assumes that an existing bridge's structural safety (as required by the Eurocodes) is guaranteed when the ultimate limit states, verified through the partial safety factors method, are satisfied. The procedure allows checking whether all vehicles approved for use on public roads in Poland can cross the existing road bridge without total weight and axle load restrictions, without risking a reduction in the structural safety level, and with a minimised likelihood of severe structural damage or loss of the bridge's basic functions.

The assessment of the structural safety of the existing bridge should follow a stepwise framework with increasing levels of detail and accuracy (Figure 1). The first level involves analysing archival documents, inventorying the bridge's geometry (if necessary), conducting a condition survey (inspection), performing basic diagnostic tests, and executing simple computational analyses. Additional information, such as historical standards, construction strategies, and operation & maintenance experiences, is also essential. The assessment must consider the current condition of the bridge. Initial material properties can be derived from historical standards for different materials. Computational analysis involves re-evaluating the existing structure based on action forces and material parameters. At the second level, advanced diagnostic tests are required to identify detailed damage, among other things. Generally, updated characteristic values of materials are defined according to EN 1990,

Appendix D [44], based on material testing. Structural safety and serviceability verifications are carried out using these updated characteristic material values and modified partial factors, in accordance with the primary design equations of prEN 1990-2 [43] within the partial factor framework. The advanced analysis (usually performed using the finite element method, FEM) considers the impact of identified failures on the bridge's structural safety. The third level involves specialised diagnostic tests of the bridge under proof or service loads to evaluate its performance. The outcomes of these tests are primarily used to update and validate the FEM model employed in the final assessment of structural safety. Throughout each assessment level, the structural safety is evaluated to determine if it is adequate for continued use or if further analysis is advisable. Based on this, the bridge authority can decide on additional maintenance actions (e.g., traffic or service load restrictions, repairs, strengthening). Any interventions should, where possible, be designed in accordance with standards (Eurocodes) for new structures.

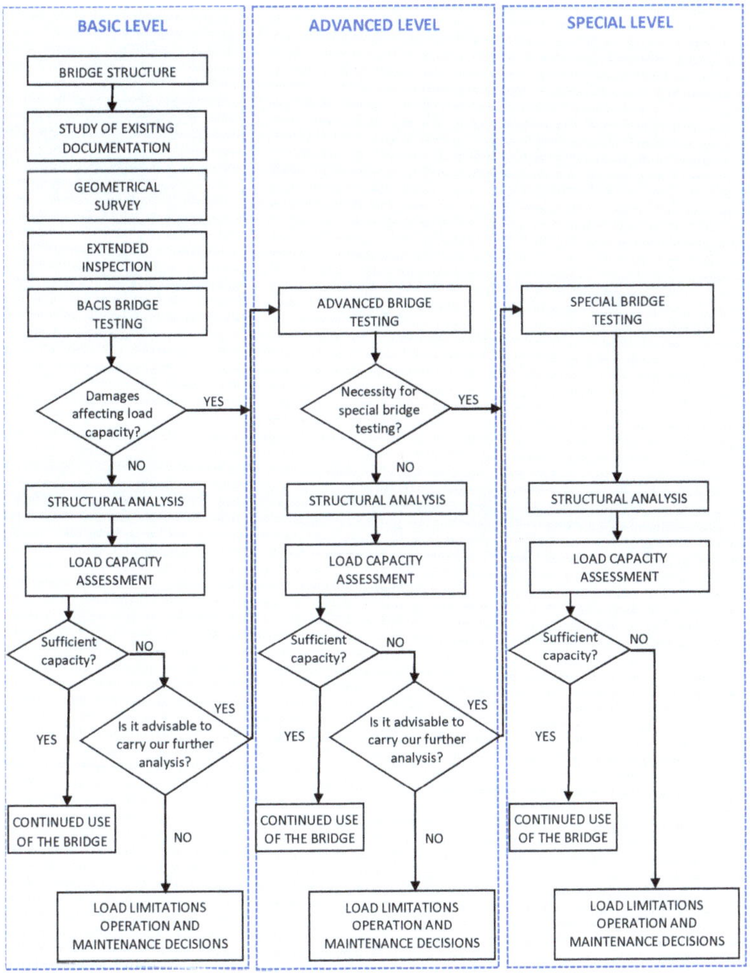

Figure 1. The framework for assessing the safety of an existing bridge according to the new Polish guidelines [47].

The criteria for selection and the scope of application for the appropriate assessment levels are as follows:

- Basic level—used in the safety assessment of bridges where no damage could compromise the structure's safety.
- Advanced level—used in safety assessments of bridges where some failures might threaten the structure's safety, and these failures can be fully detected using standard non-destructive or mildly destructive diagnostic tests, regardless of loads.
- Special level—used in the safety assessment of bridges where some failures might impact the structure's safety, and their detection requires non-standard diagnostic tests, including tests under proof and/or service loads (bridge performance tests).

The decision regarding which level of assessment to select should be based on the results of the initial condition survey of the bridge (inspection), conducted in consideration of the analysis of archival documentation and basic diagnostic tests. The recommendations for analysing the archival documentation of the bridge, as well as the methods and techniques used in diagnostic tests within the framework of assessing structural safety, complement the framework outlined in the specific guidelines [47].

3. Bridge Description

The paper discusses a 55-year-old road bridge with a superstructure made from a steel-welded shell structure. In the 1960s, it was an innovative steel shell design inspired by shipyard structural solutions. It served as a load-bearing structure for belt conveyors, supported industrial installations, and was used in road bridge construction. Between 1970 and 1989, several similar bridges were constructed in southern Poland, with two still in operation today. Despite their innovative structural concept at the time, these steel shell bridges exhibited numerous structural and quality issues, which led to cracks in welded joints and local deformations after some years of service. For example, the superstructure of the first bridge, built in 1970, exhibited numerous cracks in the welded joints of the stiffening trusses within the shell just one year after its completion. Additionally, minor local deformations were observed in some truss elements, indicating possible overloading. Analysis revealed that vibrations from the tram rolling stock on the bridge caused these cracks. After two years, a new tram track with vibration isolation was installed, and the cracks were repaired. Nevertheless, similar damages occurred later in other bridges of this type, primarily due to their location on less busy roads. The extreme form of these issues is evident in the structure of the bridge discussed in this paper.

The bridge was constructed in 1970 and designed to withstand live loads according to the old Polish bridge standard from the early 1960s. It has a total length of 288.0 m and a static scheme of a five-span continuous beam with span lengths of 48.0 m, 60.0 m, 72.0 m, 60.0 m, and 48.0 m (Figure 2). The overall width is 11.20 m, comprising a roadway with two lanes (2×3.50 m) and two sidewalks (2×2.10 m). The spans of the bridge were constructed as a steel shell structure, with a depth of 2.0 m and a width of 9.0 m (Figure 3). The shell, shaped like a basket curve, is made of steel sheets stiffened with longitudinal trapezoidal ribs. The top steel sheet with ribs forms an orthotropic steel deck. The shell superstructure is internally stiffened with steel trusses: four longitudinal trusses (arranged as two pairs) and numerous transverse trusses spaced every 2.0 m (Figure 4). The first type of trusses features a standard triangulated system, acting both as longitudinal stiffeners for the shell and as the main load-bearing girders. The other, a Vierendel truss (or rigid frame), functions as a cross-beam. Both types are made from rolled or welded I-beams, with welded joints connecting all elements of the steel superstructure. Reinforced concrete prefabricated slabs are attached atop the steel sheets on both sides to form the bridge sidewalks.

Figure 2. The general view of the bridge.

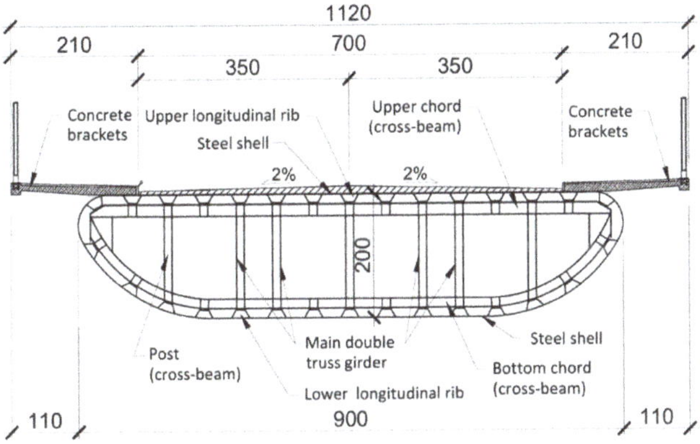

Figure 3. Cross-section of the steel shell superstructure.

Figure 4. Interior of the steel superstructure showing longitudinal and transverse stiffening trusses, with shell sheets reinforced by longitudinal trapezoidal ribs.

4. Basic Level of Assessment Framework

The initial stage of the bridge assessment involved analysing archival documents. All available information was included in the evaluation, such as the original design and construction (including workshop) documentation (i.e., calculations and drawings from 1969) and subsequent inspections and previous assessment reports conducted during the structure's service life. The inventory of the structure's geometry confirmed that it was built strictly in accordance with the archival drawings. No structural modifications were made after the original construction, nor were any documented or discovered on site. Material or product properties were determined based on relevant documents from the time of construction. The bridge was designed for live loads according to the old Polish bridge standard, dated 1966, and the basic load-bearing capacity and safety conditions for the steel structure were checked using the allowable stress method, as outlined in the Polish standard for steel structures from 1958. Therefore, the existing steel structure did not meet modern safety standards for bridges based on limit states and partial safety factors. In particular, welded joints were designed and inspected using outdated methods, and the designer at that time did not evaluate the fatigue limit state of these connections. Furthermore, the static analysis of the complex spatial superstructure was performed using simple (one-dimensional) calculation models.

The current condition of the existing steel structure, including ongoing deterioration processes and present defects, was determined based on available information from inspection reports. A condition survey was conducted to update the understanding of the structure's technical state, focusing on aspects critical to load capacity, safety, and reliability. The primary emphasis was on the structural elements that most significantly affect the safe performance of the steel structure. During the survey, the accuracy of as-built information was verified, and the superstructure's actual main dimensions and boundary conditions were confirmed. All data collected during the study were incorporated into the modelling for structural analysis.

The condition survey showed that the steel shell, some longitudinal ribs, and parts of the trusses were superficially corroded (Figures 2 and 4), mainly near equipment elements such as expansion joints, drainage inlets, and where the lighting lantern is attached to the structure, as well as at points where the prefabricated concrete slabs connect to the orthotropic bridge deck. Measurements of the thickness of these steel sheets and components indicated significant corrosion losses, reaching up to 3.0 mm in critical areas. The inspection also identified numerous (more than 100) severe plastic local deformations of the elements forming the trussed cross-beams (Figure 5). These deformations reduce both the load-bearing capacity of the cross-beams and the overall stiffness of the spans. Their high number, plastic deformation, and localisation mainly in the support zones of the spans may point to overloading as the primary cause of their occurrence.

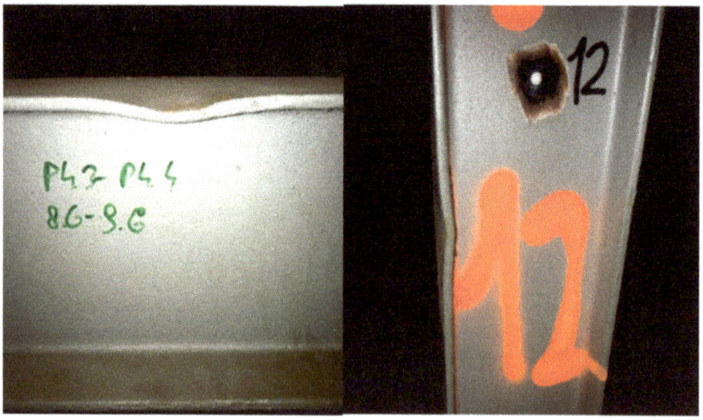

Figure 5. Local plastic deformations of the upper chord (**left**) and post (**right**) of the cross-beams.

The most serious damage to the steel structure, identified during the condition survey of the bridge, was cracks in the welds (Figure 6). At this assessment level, most superstructure welds were examined using a combination of visual inspection and penetrant tests. The inspection revealed the poor condition of the butt welds of the shell's longitudinal stiffening ribs. More than 40 cracks were discovered in the butt joints of the trapezoidal ribs, particularly located in the orthotropic deck at the support zones of the spans. These cracks disrupt the continuity of the longitudinal ribs, alter their static scheme and local internal force distribution, and negatively impact other structural elements, ultimately leading to their deterioration. For example, they caused cracks in adjacent fillet welds connecting the deck ribs to the upper chords of the cross-beams. More than 100 cracks (!) were found in these locations. The cracks varied in development, from hairline cracks that initiated weld destruction to complete joint break.

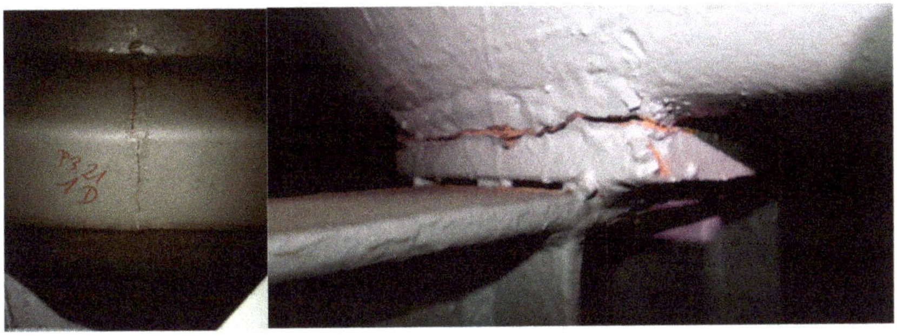

Figure 6. Typical weld cracks in the butt joints of longitudinal ribs (**left**) and an example of a crack in the fillet weld connecting the longitudinal rib to the upper chord of the cross-beam (**right**).

In most cases, the cracking of these joints was progressive, occurring 10 years after the initial repair of the welds, which revealed a permanent structural defect in this type of connection. Finally, several cracks were discovered in the welded joints between the upper cords of the main truss girders and cross-beams, particularly in the zones where the spans are supported. The primary cause of most cracks in welded joints is their poor quality, along with overloading and fatigue in some instances.

Following the assessment framework (Figure 1), it was necessary to determine whether the observed failures of the steel structure had affected the safety of individual elements or the entire superstructure of the bridge. As is well known, cracks, plastic deformations, and severe corrosion are the leading causes of performance degradation in steel bridges. Cracks threaten the safety of the bridge and can even lead to collapse. In this case, most cracks in welded joints resulted from poor and faulty quality (many initial defects) and distortion caused by overloading or restraint issues. Furthermore, steel bridges, especially those with orthotropic decks, are susceptible to fatigue cracking due to prolonged wheel loading and high cyclic stresses. The significance of the crack for the bridge's structural safety is of primary importance. If the crack is perpendicular to the tension direction induced by the service load, as shown in Figure 6, it may compromise the bridge's load-bearing capacity. Additionally, extensive plastic deformations in the cross-beams and local corrosion damage to the shells can substantially reduce the load-carrying ability. In summary, the damages observed in the steel superstructure, particularly their type and extent, may impact the structural safety of the bridge. Therefore, advancing to the second (more detailed) level of the assessment framework was necessary, as shown in Figure 1.

5. An Advanced Level of Assessment Framework

An advanced assessment should be undertaken if the basic assessment has not demonstrated the bridge's adequacy. According to the standard prEN-1990-2 [43], an advanced assessment may include: detailed documentary searches and reviews; further structural investigations to update information about the structural system, actions and influences—including possible damage and deterioration mechanisms, material properties, ground conditions, etc.; structural analysis and verifications; recommendations for immediate interventions; and other suggestions such as additional assessment stages, potential future interventions, and future use of the structure. All these measures are encompassed in the new national framework (Figure 1), with nearly all being applied in the case of the bridge in question. Since all archival documents found were used in the basic assessment, no detailed documentary search or review was necessary. Using archival static calculations, it was identified which elements and connections bear the most significant loads, and the structure's load-bearing capacity was established. Identifying these critical points limited the scope of advanced inspections. The advanced assessment started with further structural investigations (such as measurements or tests) to update information about the structural system, including possible damage and deterioration mechanisms, covering aspects like structural conditions, behaviour, actions, and influences.

5.1. Advanced Bridge Testing

The purpose of the material tests was to determine the grade of structural steel (confirming compliance with the design documentation in this respect), its mechanical parameters necessary for structural analysis, and to assess the degree of steel weldability in case it would be required to strengthen the structure. The steel grade was determined using a spectral examination of the material composition with the Spectroport F emission spectrometer. The tests revealed that the structure was made of low-carbon steel type St3m (formerly the national steel designation), corresponding to the modern steel type S235 according to the standard EN 10025-2:2017-03 [52]. Based on this, the characteristic yield strength of steel was assumed to be 235 MPa. The maximum carbon equivalent for the tested steel was calculated at 0.323%, which is below the threshold of 0.4%, indicating that the structural steel used to construct the bridge is easily weldable.

The advanced testing of welds aimed to identify damage that could affect the bridge's load-carrying capacity. Welds without visible defects were selected for testing. Two types of welds were fully identified using NDT methods: the transverse butt welds in the longitudinal trapezoidal ribs of the orthotropic deck and the fillet welds connecting the longitudinal ribs of the deck with the upper chords of transverse cross-beams. Based on weld type and size, the first were tested using ultrasound, while the latter were tested using the magnetic-particle method. Ultrasonic testing of the selected transverse butt welds was conducted using an ultrasonic flaw detector, type Epoch 600, in accordance with the standard EN ISO 1764:2018-01 [53]. The purpose was to detect subsurface discontinuities in the welds. The entire surface of the transverse joints was examined along their length. A total of 23 welds were tested, and no cracks were observed during visual inspection. However, all tested welds revealed severe discontinuities such as cracks and/or lack of penetration, indicating very low quality. The most extensive cracks reached up to three-quarters of the weld's length. The poor quality of the welds was mainly responsible for the large cracks. Additionally, numerous structural notches and cyclic loading of the bridge decks contributed to fatigue, which accelerated the growth of cracks.

Magnetic particle tests on selected welds were performed using a magnetic jet flaw detector AC42 K+D Flux Technic type, following standards EN ISO 9934-1:2017-02 [54] and EN ISO 17638:2017-01 [55]. The purpose was to detect surface and subsurface discontinuities in the fillet welds connecting the longitudinal ribs of the deck to the upper chords of the transverse cross-beams. The examination covered 22 welds, which showed no cracks based on visual inspection. The surface and subsurface of the joints were inspected over an approximately 20 mm area. On all tested welds, several inadmissible indications (i.e., flaws with $l > 2$ mm and/or $d > 1$ mm) were found on the surface, indicating very low weld quality (Figure 7).

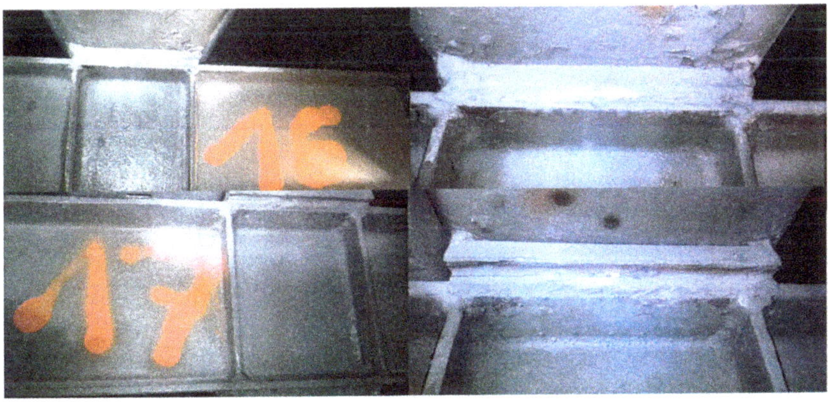

Figure 7. Surface discontinuities of the fillet welds connecting the longitudinal ribs of the deck with the upper chords of transverse cross-beams.

The observed discontinuities in the fillet-welded joints connecting the orthotropic deck and the cross-beams may lead to a lack of composite action between these elements, as assumed in the original design. Consequently, it was also investigated whether these discontinuities had a significant impact on the overall stiffness of the superstructure. The potential change in stiffness was evaluated by measuring the permanent deflection at the midpoint of the longest span (L = 72.0 m) using precision levelling. The average measured permanent deflection was 105 mm (L/686), which is considered substantial after 55 years of bridge operation, especially since the superstructure was initially built with reverse deflection. This indicates that the flaws found in the welded joints of the steel structure considerably reduce the overall stiffness of the spans.

5.2. Structural Analysis

The next step in the advanced-level assessment framework is the structural analysis (assessment by calculation). If an evaluation by calculation is performed, the values of actions and resistances should be considered for the limit states in the calculation situations being assessed. The rules for action combinations should also be consistent with the assessment situations relevant to the structure. The combination rules were taken from EN 1990 [44]. The identified damage and deterioration rates, along with their impact on structural performance and load-carrying capacity, are incorporated into the calculation models based on the available information obtained to date.

The structural analysis of the steel superstructure was conducted using the finite element method with a three-dimensional numerical model created in SOFISTIK version 2024 software. Information related to the structural system—such as the layout of the structure, arrangement of individual structural elements, and overall dimensions—was obtained from design documents. The numerical model accurately reflected the structure's geometry and components based on these documents, which, in some cases, were verified through inspection data (e.g., thickness of steel sheets). If the records did not contain the necessary information or if there were doubts about their accuracy, the structural model was updated during the initial inspection of the inventory. The assumptions regarding the static and kinematic boundary conditions of the structural system, including those related to support conditions, were confirmed to be valid.

Shell (quad) and beam finite elements were used to model the superstructure. Depending on the structure's geometry, the appropriate type of finite elements and meshing techniques were selected to generate the mesh. A general meshing approach was employed to ensure all nodes in contact areas converged. An ordered reticulation method was applied. The four-node shell elements discretised the superstructure shell with a maximum size of 300 × 500 mm. Mesh refinement using shell elements of size 150 × 500 mm was carried out for the longitudinal trapezoidal ribs. Various shell sheet thicknesses, as well as different types and sizes of ribs and truss sections, were considered. Two-node beam elements were used to model the internal trusses of girders (posts, diagonals, and both transverse chords). The actual thicknesses of individual truss elements, which vary across the cross-section of the bridge span and along its length, were incorporated into the model. The truss connections were modelled as rigid, transmitting the bending moment, on a standard grid of nodes. In the key connections between the transverse upper chords and the trapezoidal longitudinal ribs, restrictions were introduced using spring elements (spri). The spring elements connecting the nodes of the upper chord with those of the longitudinal ribs were given maximum rigidity by modelling the welded joint. Axial stiffnesses of 10^{16} kN/m^3 were assumed in three orthogonal directions. Additionally, the spring elements in the three orthogonal bending planes have flexible stiffnesses of 10^{16} kN/rad/m in each plane. This node model represented a welded joint with full rigidity, free from damage. The visualisation of the numerical model of the steel superstructure is shown in Figure 8.

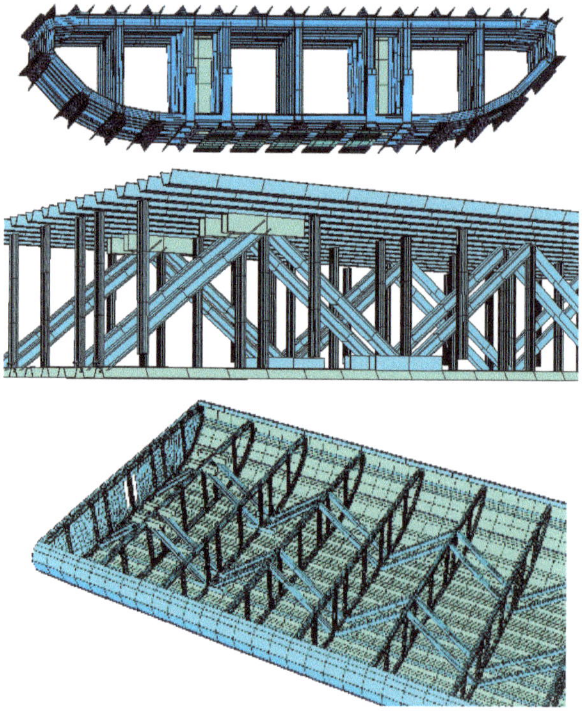

Figure 8. Numerical model of the superstructure.

The results of the condition survey and advanced bridge testing were integrated into the model, including corrosion losses of deteriorated elements and cracks in the fillet welds between the longitudinal ribs of the deck and the upper chord of the cross-beams. The latter failures were simulated by removing the axial stiffness in the (x) and (y) directions, along with the corresponding bending stiffnesses in the same planes. An axial stiffness of 1 kN/m^3 in these two directions and a matching bending stiffness of 1 kN/rad/m in the same planes were introduced. In the (z) direction, the full axial stiffness, namely

10^{16} kN/m^3, and the corresponding full bending stiffness of 10^{16} kN/rad/m were retained, thereby modelling the pressure in the gravitational direction.

The material properties of the finite elements were assumed to be derived from and updated based on analysis of design specifications and construction records, corroborated by the condition survey and material testing results. Based on the material tests conducted, the parameters of the structural steel were found to align with those of modern S-235 steel. At this stage of the structural analysis, steel was modelled as a linear-elastic isotropic material using two engineering constants: E and ν. The material constants for steel and strength parameters were assumed to be E = 210 GPa, ν = 0.3, and f_y = 235 MPa.

According to prEN-1990-2 [43], traffic loads on existing bridges should be assumed using the same live load models established in EN 1991-2 [48] for designing new bridges. However, case-specific traffic load models may be specified by the relevant authority, based on the actual traffic characteristics and the traffic forecast over the remaining service life. In our case, the traffic load models used to evaluate the bridge's load-bearing capacity were adopted in accordance with the requirements of the Polish road administration [9]. The vehicle schemes, loads, and markings, as adopted as actual traffic loads in the structural analysis, are shown in Figure 9.

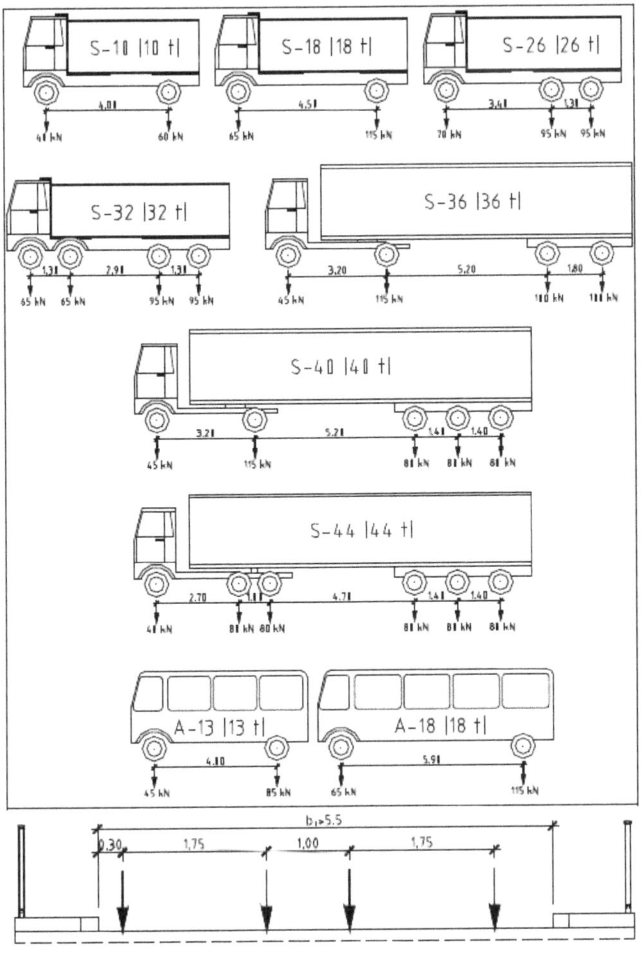

Figure 9. Models of traffic vehicles for assessing the structural safety of existing bridges, as per [9].

The single vehicles, as shown in Figure 9, were used in the numerical analysis as a surface load (pressure) of a single wheel, acting on the actual surface of the upper deck plate, considering a wheel load area of 0.30 × 0.40 m. An example of the load position of the traffic vehicle, marked as 2×S10, is shown in Figure 10.

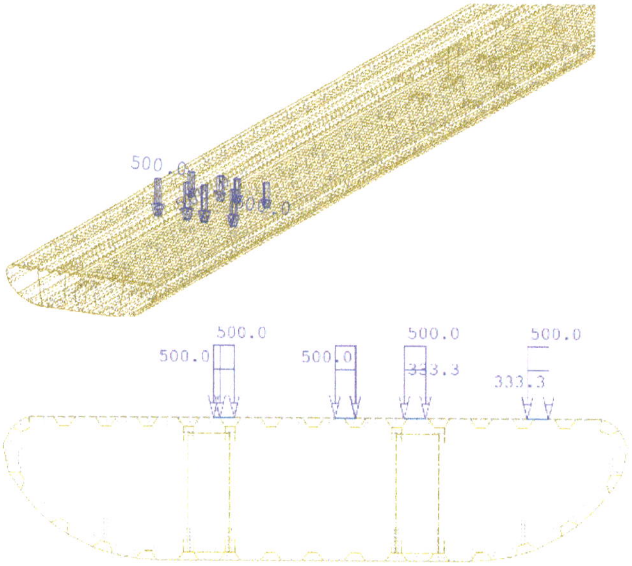

Figure 10. Wheel load of the 2×S-10 single traffic vehicle [kN/m^2].

The following traffic vehicles, according to Figure 9, were set to move along the length of the bridge spans, assuming two possible combinations:

1. Unrestricted traffic: single vehicles moving individually in both directions on each lane of the bridge roadway, occupying the most disadvantageous position from the perspective of the calculated static value.
2. Shuttle traffic: individual vehicles moving separately in one direction on a single lane of the bridge road, occupying the least favourable position from the calculated static value's perspective.

The analysis also assumed that no other vehicles were on the bridge roadway apart from the single vehicles of each type.

Other actions were implemented in accordance with the relevant Eurocodes. The self-weight of the structure or its elements, as well as other permanent actions, was determined based on the relevant geometrical properties and material densities. The pedestrian load, with a characteristic value of 5.00 kN/m^2, was applied to the actual contact area between the prefabricated concrete slabs and the orthotropic deck, spanning a width of 0.50 m and the entire length of the bridge. The pedestrian load was modelled as a surface load on the top plate of the steel deck, acting in the gravitational direction with the corresponding bending moment related to the transfer of the load's centre of gravity (Figure 11). The load was modelled separately for both sidewalks, varying along the length of the bridge. The analysis also included the effect of temperature changes at the depth of the steel structure. The temperature difference for the simplified model at the extreme fibres was assumed to be 18 °C for heating and 13 °C for cooling (Figure 12). The scenario of uniform heating and cooling actions on the structure was not considered in the calculation.

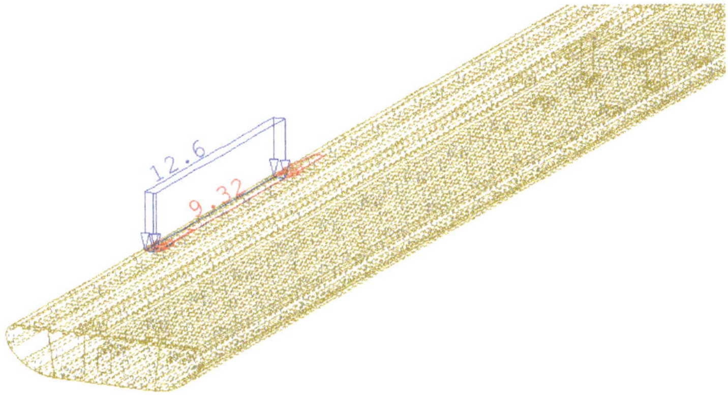

Figure 11. Pedestrian load on sidewalks.

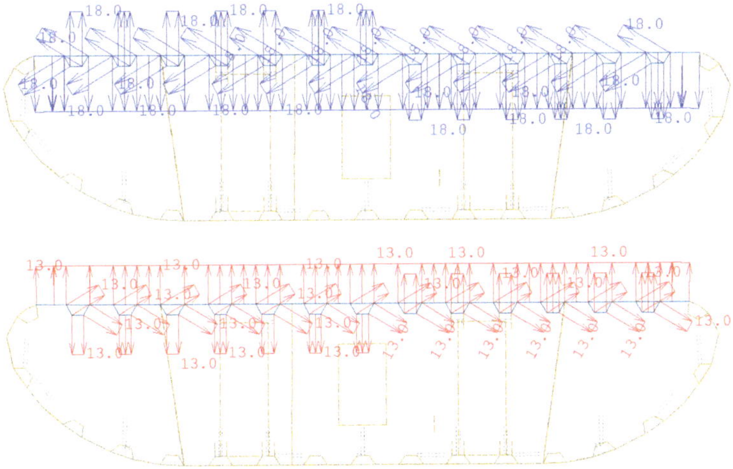

Figure 12. Temperature actions: heating (**top**) and cooling (**bottom**).

The action combinations followed EN 1990 [44], using the partial safety and combination factors shown in Table 4. These combinations were created to simulate the most severe conditions of the considered actions.

Table 4. Partial safety factors and combination factors applied in structural analysis.

Type of Action	$\gamma_{G,sup}$	$\gamma_{G,inf}$	Ψ_0	Ψ_1	Ψ_2
Self-weight	1.35	1.35	1.00	1.00	1.00
Permanent actions	1.35	1.35	1.00	1.00	1.00
Traffic live load	1.35 [1]	0.00 [1]	0.75 [1]	0.75 [1]	0.00 [1]
Pedestrian load	1.35	0.00	0.40	0.40	0.00
Thermal actions	1.50	0.00	0.60	0.60	0.50

[1] taken as for tandem TS of the LM1 model in Eurocode.

5.3. Structural Safety Verification

The ultimate limit state (ULS) verification (or structural safety verification) was performed using the limit state method with the inequality given by Formula (1):

$$Ea \leq Ra, \tag{1}$$

where

*E*a is the assessment value of the effect of actions;

*R*a is the assessment value of the resistance.

*E*a and *R*a can be expressed as functions of the assessment values of the basic variables, including the relevant partial safety factors and combination factors. In our case, the equivalent stress according to the Huber-Mises-Hencky strength criterion (H-M-H) was taken as the assessment value of the effect of actions *E*a, while the design strength of structural steel, determined based on material tests and archival documents, was used as the assessment value of the resistance *R*a. The relevant load and material partial safety factors, as well as combination factors, were applied to both assessment values in accordance with the Eurocodes. Consequently, using Formula (1), the ultimate limit state (ULS) of individual superstructure elements was verified. Since the structure lacked sufficient ductility for the required deformations (due to cracks), plastic redistribution was not considered for assessment at the ultimate limit state. The selected results of the ULS analysis, presented as H-M-H stress plots, are shown in Figures 13–15.

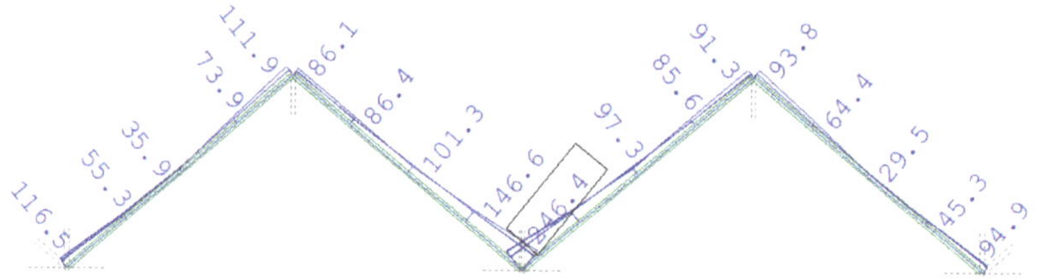

Figure 13. Maximum H-M-H stresses for the diagonals of the longitudinal stiffening truss (under 2 × S-10 vehicle load).

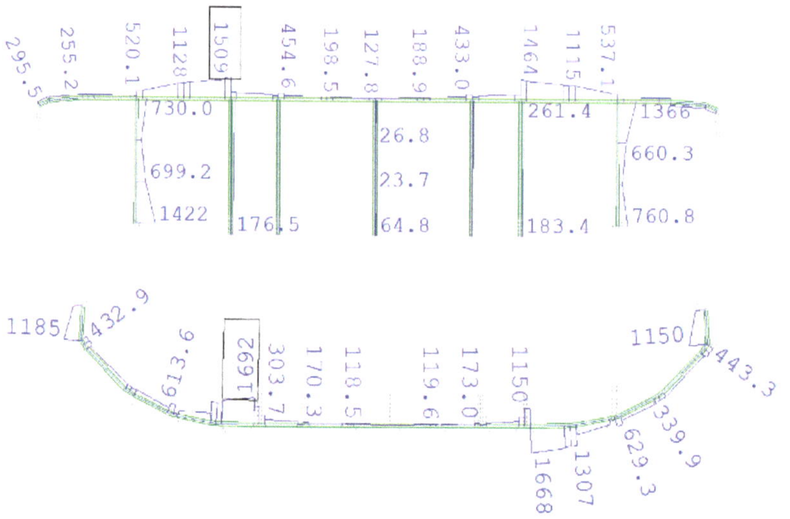

Figure 14. Maximum H-M-H stresses for the cross-beam elements (under 2 × S-10 vehicle load).

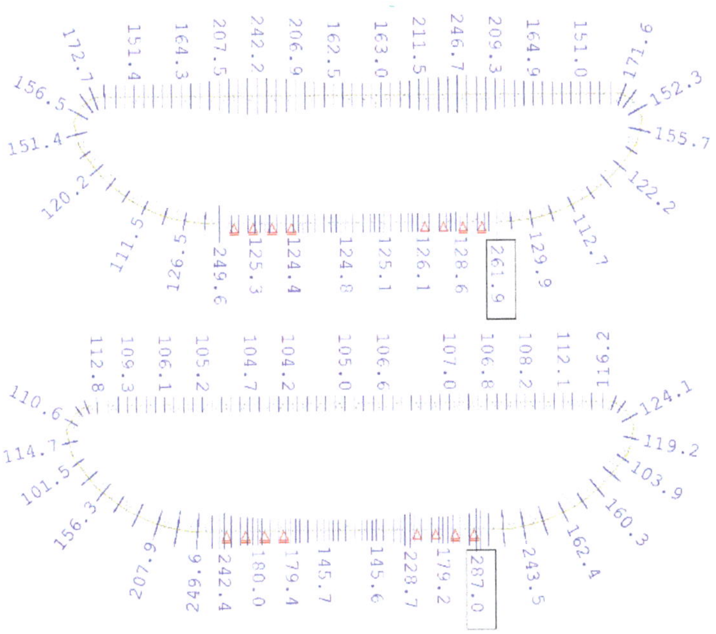

Figure 15. Maximum H-M-H stresses for the shell sheets at the P-3 support section of the span (under 2 × S-10 vehicle load).

Although the FEA results are displayed visually in the figures above, tabulated stress values or safety margins for key elements would enhance transparency and clarity. Therefore, in Table 5, these values are additionally provided for selected elements of the steel superstructure.

Table 5. H-M-H stress values and safety margins for key elements of superstructure.

Superstructure Element	Location	H-M-H Stresses [1] [MPa]	Steel Design Strength [MPa]	Exceeding the ULS [%]
Upper chord	P3 support	1509.0		542.1
Bottom chord	P3 support	1692.0		620.0
External post	P3 support	1422.0		505.1
Internal post	P3-P4 span	446.8	235	90.1
Shell	P3 support	261.9		11.4
Upper rib	P3 support	198.1		----
Bottom rib	P3 support	306.6		30.5
Diagonal	All supports	246.4		4.9

[1] under 2 × S-10 vehicle load.

The assessment of the steel superstructure revealed that the structure lacks a sufficient safety margin regarding the ultimate limit state (ULS), even when subjected to the weight of a traffic vehicle of the lowest class, 2 × S-10 (a vehicle with a total weight of 10 tonnes). The main steel elements responsible for the superstructure's insufficient safety are as follows:

- Longitudinal truss diagonals—the assessment value of the resistance (*R*a) is exceeded by up to 5%; number of deficient elements—4;
- Upper chord of cross-beams—the assessment value of the resistance (*R*a) is exceeded by up to 550% (!!!); number of deficient elements—20;

- Lower chord of cross-beams—the assessment value of the resistance (*R*a) is exceeded by up to 620% (!!!); number of deficient elements—20;
- Cross-beam posts—the assessment value of the resistance (*R*a) is exceeded by up to 505% (!!!); the number of deficient elements is 120.

The shell sheets of the superstructure, its longitudinal stiffening ribs, and its upper part, the orthotropic deck, have sufficient safety regarding ULS; however, only for the lowest class 2 × S-10 traffic vehicle. Additionally, the analysis showed that the resistance values of some of the elements mentioned above were exceeded under self-weight and permanent load (by up to 200%). A comparison between analysis results and the condition survey indicated the same elements, where local plastic deformations were observed (Figure 5). The structural analysis revealed that more than 180 superstructure elements are insufficiently safe to carry the actual minimum traffic loads with the safety level required by the new national guidelines [47] and the European standard prEN 1990-2 [43].

5.4. Maintenance Decisions

According to the European standard EN 13306:2010 [56], the general definition of maintenance is given as the *"combination of all technical, administrative, and managerial actions throughout the life cycle of an item intended to keep it in, or restore it to, a state in which it can perform the required function"*. It is clear that determining appropriate maintenance actions is a complex decision-making process; that is, several decisions must be made regarding which actions are relevant, how they should be implemented, when they should be executed, who should be responsible, and how many resources should be allocated. To optimise these decisions, they need to be supported by information about the actual technical condition (including load-carrying capacity) of the specific asset.

Considering the poor technical condition of the steel superstructure (corrosion, numerous cracks in welds, and local deformations) and the results of the structural safety assessment, which essentially disqualified the continued use of the steel structure for the bridge's operation, the fatigue life analysis (typically conducted in such cases, see [57]) and the special bridge testing were deemed unnecessary. Because of the very low structural safety of the superstructure (even under self-weight and permanent load) revealed during the assessment, the bridge authority decided not to impose any traffic restrictions (such as vehicle load or speed limits, or operational constraints). Instead, they ordered the immediate closure of the bridge until a decision was made on whether to repair/strengthen or replace the superstructure. However, repairing or strengthening the steel structure would involve extensive work and high direct costs. Furthermore, the life cycle cost analysis comparing repair versus replacement convinced the road authority to replace the superstructure completely. Over the following two years, the old steel superstructure will be replaced with a modern steel–concrete composite structure supported on new piers with reinforced existing foundations.

6. Conclusions

The structural safety assessment of the existing steel bridge, considering the technical condition of the structure and current traffic loads, showed that the steel superstructure did not meet the required safety standards according to the new national Polish guidelines, based on the prEN-1990-2 standard for assessing existing structures. The following detailed conclusions were drawn from the assessment:

- More than 180 superstructure elements are insufficiently safe to support the actual minimum traffic loads with the safety level required by the new national guidelines and the European standard prEN 1990-2.

- The resistance values of some structural elements were exceeded even under self-weight and permanent load (by up to 200%).
- The condition survey indicated local plastic deformations in the same elements, demonstrating a lack of load-bearing capacity based on the assessment.
- The remaining superstructure elements have sufficient safety concerning ULS; however, only for the lowest class 2 × S-10 traffic vehicle.

The structural analysis revealed a very low level of structural safety, significantly below the standards established by the Eurocodes. This is primarily due to the inadequate structural analysis of the highly complex steel structure conducted during the bridge's design in 1969. Additionally, observed failures such as cracks in welds and local deformations significantly compromise the bridge's safety. In this situation, permitting any traffic load on the bridge would pose a direct threat to its safety and that of its users. Therefore, it was recommended that the bridge be closed to public traffic and replaced with a new structure.

The new guidelines for the structural safety assessment of existing road bridges act as a tool to support decision-making by road authorities. According to the initial assumptions of the Ministry of Infrastructure, the guidelines may be utilised, among other things, in cases such as (needs): lack of maintenance information from the previous operational period of the bridge; doubts or irregularities observed during bridge monitoring; an increase in service load due to changes in road classification; determining the military load class (MLC) for bridges on public roads; evaluating the extent of structural modifications needed for rehabilitation or strengthening of existing structures; and so on.

The application of the new guidelines has eased the assessment of the complex steel bridge in accordance with the unified principles introduced across the European Union through the implementation of the prEN 1990-2 standard. This new standard represents a significant step forward, as it is among the first international design codes to specifically focus on assessment. The new Polish guidelines standardise, at a European level, methods for evaluating existing structures and outline the criteria to consider when investigating and appraising existing bridges, while also establishing differences compared to conventional design for new structures. The authors emphasise that particular importance was given to understanding the structural configuration and the most probable failure modes in existing structures, as this provides a crucial basis for planning and conducting verifications most suitable and necessary for each case to evaluate the safety, durability, and functionality of the structure.

The application of these new national guidelines to assess the structural safety of an existing 55-year-old bridge, as discussed in the paper, highlights their practical significance. The scientific significance of the paper lies in presenting new Polish national guidelines for assessing existing bridges, developed with the authors' participation as a national implementation of the new Eurocode prEN 1990-2. The paper illustrates the application of these guidelines using a complex steel bridge structure with poor technical condition. This example could serve as a benchmark for similar assessments carried out in accordance with the new Eurocode, making it of considerable practical importance. However, it should be noted that the assessment procedure was not completed (i.e., the special level was not used) due to the critical condition of the bridge and the urgent need to close it.

Author Contributions: Conceptualisation, T.S.; methodology, T.S.; software, M.R.; validation, T.S.; investigation, M.R.; resources, T.S.; data curation, M.R.; writing—original draft preparation, M.R.; writing—review and editing, T.S.; visualisation, M.R.; supervision, T.S.; project administration, T.S.; funding acquisition, T.S. All authors have read and agreed to the published version of the manuscript.

Funding: This work is based on research partially supported by the Bridge Administrator—DISTRICT ROAD AUTHORITY IN TARNÓW, under Commercial Agreement No. PZD.1131.26.2019.

Data Availability Statement: The data presented in this study are available upon request from the corresponding author, subject to approval by the District Road Authority in Tarnów, Poland.

Acknowledgments: The research team also thanks Grupa AZOTY and PROReM Sp. z o.o. Laboratorium Metaloznawstwa, Tarnów, for sharing material testing data.

Conflicts of Interest: The authors declare no conflicts of interest. The funders had no role in the design of the study, in the collection, analysis, or interpretation of data, in the writing of the manuscript, or in the decision to publish the results.

Abbreviations

The following abbreviations are used in this manuscript:

AHP	Analytical hierarchy process
NDT	Non-destructive testing
FEA	Finite element analysis
SHM	Structural health monitoring
NBI	National Bridge Inventory
CNN	Convolutional Neural Network
GDDKiA	General Directorate for National Roads and Motorways, Poland
H-M-H	Huber-Mises-Hencky strength criterion
ULS	Ultimate Limit State (load-bearing capacity)
MLC	Military load class

References

1. Maljaars, J.; Steenbergen, R.; Abspoel, L.; Kolstein, H. Safety assessment of existing highway bridges and viaducts. *Struct. Eng. Int.* **2012**, *22*, 112–120. [CrossRef]
2. Björnsson, I.; Larsson Ivanov, O.; Honfi, D.; Leander, J. Decision support framework for bridge condition assessments. *Struct. Saf.* **2019**, *81*, 101874. [CrossRef]
3. Bień, J.; Salamak, M. The management of bridge structures—Challenges and possibilities. *Arch. Civ. Eng.* **2022**, *2*, 5–35. [CrossRef]
4. Matos, J.C.; Nicoletti, V.; Kralovanec, J.; Sousa, H.S.; Gara, F.; Moravcik, M.; Morais, M.J. Comparison of condition rating systems for bridges in three European countries. *Appl. Sci.* **2023**, *13*, 12343. [CrossRef]
5. Casas, J.R.; Frangopol, D.M.; Turmo, J.; Tsompanakis, Y. Bridge safety assessment, maintenance, monitoring, and life-cycle performance. *Struct. Infrastruct. Eng.* **2024**, *20*, 957–959. [CrossRef]
6. Gara, F.; Carbonari, S.; Nicoletti, V.; Martini, R.; Brunetti, A.; Torsani, A.; Dall'Asta, A. Assessment and management of existing bridges following the innovative Italian guidelines: A pilot study. *Int. J. Bridge Eng. Manag. Res.* **2024**, *2*, 21425008–1:18. [CrossRef]
7. Lu, P.; Easa, S.M.; Wu, Y.; Qi, Z.; Zhuang, Y. Hybrid assessment method of bridge bearing capacity. *Balt. J. Road Bridge Eng.* **2024**, *19*, 69–101. [CrossRef]
8. Ferrara, M.; Gino, D.; Miceli, E.; Giordano, L.; Malavisi, M.; Bertagnoli, G. Safety assessment of existing prestressed reinforced concrete bridge decks through different approaches. *Struct. Concr.* **2024**, *25*, 1637–1657. [CrossRef]
9. *Manual for Assessing the Structural Safety of Road Bridge Structures*; General Directorate for National Roads and Motorways: Warsaw, Poland, 2004.
10. Siwowski, T. Comparison of standard and actual structural safety of a typical truss bridge. *J. Civ. Eng. Environ. Archit.* **2009**, *51*, 91–106. Available online: https://yadda.icm.edu.pl/baztech/element/bwmeta1.element.baztech-article-BTB2-0055-0108 (accessed on 4 September 2025). (In Polish)
11. Wiśniewski, D.; Majka, M.; Bień, J. Capacity assessment of existing bridges during their service life—National and international experience. *Inżynieria I Bud.* **2013**, *69*, 436–441. (In Polish)
12. Ambroziak, A.; Malinowski, M. Technical assessment of 120-year-old railway riveted truss bridge. *Balt. J. Road Bridge Eng.* **2025**, *20*, 70–93. [CrossRef]
13. Tanner, P.; Hingorani, R.; Lara, C. New or existing, does it matter? *Acta Polytech. Czech Tech. Univ. Prague—CTU Proc.* **2022**, *36*, 286–295. [CrossRef]
14. Xia, P.-Q.; Brownjohn, J.M.W. Bridge structural condition assessment using systematically validated finite-element model. *J. Bridge Eng.* **2004**, *9*, 418–423. [CrossRef]
15. Ding, L.; Hao, H.; Xia, Y.; Deeks, A.J. Assessment of bridge load carrying capacity using updated finite element model and nonlinear analysis. *Adv. Struct. Eng.* **2012**, *15*, 1739–1750. [CrossRef]

16. Owerko, P. The stochastic finite element method in the assessment of bridge infrastructure objects—Review. *E3S Web Conf.* **2018**, *45*, 00062. [CrossRef]
17. Islam, N.; Miyashita, T.; Shill, S.K.; Takeda, K.; Fukada, S.; Takasu, A.; Al-Deen, S.; Subhani, M. Assessment of structural health of an existing prestressed concrete bridge by finite element analysis. *Aust. J. Civ. Eng.* **2022**, *21*, 265–278. [CrossRef]
18. Frangopol, D.M.; Kallen, M.-J.; van Noortwijk, J.M. Probabilistic models for life-cycle performance of deteriorating structures: Review and future directions. *Prog. Struct. Eng. Mater.* **2004**, *6*, 197–212. [CrossRef]
19. Casas, J.R.; Wisniewski, D. Safety requirements and probabilistic models of resistance in the assessment of existing railway bridges. *Struct. Infrastruct. Eng.* **2013**, *9*, 529–545. [CrossRef]
20. Obrien, E.J.; Schmidt, F.; Hajializadeh, D.; Zhou, X.-Y.; Enright, B.; Caprani, C.C.; Wilson, S.; Sheils, E. A review of probabilistic methods of assessment of load effects in bridges. *Struct. Saf.* **2015**, *53*, 44–56. [CrossRef]
21. Matos, J.C.; Cruz, P.J.S.; Valente, I.B.; Neves, L.C.; Moreira, V.N. An innovative framework for probabilistic-based structural assessment with an application to existing reinforced concrete structures. *Eng. Struct.* **2016**, *111*, 552–564. [CrossRef]
22. Olaszek, P.; Lagoda, M.; Casas, J.R. Diagnostic load testing and assessment of existing bridges: Examples of application. *Struct. Infrastruct. Eng.* **2014**, *10*, 834–842. [CrossRef]
23. Alampalli, S.; Frangopol, D.M.; Grimson, J.; Halling, M.W.; Kosnik, D.E.; Lantsoght, E.O.L.; Yang, D.; Zhou, Y.E. Bridge load testing: State-of-the-practice. *J. Bridge Eng.* **2020**, *26*, 03120002. [CrossRef]
24. Lantsoght, E.O.; De Boer, A.; Van der Veen, C.; Hordijk, D.A. Optimizing finite element models for concrete bridge assessment with proof load testing. *Front. Built Env.* **2019**, *5*, 99. [CrossRef]
25. Nguyen, D.C.; Salamak, M.; Katunin, A.; Gerges, M. Finite element model updating of RC bridge structure with static load testing: A case study of Vietnamese Thithac bridge in coastal and marine environment. *Sensors* **2022**, *22*, 8884. [CrossRef]
26. De Domenico, D.; Messina, D.; Recupero, A. Quality control and safety assessment of prestressed concrete bridge decks through combined field tests and numerical simulation. *Structures* **2022**, *39*, 1135–1157. [CrossRef]
27. Wenzel, H. *Health Monitoring of Bridges*, 1st ed.; John Wiley & Sons: Hoboken, NJ, USA, 2009.
28. Bell, E.S.; Lefebvre, P.J.; Sanayei, M.; Brenner, B.; Sipple, J.D.; Peddle, J. Objective load rating of a steel-girder bridge using structural modelling and health monitoring. *J. Struct. Eng.* **2013**, *139*, 1771–1779. [CrossRef]
29. Sousa, H.; Bento, J.; Figueiras, J. Assessment and management of concrete bridges supported by monitoring data-based finite-element modelling. *J. Bridge Eng.* **2014**, *19*, 05014002. [CrossRef]
30. Seo, J.; Czaplewski, T.M.; Kimn, J.H.; Hatfield, G. Integrated structural health monitoring system and multi-regression models for determining load ratings for complex steel bridges. *Measurement* **2015**, *75*, 308–319. [CrossRef]
31. Jamali, S.; Chan, T.H.T.; Nguyen, A.; Thambiratnam, D.P. Reliability-based load-carrying capacity assessment of bridges using structural health monitoring and nonlinear analysis. *Struct. Health Monit.* **2019**, *18*, 20–34. [CrossRef]
32. Ko, S.-W.; Kim, J.-K. A framework for assessment the load-carrying capacity of bridges without design document using an AI technique. *Appl. Sci.* **2023**, *13*, 1283. [CrossRef]
33. Liu, H.; Zhang, Y. Bridge condition rating data modelling using deep learning algorithm. *Struct. Infrastruct. Eng.* **2020**, *16*, 1447–1460. [CrossRef]
34. *The Manual for Bridge Assessment*, 3rd ed.; American Association of State Highway and Transportation Officials (AASHTO): Washington, DC, USA, 2018.
35. *Canadian Highway Bridge Design Standard CAN/CSA-S6-06*; Canadian Standards Association: Toronto, ON, Canada, 2006.
36. *Specification for Inspection and Assessment of Load-Bearing Capacity of Highway Bridges JTG/T J21–2011*; Ministry of Communications and Transportation: Beijing, China, 2011.
37. *Guidelines on Risk Classification and Management, Safety Assessment and Monitoring of Existing Bridges*; Ministry of Infrastructure: Rome, Italy, 2020.
38. *Bases pour la Maintenance Des Structures Porteuses SIA-269*; Société Suisse des ingénieurs et des Architects: Zurich, Switzerland, 2011.
39. *Design Manual for Road and Bridges, Highway Structures: Inspection and Maintenance—Assessment*; Vol. 3, Sec. 4; Highways Agency: London, UK, 2006.
40. *EN ISO 13822: 2010*; Basis for Design of Structures—Assessment of Existing Structures. International Organization for Standardization: Geneva, Switzerland, 2010.
41. *Fib Bulletin 80: 2016*; Assessment of an Existing Prestressed Concrete Bridge. Fédération Internationale du Béton: Lausanne, Suisse, 2016.
42. Wiśniewski, D.F.; Casas, J.R.; Ghosn, M. Codes for safety assessment of existing bridges—Current state and further development. *Struct. Eng. Int.* **2012**, *22*, 552–561. [CrossRef]
43. Allaix, D.L.; Bigaj-van Vliet, A. Existing standardization on monitoring, safety assessment and maintenance of bridges and tunnels. *Ce/Papers* **2023**, *6*, 498–504. [CrossRef]

44. *prEN 1990-2:2024*; Eurocode 0: Basis of Structural and Geotechnical Design—Part 2: Assessment of Existing Structures. Comité Européen de Normalisation: Brussels, Belgium, 2024.
45. *EN 1990:2002*; Eurocode 0: Basis of Structural Design. Comité Européen de Normalisation: Brussels, Belgium, 2002.
46. *The Instructions to Conduct the Inspections of Road Bridges*, 3rd ed.; General Directorate for National Roads and Motorways: Warsaw, Poland, 2020.
47. Siwowski, T.; Bień, J.; Wiśniewski, D.; Żółtowski, K. *Guidelines for Determining the Structural Safety of Existing Road Bridges—WR-M-82*; Guidelines and standards recommended by the Minister of Infrastructure; Ministry of Infrastructure: Warsaw, Poland, 2024.
48. *EN 1991-2:2003*; Eurocode 1: Actions on Structures—Part 2: Traffic Loads on Bridges. Comité Européen de Normalisation: Brussels, Belgium, 2003.
49. *EN 1992-2:2005*; Eurocode 2: Design of Concrete Structures—Part 2: Concrete Bridges—Design and Detailing Rules. Comité Européen de Normalisation: Brussels, Belgium, 2005.
50. *EN 1993-2:2006*; Eurocode 3: Design of Steel Structures—Part 2: Steel Bridges. Comité Européen de Normalisation: Brussels, Belgium, 2006.
51. *EN 1994-2:2005*; Eurocode 4: Design of Composite Steel and Concrete Structures—Part 2: General Rules and Rules for Bridges. Comité Européen de Normalisation: Brussels, Belgium, 2005.
52. *EN 10025-2:2017-03*; Hot Rolled Products of Structural Steels—Part 2: Technical Delivery Conditions for Non-Alloy Structural Steels. Comité Européen de Normalisation: Brussels, Belgium, 2017.
53. *EN ISO 17640:2018-01*; Non-Destructive Testing of Welds—Ultrasonic Testing—Techniques, Testing Levels, and Assessment. International Organization for Standardization: Geneva, Switzerland, 2018.
54. *EN ISO 9934-1:2017-02*; Non-Destructive Testing—Magnetic Particle Testing—Part 1: General Principles. International Organization for Standardization: Geneva, Switzerland, 2017.
55. *EN ISO 17638:2017-01*; Non-Destructive Testing of Welds—Magnetic Particle Testing. International Organization for Standardization: Geneva, Switzerland, 2017.
56. *EN 13306:2010*; Maintenance—Maintenance Terminology. Comité Européen de Normalisation: Brussels, Belgium, 2010.
57. Rajchel, M.; Siwowski, T. Fatigue assessment of a 100-year-old riveted truss railway bridge. *J. Constr. Steel Res.* **2024**, *217*, 108662. [CrossRef]

applied sciences

MDPI

Article

Experimental and Parametric Analysis of Pull-Out Resistance of Notched T-Perfobond Connectors

Zifa Dong [1], Shuangjie Zheng [1,2,*], Lizhe Jiao [1], Xiaoqing Xu [3], Yao Yao [1], Zhuoru Gao [1] and Haifeng Li [1]

[1] College of Civil Engineering, Huaqiao University, Xiamen 361021, China; zfdong@stu.hqu.edu.cn (Z.D.); lz.jiao@stu.hqu.edu.cn (L.J.); 23013086028@stu.hqu.edu.cn (Y.Y.); gaozhuoru@163.com (Z.G.); lihai_feng@126.com (H.L.)

[2] Key Laboratory for Intelligent Infrastructure and Monitoring of Fujian Province, Huaqiao University, Xiamen 361021, China

[3] Department of Bridge Engineering, Tongji University, Shanghai 200092, China; xq.xu@foxmail.com

* Correspondence: sjzheng@hqu.edu.cn; Tel.: +86-188-5016-4929

Abstract: To enhance the uplift capacity and facilitate the installation of multi-row perfobond connectors at shallow burial depths, this study puts forward a novel notched T-perfobond connector. The design incorporates an integrated flange at the bottom of the connector and a notch at the edge of the hole. Through pull-out model tests on four notched T-perfobond connectors, this research investigates their failure mechanisms and pull-out capacities. Utilizing the explicit dynamics method in ABAQUS, a finite-element model of the pull-out resistance test for notched T-perfobond connectors is established and verified against experimental data. Furthermore, a detailed parametric analysis involving 54 models is conducted, examining crucial parameters such as rib dimensions, hole geometry, flange size, notch width, bar diameter, and material properties. Based on the combined experimental and numerical results, this paper assesses the suitability of current formulas for calculating the pull-out capacity of perfobond connectors and proposes a refined calculation method specifically for notched T-perfobond connectors. All the findings reported in this paper can serve as a reference in the design and construction of composite structures.

Keywords: composite structures; perfobond connectors; flange; notch; pull-out performance; finite-element analysis

1. Introduction

As pivotal interfacing elements in composite structures, perfobond connectors have been widely utilized in various structural systems, such as composite beams, hybrid beams, composite towers, bridge decks, and composite anchorages [1–4], due to their outstanding shear resistance and fatigue resistance, as well as ease of installation.

In recent years, extensive research has been conducted on perfobond connectors, employing a variety of methodologies. Physical modeling, the first of these methodologies, involves the creation of scaled or full-sized replicas of the connectors and the structures they are a part of. In particular, push-in, push-out, and pull-out model tests are usually used to obtain the shear and tensile behavior of perfobond connectors [5–10]. This hands-on approach allows researchers to observe, firsthand, the mechanical interactions that occur under real-world conditions, such as the distribution of forces and the development of stresses within the composite system. Physical models are particularly useful for studying complex interactions that cannot be fully captured by computational methods alone.

The second methodology is computational simulation, which leverages advanced software tools, such as ABAQUS (https://www.3ds.com/products/simulia/abaqus accessed date 10 August 2024) and ANSYS (https://www.ansys.com/ accessed date 10 August 2024), to create virtual models of perfobond connectors and the composite structures they are used in [11–16]. These simulations can account for a wide range of variables, including material properties, loading conditions, and environmental factors. By simulating various scenarios, researchers can predict the behavior of perfobond connectors under different circumstances, which is invaluable for optimizing their design and application.

Theoretical analysis, the third methodology, involves the development of mathematical models to describe the behavior of perfobond connectors [17–22]. This approach is based on established principles of mechanics and materials science and can provide deep insights into the fundamental mechanisms at play. Theoretical models can be used to derive equations that predict shear performance and pull-out characteristics, offering a basis for the design and evaluation of perfobond connectors.

Substantial research has been devoted to the fundamental mechanical attributes of perfobond connectors within conventional concrete slabs, yet some knowledge deficits persist [23–25]. For example, specific composite structures such as ultra-thin bridge decks possess an exceptionally thin concrete layer, raising the slab's out-of-plane uplift resistance to a critical concern. Nevertheless, current literature mainly focuses on shear performance, frequently neglecting pull-out behavior.

Moreover, the insertion of steel bars to enhance connector performance is highly advantageous, but the closed circular holes of conventional perfobond connectors impede this procedure, necessitating the exploration of the mechanics of notched perfobond connectors. Enlarging the notch to align with the hole size might compromise the pull-out capacity, resulting in our proposal to incorporate a flange into the design of the notched perfobond connector.

In addition, although thin concrete components are prone to tensile cracking, the utilization of ultra-high-performance concrete (UHPC) could significantly enhance crack resistance. In spite of this, there is a marked scarcity of studies investigating the mechanical properties of perfobond connectors in UHPC settings.

Our research presents a novel notched T-perfobond connector devised to surmount the obstacles of insufficient pull-out bearing capacity in shallow-embedded multi-row connectors and the difficulty of steel bar penetration. The model test provides the pull-out bearing capacity of this new connector, and an ABAQUS finite-element model is constructed and validated with experimental data. Subsequently, 54 parametric models are formulated to analyze the key factors influencing performance. The study concludes with the proposition of a calculation method for the pull-out bearing capacity of the redesigned connector.

2. Pull-Out Test

2.1. Test Program

As shown in Table 1, four pull-out tests were conducted on the T-perfobond connectors and notched T-perfobond connectors. The aims were to obtain the failure mode, ultimate pull-out resistance, and load-separation behavior of the notched T-perfobond connectors and to verify the reliability of the finite-element analysis. Depending on the type of connection, these specimens can be evenly divided into the following two groups: PFR specimens with a flange of the same material at the bottom of the perfobond connector to enhance the pull-out resistance of the specimens and PMFR specimens with a notch at the bottom edge of the perfobond including the flange. The main parameters of the pull-out test specimens are the hole length (d_l), hole height (d_h), rib thickness (t_p), rib length (l_p), the width of the flange of the perfobond connector (b_f), the thickness of the flange (t_f), the length of the flange ($l_f = l_p$), the width of the notch (c_w), and the diameter of the rebar (d_r). P denotes the connecting piece, R represents for the penetrating rebar, F corresponds to the flange, and M represents the notch.

Table 1. Model grouping and experimental results.

Specimen	d_l (mm)	d_h (mm)	d_r (mm)	t_p (mm)	t_f (mm)	b_f (mm)	l_p (mm)	c_w (mm)	T_u (kN)	$T_{u,avg}$ (kN)
PFR–1	40	20	8	6	6	26	80	0	140	132.0
PFR–2	40	20	8	6	6	26	80	0	124	
PMFR–1	40	20	8	6	6	26	80	16	102	100.5
PMFR–2	40	20	8	6	6	26	80	16	99	

2.2. Specimen Layout

Figure 1 presents the constructional forms of the pull-out model specimens for the perfobond connector. The standard-model UHPC cube has a size of 500 mm × 500 mm × 500 mm. The T-perfobond rib is physically attached to the steel beam through welding. In the pull-out test, the prototype steel beam only needs to retain its steel flange and a portion of the web, forming a T-beam. Perfobond shear connectors are vertically welded to the T-beam, and the actuator is connected to the T-beam to transfer the force to the perfobond rib. Considering that the penetrating rebar is not truncated for its full length in the actual project, it is designed to be equal to the side length of the UHPC cube at the geometric center of the hole. The buried depth (h) of the notched T-perfobond shear connectors is 40 mm. A 6 mm-thick flange is added to the bottom of the connectors to increase the thickness, with an actual buried depth of 46 mm. A 30 mm space is left between the T-beam and the concrete surface to facilitate the observation of damage to the concrete and the perfobond rib. Four specimens are embedded along all four sides of the UHPC cube. Due to the shallow buried depth of the specimens, they do not affect each other, and the design can reduce the use of UHPC materials by approximately 40%.

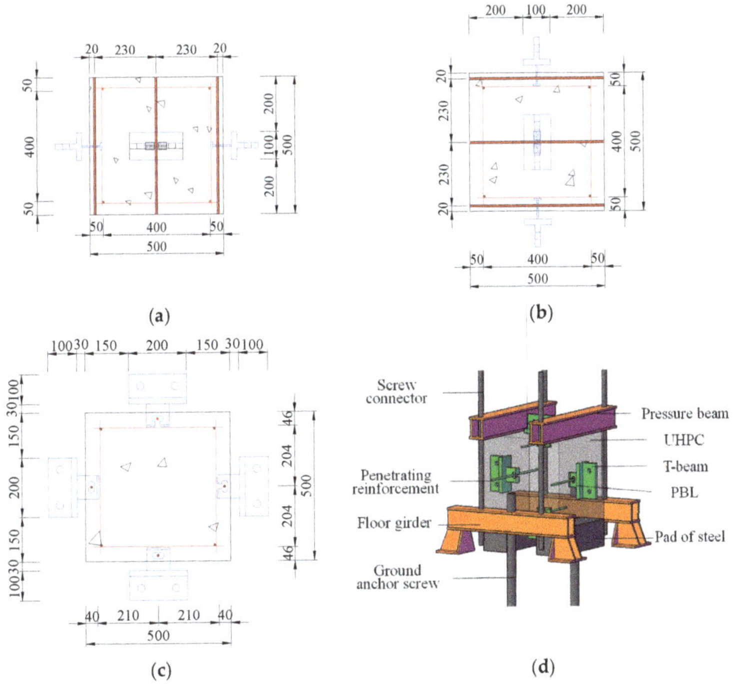

Figure 1. Layout of pull-out test specimens. (**a**) Front view; (**b**) side view; (**c**) top view; (**d**) 3D view. (Note: The black lines indicate the concrete cube, the black triangles indicate the concrete fill; the blue lines indicate the steel plate; and the red lines indicate the rebars.)

2.3. Details of Perfobond Ribs

Figure 2 shows the detailed structure and dimensions of the pull-out specimens of perfobond connectors. The material properties and construction dimensions of the perfobond connectors are the same, except for the presence or absence of notches, and the two groups of pull-out specimens are provided with penetrating rebars. Since the embedment depth of the perfobond connectors is only 40 mm, to increase the pull-out resistance of the members, all specimen holes are long and transverse.

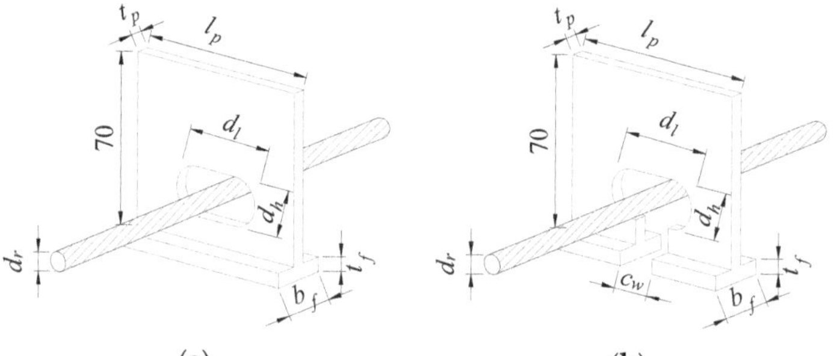

Figure 2. Details of perfobond ribs. (**a**) T-perfobond rib; (**b**) notched T-perfobond rib. (Note: The blue lines indicate the perfobond connectors, and the red lines indicate the rebar.)

2.4. Material Properties

As depicted in Figure 3, steel fibers, accounting for approximately 6.78% of the mass fraction, are integrated into the ultra-high-performance concrete (UHPC) to enhance its ultimate tensile resistance. These steel fibers are in the form of straight lengths with a length of 14 mm, a diameter of 0.2 mm, and a tensile strength of 2200 MPa. The concrete's cubic compressive strength, denoted by f_{cu}, is measured at 118.75 MPa. Correspondingly, the uniaxial compressive strength, represented by f_c, stands at 95.00 MPa, which is 80% of f_{cu}. The tensile strength of the concrete, determined using a specimen shaped like a dog bone and labeled as f_t, is recorded at 10.38 MPa. For the reinforcing bars, the yield strength (f_{ry}) and ultimate tensile strength (f_{ru}) are 351.30 MPa and 691.50 MPa, respectively. In the case of the perfobond rib and its flange, the yield strength (f_{sy}) and tensile strength (f_{su}) are 384.67 MPa and 531.33 MPa, respectively.

Figure 3. Steel fibers.

2.5. Test Setup and Instrumentation

Figure 4 illustrates the loading apparatus used for the pull-out test. A vertical pull-out force was exerted on the upper section of the specimens utilizing an MTS 50t push–pull actuator. The UHPC base of each specimen was firmly anchored to the floor girder through two short steel girders affixed to the floor girder by four screw connectors. A specially designed π-connector plate facilitated the linkage between the T-beam and the actuator, with the assembly secured using high-strength tension bolts. To ascertain the separation between the perfobond rib and the concrete, two displacement transducers were symmetrically positioned on the lateral surfaces of the specimens—one on each side. The pull-out load of the actuator increased monotonically until the T-perfobond connector and the notched T-perfobond connector were completely damaged and lost their bearing capacity. During the loading process, the pull-out load was continuously recorded by the actuator, and the separation displacement was collected by LVDT displacement gauges.

Figure 4. Test setup and instrumentation.

3. Test Results and Analysis

3.1. Failure Mode

The pull-out resistance of the specimens in pull-out tests is shown in Table 1. The PFR specimens were not notched at the hole edge. With an increase in vertical displacement, the constraint of concrete on the flange of the perfobond ribs was enhanced. The damage mode consists of the perfobond ribs on both sides of the hole of the perfobond connector being pulled off, which results in the loss of bearing capacity of the connectors. The surface of the UHPC was intact, without obvious bulges and cracks. The PMFR specimens were notched at the hole edge. Compared with the specimens without notches, the restraining area of the concrete on the flange of the perfobond connector was small. The UHPC was lifted upward diagonally along the edge of the flange at a certain angle to form a conical damage surface, which then pulled the perforated rebar out from the UHPC. The failure mode of the specimens is shown in Figure 5.

(a) (b)

Figure 5. Uplift failures of specimens. (**a**) Fracture damage of perfobond ribs of PFR; (**b**) concrete punching damage of PMFR.

3.2. Load–Separation Behavior

Figure 6 presents the load—separation curves, illustrating the progression of separation between the steel beam and the concrete slab in response to the applied pull-out force. The curves delineate three distinct stages. Initially, upon specimen loading, the load increased linearly with separation, indicating a period of elastic behavior. As the specimens approached the damage stage, a noticeable reduction in stiffness was observed, reflecting the initiation of inelastic deformation. Post-ultimate load, separation persisted, augmenting while the pull-out force diminished, signifying the onset of the post-peak softening phase.

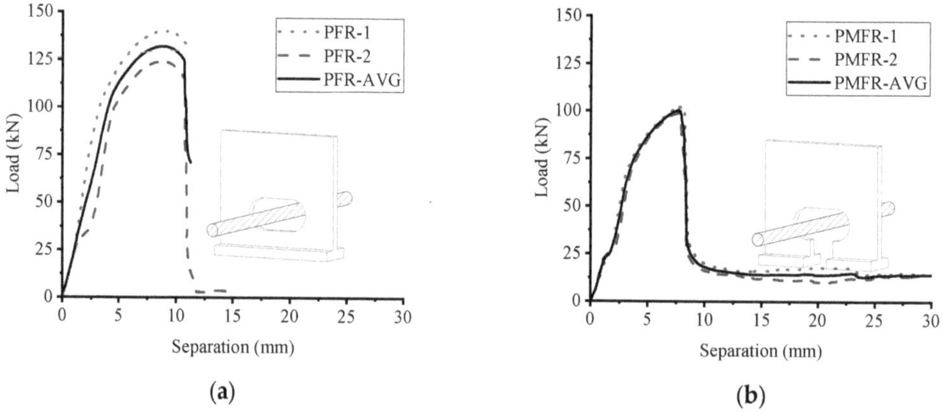

Figure 6. Load–separation curves. (**a**) PFR; (**b**) PMFR.

In this experiment, the average pull-out bearing capacity for the T-perfobond (PFR) and notched T-perfobond (PMFR) connectors was 132.0 kN and 100.5 kN, respectively, while the average peak separation was 8.83 mm and 7.73 mm, respectively. The introduction of notches decreased the pull-out bearing capacity by 23.86% and the peak separation by 12.46%. Nevertheless, due to the limited number of model specimens, additional parametric analysis is necessary to affirm the influence of notches on pull-out bearing performance.

3.3. Pull-Out Mechanism

The tensile strength of the perfobond connectors is a collective result of the perfobond rib, the flange, the concrete, and the rebar that passes through the perforations. An external force applied to the perfobond rib was transmitted from the rib, through the concrete, and into the rebar and flange. The force exerted on the perfobond rib and its immediate vicinity was the most significant and diminished axially towards the rebar ends. The concrete dowel effect and the rebar's role are analogous to that of an elastic foundation beam. The surrounding concrete serves as a foundation, providing the reaction force for this composite elastic beam. When the perfobond rib's strength is less than that of the concrete dowel within the hole, the constraining effect of the rebar on the flange becomes critical. The fracture of the perfobond rib is the decisive factor that leads to the specimen's loss of load-bearing capacity. On the other hand, if the concrete dowel's strength predominates, an inverted pyramid-shaped wedge of concrete is formed diagonally upwards from the flange's edge. In this scenario, the concrete's punching shear failure is the determining factor that causes the ultimate loss of the specimen's load-bearing capacity, as depicted in Figure 7.

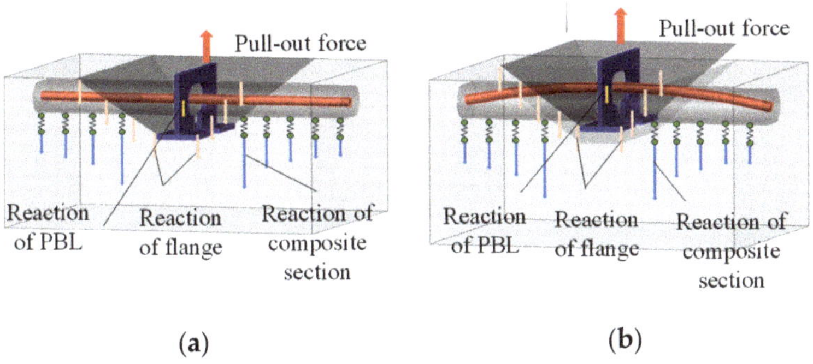

Figure 7. Pull-out mechanism. (**a**) T-perfobond; (**b**) notched T-perfobond.

4. Numerical Simulation

4.1. General

Figure 8 displays the simulated pull-out test for the notched T-perfobond shear connector, which was conducted utilizing the finite-element method. The numerical analysis aimed to elucidate the damage and pull-out resistance mechanisms inherent to these connectors. The simulation was executed using ABAQUS versatile analysis software [26], which served as a cost-effective and time-efficient alternative to physical pull-out tests. The analysis employed a dynamic explicit method, accounting for both material and geometric nonlinearities. Given the slow loading rate of the test, the loading was treated as quasi-static.

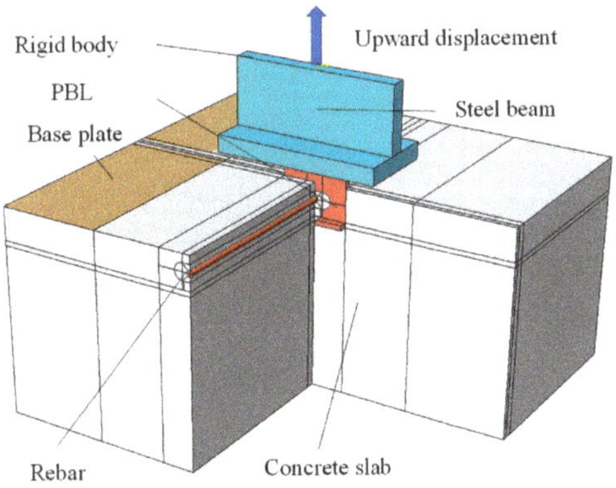

Figure 8. Finite-element model.

4.2. Finite-Element Type and Mesh

Figure 9 illustrates the meshing strategy for the pull-out model of the notched T-perfobond connectors, which incorporated both solid and rigid elements. The concrete slab, T-shaped steel plate, perforating rebar, and perfobond rib were represented using eight-node reduced integration hexahedral elements (C3D8R). A rigid body element (R3D4) discretized the upper surface of the T-shaped steel plate to simulate perfect bonding. To optimize analysis time, a global mesh with an average element size of 20 mm was implemented. However, for enhanced accuracy, a locally refined mesh was applied to critical areas. Specifically, the bond region of the UHPC was meshed with a minimum size of approximately 10 mm, while the perfobond ribs were further refined with a minimum mesh size of about 5 mm to accurately capture the detailed stress distribution.

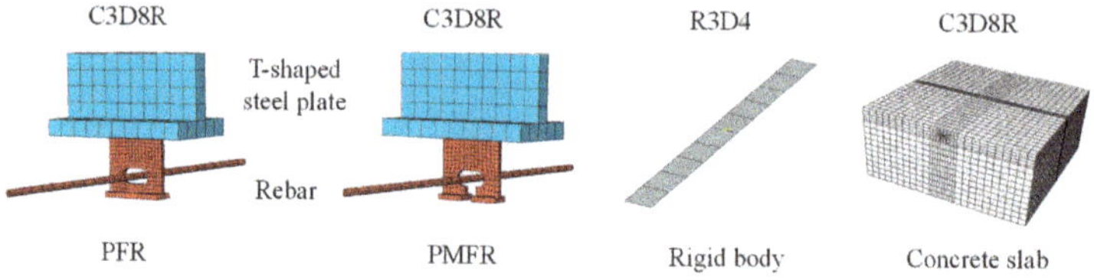

Figure 9. Finite-element type and mesh.

4.3. Interaction and Boundary Conditions

Figure 10 depicts the setup where the reference point of the rigid body is constrained against all translational and rotational movements. An upward displacement was imposed on this reference point to simulate the pull-out action. To precisely simulate the connection between the steel beam and perfobond connectors in the finite-element model, tie-binding constraints were applied. At the interfaces between the concrete and both the perfobond connectors and the rebar, generalized contact interactions were established. To prevent interpenetration, a "hard" contact condition was specified in the normal direction. In the tangential direction, a "penalized" friction model was implemented to account for the sliding resistance. To simplify the model's actual loading boundary conditions, fixed constraints were applied to the upper surface of the concrete slab, creating a base rib. This approach effectively replicates the experimental setup within the numerical simulation.

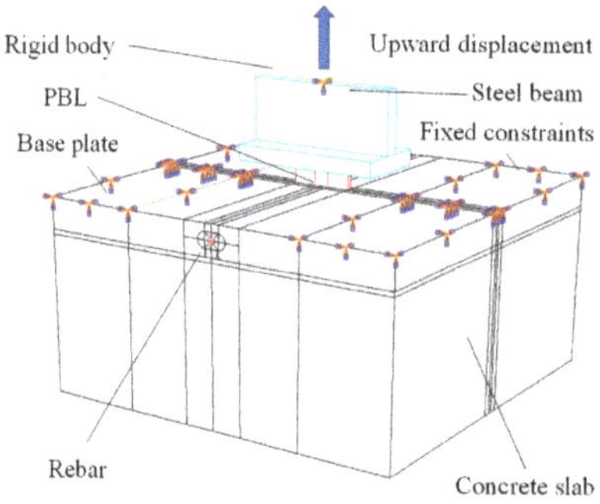

Figure 10. Interaction and boundary conditions.

4.4. Material Constitution of Concrete

Figure 11 illustrates the nonlinear response of concrete under compression and tension, characterized by the uniaxial compressive stress–strain curves and the relationships between tensile stress and crack width.

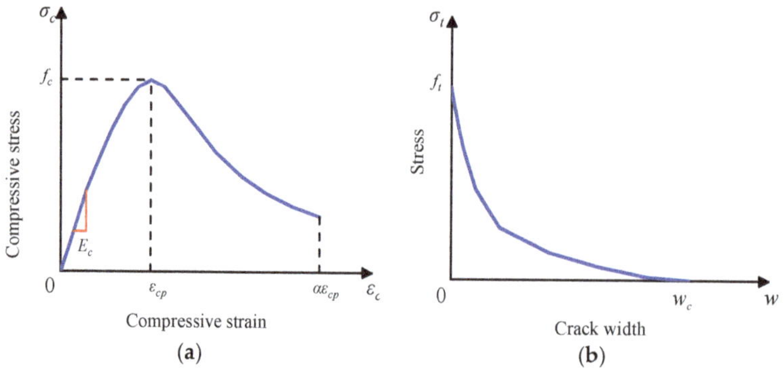

Figure 11. Material constitution of concrete. (**a**) Compression; (**b**) tension.

The intrinsic compressive behavior of concrete is articulated by Equation (1) [27]. As depicted in Figure 11a, the stress–strain curve commences with an initial linear elastic branch. Subsequently, the curve transitions into a nonlinear parabolic segment, followed by a descending branch, each reflecting different phases of the material's response under compression.

$$\begin{cases} \sigma_c = f_c \cdot \frac{n \cdot \xi - \xi^2}{1+(n-2)\cdot\xi} & \varepsilon_c \leq \varepsilon_{cp} \\ \sigma_c = f_c \cdot \frac{\xi}{2\cdot(\xi-1)^2+\xi} & \varepsilon_c > \varepsilon_{cp} \end{cases} \quad (1)$$

The model outlined by CEB-FIP (1993) governs the ascending segment of the curve, where σ_c denotes the compressive stress in MPa and f_c represents the concrete's axial compressive strength, also in MPa. The strain at the peak stress is defined as ε_{cp} = 0.0035, and ξ quantifies the strain relative to the peak strain, expressed as $\xi = \varepsilon_c / \varepsilon_{cp}$. Furthermore, n is the ratio of the initial modulus of elasticity (E_c) to the modulus of the descending branch at the peak point (E_{cp}), calculated as $n = E_c / E_{cp}$ [27].For uncracked concrete, a linear stress–strain relationship captures the material's elongation characteristics. In contrast, the cracked section, as illustrated in Figure 11b, necessitates a nonlinear description of the stress–crack width relationship. This is articulated through Equation (2) [27].

$$\frac{\sigma_t}{f_t} = \left[1 + \left(c_1 \cdot \frac{w}{w_c}\right)^3\right] \cdot \exp\left(-c_2 \cdot \frac{w}{w_c}\right) - \frac{w}{w_c} \cdot \left(1 + c_1^3\right) \cdot \exp(-c_2) \quad (2)$$

In the provided model, σ_t signifies the tensile stress in MPa, while f_t is the concrete's tensile strength, also expressed in MPa. The crack width is denoted by w in mm, and w_c represents the crack width at complete stress relief, defined as $w_c = 5.14\, G_F / f_t$ (mm), where G_F is the fracture energy required to generate a unit area of stress-free crack, as expressed by $G_F = 0.073 {f_c}^{0.18}$ (N/mm). The model incorporates constants of $c_1 = 3$ and $c_2 = 6.93$ [27].

The concrete damage plasticity model encapsulates the degradation response of concrete under stress. This model employs two independent uniaxial damage variables, namely d_c and d_t, to quantify concrete damage resulting from compressive crushing and tensile cracking, respectively [27].

For concrete under compression, the progression of d_c is linked to the plastic strain (${\varepsilon_l}^{cp}$). It is determined to be proportional to the inelastic strain (${\varepsilon_c}^{in} = \varepsilon_c - \sigma_c / E_c$) with a constant factor of b_c ($0 < b_c < 1$), as delineated in Equation (3) [27].

$$d_c = 1 - \frac{\sigma_c}{E_c \cdot \varepsilon_c^{pl} \cdot (1/b_c - 1) + \sigma_c} \quad (3)$$

In this context, d_c represents the compressive damage component of concrete. The ratio of plastic to inelastic strain is denoted by b_c, as defined by $b_c = {\varepsilon_c}^{pl} / {\varepsilon_c}^{in}$, and b_c is assigned a value of 0.7 according to [27].

For concrete under tension, the damage evolution component (d) is correlated with the "plastic" crack width (w^{pl}). This component is proportional to the actual crack width (w) with the proportionality governed by the constant factor (b_t, where $0 < b_t < 1$), as described by Equation (4) [27].

$$d_t = 1 - \frac{\sigma_t \cdot l_0}{E_c \cdot w^{pl} \cdot (1/b_t - 1) + \sigma_t \cdot l_0} \quad (4)$$

Here, d_t signifies the tensile damage component of the concrete. The term l_0 is considered a reference unit length for normalization purposes. Additionally, b_t denotes the ratio of the 'plastic' crack width to the actual crack width, expressed as $b_t = w^{pl} / w$, with a designated value of 0.1, in accordance with [27].

4.5. Material Constitution of Steel

Figure 12 illustrates the stress–strain relationship for perfobond ribs and bars, modeled with three linear segments. Initially, the relationship is characterized by an elastic phase with a Young's modulus denoted as E_S. This is succeeded by a yielding phase, culminating in a strain-hardening phase. It is presumed that the stress–strain behavior mirrors that of steel under compression.

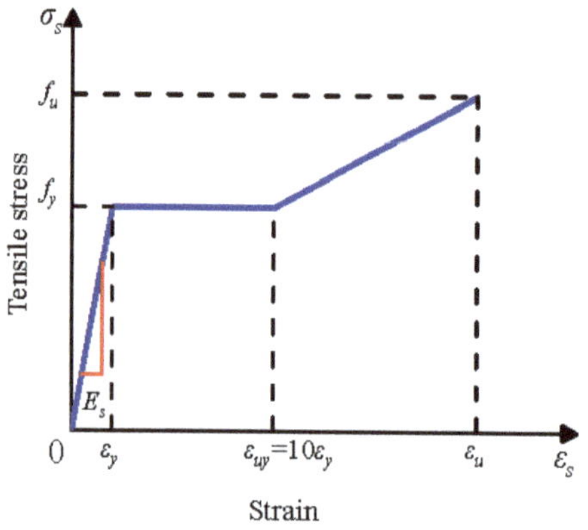

Figure 12. Material constitution of steel.

4.6. Verification of Finite-Element Model

The numerical simulation outcomes closely resemble the pull-out damage patterns observed in perfobond connectors, as depicted in Figure 13. In the case of notched T-perfobond connectors, an increase in pull-out force intensifies the concrete's constraint on the flange, predisposing the perfobond rib and flange to a wedge-shaped inverted pyramid damage mechanism, as illustrated in Figure 13a. Conversely, for T-perfobond connectors, the larger area of the flange in contact with the concrete during upward pull-out results in enhanced anchorage. This robust anchorage ultimately culminates in the fracture of the perfobond connectors, as shown in Figure 13b. Specifically, the peak stress in the perfobond rib reached 667.99 MPa, a threshold sufficient to confirm the rib's fracture. In the notched T-perfobond connectors, the shallow embedment depth coupled with the flange's influence means that the perforated rebar's loss of bearing capacity plays a limited role, predominantly in anchoring.

In Figure 14, the load–separation curves derived from the finite-element model (FEM) are juxtaposed with the outcomes of the pull-out tests. The experimental curves for T-perfobond connectors comprise three distinct phases, namely the elastic phase, the nonlinear parabolic phase, and the descending branch. With the concrete dilation angle set at 30°, the finite-element analysis predicted pull-out forces of 106.51 kN for the notched T-perfobond connectors and 111.52 kN for the T-perfobond connectors. These values correspond to 106.0% and 107.5% of the average experimental results, respectively. The initial stiffness observed in the numerical simulation's load–separation curves exceeds the experimental values, possibly due to the idealized boundary conditions in the simulation, which do not account for the deformations present in the experimental setup. Additionally, discrepancies may arise from the inherent variability in the mechanical properties of the steel fiber-reinforced UHPC (ultra-high-performance concrete) material used in the tests. Therefore, the stiffness of the loaded steel beam was adjusted between 4 and 800 kN/mm by trial calculation so that the load–separation curves of the simulations were in better agreement with the model test results.

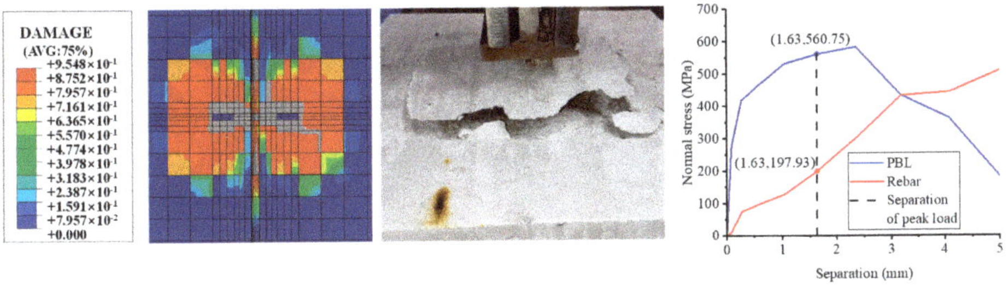

(**a**)

(**b**)

Figure 13. Tested and analyzed failure modes. (**a**) Concrete damage verification; (**b**) perfobond rib fracture verification.

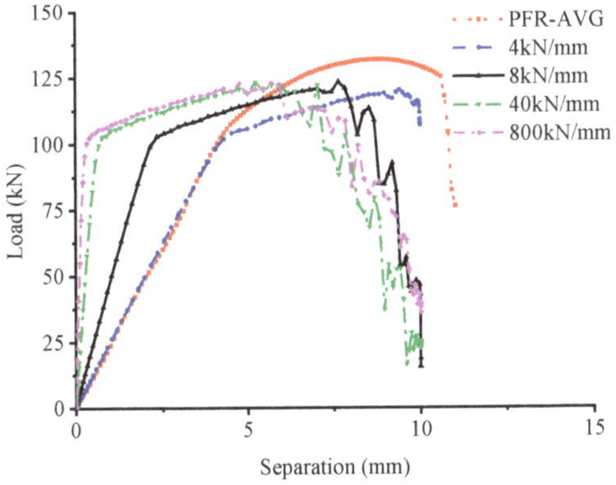

(**a**)

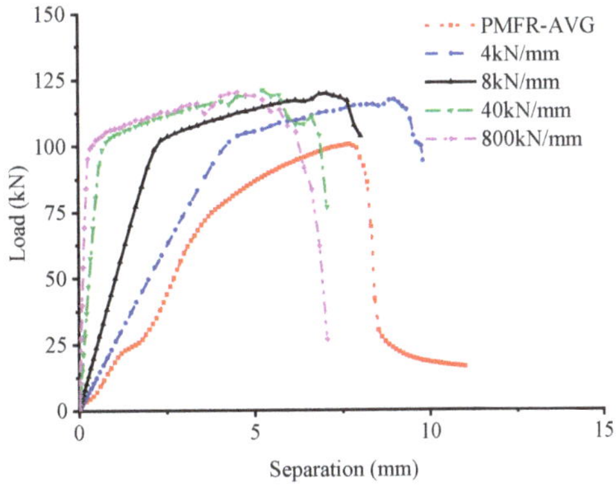

(**b**)

Figure 14. Tested and analyzed load–slip curves. (**a**) T-perfobond; (**b**) notched T-perfobond.

5. Parametric Study

As shown in Table 2, a total of 54 simulations were conducted to further investigate the effects of the connectors' dimensions and material properties on the pull-out resistance of notched T-perfobond connectors. The considered parameters encompass the perfobond connector flange thickness (t_f), width (b_f), and cut width (c_w) and perfobond rib-hole height (d_h), hole length (d_l), thickness (t_p), width (l_p), rebar diameter (d_r), concrete axial compressive strength (f_c), rebar yield strength (f_{ry}), and perfobond rib yield strength (f_{sy}).

Table 2. Generalization of parametric models.

Model	t_f (mm)	b_f (mm)	c_w (mm)	d_h (mm)	d_l (mm)	t_p (mm)	l_p (mm)	d_r (mm)	f_c (MPa)	f_{ry} (MPa)	f_{sy} (MPa)	T_u (kN)
TF-02	2	26	16	20	40	6	80	8	95	400	390	103.44
TF-04	4	26	16	20	40	6	80	8	95	400	390	98.39
TF-06	6	26	16	20	40	6	80	8	95	400	390	106.51
TF-08	8	26	16	20	40	6	80	8	95	400	390	117.64
TF-10	10	26	16	20	40	6	80	8	95	400	390	117.82
BF-06	6	6	16	20	40	6	80	8	95	400	390	66.71
BF-12	6	12	16	20	40	6	80	8	95	400	390	114.31
BF-18	6	18	16	20	40	6	80	8	95	400	390	112.2
BF-26	6	26	16	20	40	6	80	8	95	400	390	106.51
BF-34	6	34	16	20	40	6	80	8	95	400	390	99.43
CW-00	6	26	0	20	40	6	80	8	95	400	390	111.52
CW-08	6	26	8	20	40	6	80	8	95	400	390	109.01
CW-16	6	26	16	20	40	6	80	8	95	400	390	106.51
CW-24	6	26	24	20	40	6	80	8	95	400	390	108.45
CW-40	6	26	40	20	40	6	80	8	95	400	390	107.27
DH-10	6	26	16	10	40	6	80	8	95	400	390	108.3
DH-15	6	26	16	15	40	6	80	8	95	400	390	110.02
DH-20	6	26	16	20	40	6	80	8	95	400	390	106.51
DH-25	6	26	16	25	40	6	80	8	95	400	390	105.19
DH-30	6	26	16	30	40	6	80	8	95	400	390	88.66
DL-20	6	26	16	20	20	6	80	8	95	400	390	103.98
DL-30	6	26	16	20	30	6	80	8	95	400	390	107.93
DL-40	6	26	16	20	40	6	80	8	95	400	390	106.51
DL-50	6	26	16	20	50	6	80	8	95	400	390	106.51
DL-60	6	26	16	20	60	6	80	8	95	400	390	65.55
TP-02	6	26	16	20	40	2	80	8	95	400	390	43.22
TP-04	6	26	16	20	40	4	80	8	95	400	390	83.51
TP-06	6	26	16	20	40	6	80	8	95	400	390	106.51
TP-08	6	26	16	20	40	8	80	8	95	400	390	110.95
TP-10	6	26	16	20	40	10	80	8	95	400	390	107.13
LP-60	6	26	16	20	40	6	60	8	95	400	390	63.1
LP-70	6	26	16	20	40	6	70	8	95	400	390	91.62
LP-80	6	26	16	20	40	6	80	8	95	400	390	106.51
LP-90	6	26	16	20	40	6	90	8	95	400	390	117.26
LP-100	6	26	16	20	40	6	100	8	95	400	390	115.54
DR-04	6	26	16	20	40	6	80	4	95	400	390	109.9
DR-06	6	26	16	20	40	6	80	6	95	400	390	109.77
DR-08	6	26	16	20	40	6	80	8	95	400	390	106.51
DR-10	6	26	16	20	40	6	80	10	95	400	390	108.29
DR-12	6	26	16	20	40	6	80	12	95	400	390	109.53

Table 2. *Cont.*

Model	t_f (mm)	b_f (mm)	c_w (mm)	d_h (mm)	d_l (mm)	t_p (mm)	l_p (mm)	d_r (mm)	f_c (MPa)	f_{ry} (MPa)	f_{sy} (MPa)	T_u (kN)
FC-80	6	26	16	20	40	6	80	8	80	400	390	97.83
FC-95	6	26	16	20	40	6	80	8	95	400	390	106.51
FC-110	6	26	16	20	40	6	80	8	110	400	390	120.13
FC-130	6	26	16	20	40	6	80	8	130	400	390	122.59
FC-140	6	26	16	20	40	6	80	8	140	400	390	122.57
RY-235	6	26	16	20	40	6	80	8	95	235	390	106.51
RY-335	6	26	16	20	40	6	80	8	95	335	390	106.51
RY-400	6	26	16	20	40	6	80	8	95	400	390	106.51
RY-500	6	26	16	20	40	6	80	8	95	500	390	106.51
SY-235	6	26	16	20	40	6	80	8	95	400	235	75.1
SY-345	6	26	16	20	40	6	80	8	95	400	345	106.38
SY-390	6	26	16	20	40	6	80	8	95	400	390	106.51
SY-420	6	26	16	20	40	6	80	8	95	400	420	106.11
SY-460	6	26	16	20	40	6	80	8	95	400	460	114.61

Note: The datas in the gray background represent the same set of parameters.

5.1. Influence of Flange Size

Figure 15 demonstrates the significant impact of flange thickness on the pull-out performance of notched T-perfobond connectors. Increases in flange thickness from 2 mm to 4 mm, 6 mm, 8 mm, and 10 mm resulted in corresponding pull-out resistance changes of −4.88%, 2.97%, 13.73%, and 13.90%, respectively. The findings indicate that enhancing flange thickness augments the pull-out resistance of these connectors. The primary rationale is the pronounced anchorage effect of the notched perfobond rib's bottom flange. Between 2 mm and 8 mm flange thickness, the flange is the weaker constraint, and the pull-out resistance increases linearly with flange thickness. However, when the flange thickness exceeds 8 mm, the notched perfobond rib or the concrete becomes the limiting constraint and further increases in flange thickness yield minimal changes in pull-out resistance.

Figure 16 illustrates the substantial effect of flange width on the pull-out performance of notched T-perfobond connectors. Increasing the flange width from 6 mm, which matches the connector's thickness, to 12 mm, 18 mm, 26 mm, and 34 mm, resulted in increases in pull-out resistance of 71.35%, 68.19%, 59.66%, and 49.05%, respectively. The data indicate that broadening the flange width from 6 mm to 12 mm significantly enhances the pull-out resistance due to the robust anchorage provided by the bottom flange of the connectors. However, as the flange width extends beyond 12 mm, there is a gradual decline in the ultimate pull-out resistance. This decrease may be attributed to the excessive length of the flange, which leads to a more pronounced upward force on the concrete at the junction of the notched perfobond rib and the flange. The resultant moment at the flange's outer edge due to this force is larger and may cause stress concentration, leading to premature concrete failure at the flange's outer edge and a consequent reduction in pull-out resistance.

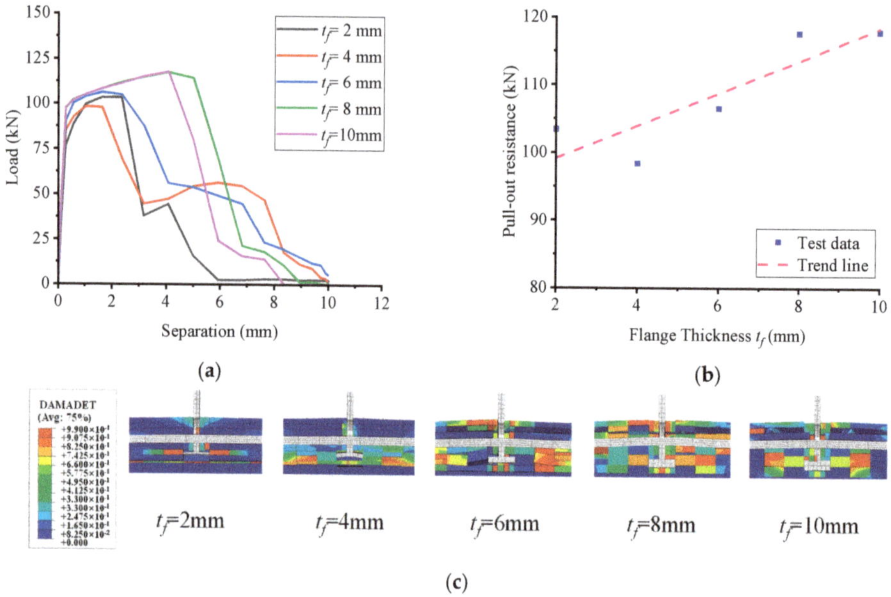

Figure 15. Influence of flange thickness. (**a**) Load–separation curves; (**b**) influence analysis; (**c**) failure mode.

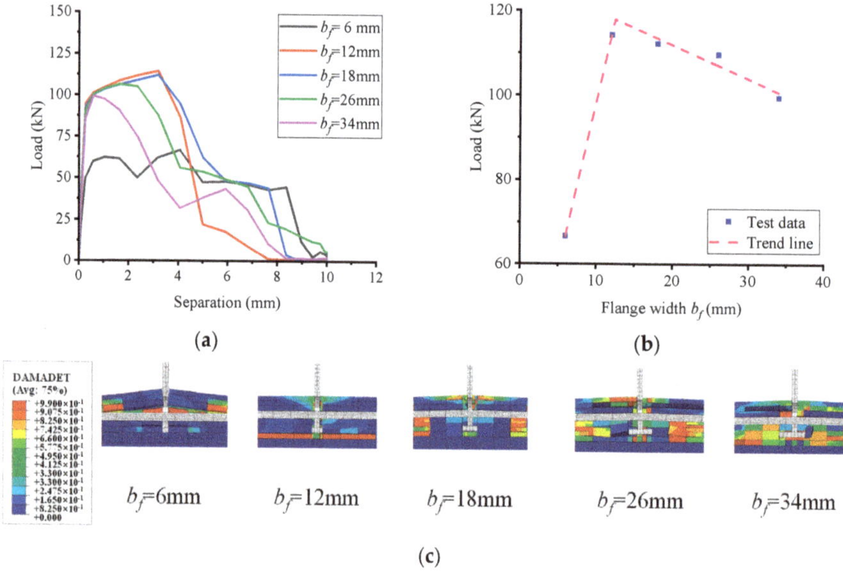

Figure 16. Influence of flange width. (**a**) Load–separation curves; (**b**) influence analysis; (**c**) failure mode.

5.2. Influence of Cut Width

Figure 17 displays the impact of notch width on the pull-out performance of notched T-perfobond connectors. In this study, the notched ratio (c_w/l_f) was varied from 0 to 0.50 across the specimens. The conventional T-perfobond connectors featured a notched width of 0 mm, corresponding to a notched ratio of 0. The notched T-perfobond connectors had notched widths of 8 mm, 16 mm, 24 mm, and 40 mm, with notched ratios of 0.10, 0.20, 0.30, and 0.50, respectively. The findings indicate that an increase in notch width has a minor influence on the pull-out resistance of notched T-perfobond connectors. When the notch width is less than 16 mm, corresponding to a notch ratio below 0.20, the reduction in pull-out resistance does not exceed 4.50%. Beyond a notch width of 16 mm up to 40 mm, the ultimate pull-out resistance of the specimens remains relatively stable. This constancy could be due to the flange providing the necessary pull-out resistance to counteract the effects of the notch.

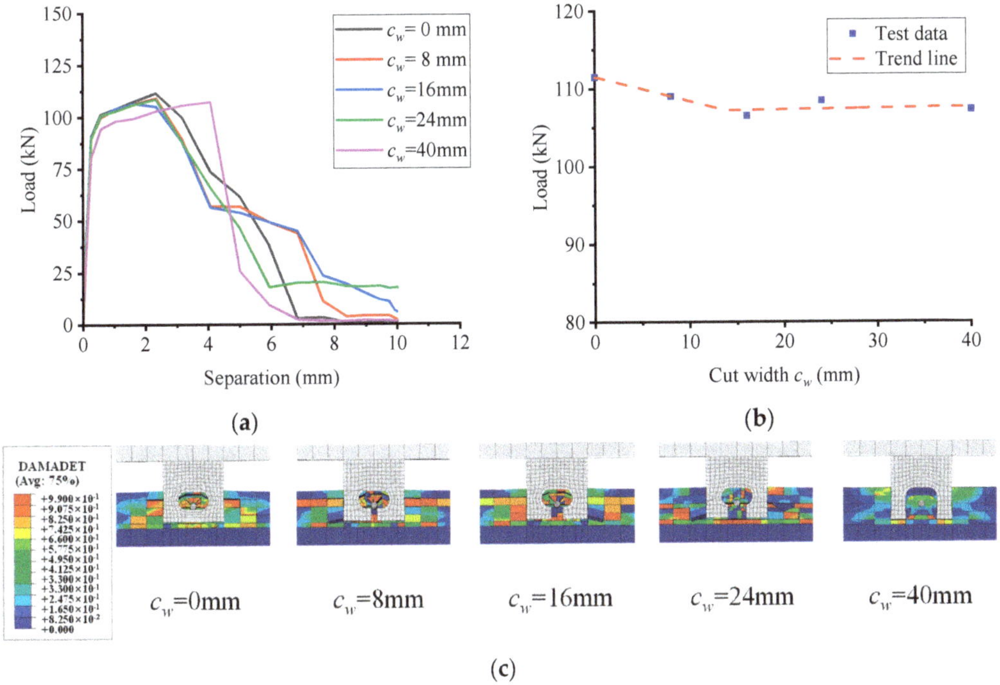

Figure 17. Influence of cut width. (**a**) Load–separation curves; (**b**) influence analysis; (**c**) failure mode.

5.3. Influence of Hole Size

Figure 18 illustrates the impact of hole height on the pull-out performance of notched T-perfobond connectors. When the hole height was increased from 10 mm to 15 mm, 20 mm, 25 mm, and 30 mm, the corresponding changes in pull-out resistance were 1.59%, −1.65%, −2.87%, and −18.13%, respectively. The data reveal that while an increase in hole height from 10 mm to 15 mm slightly enhances pull-out resistance, further increases result in a decline. This trend is primarily attributed to the reduction in the perfobond rib's thickness at the lower edge of the hole as the hole height increases. Concurrently, the moment at the notch of the perfobond rib and the inner edge of the hole at its maximum length intensifies. Despite this, the projected area of the concrete dowel and the contact area of the perfobond connectors remain constant, potentially causing earlier yielding and damage at the inner edge of the hole, which, in turn, diminishes pull-out resistance. Examination of the load–separation curve indicates that a larger hole height correlates with an extended ductile section.

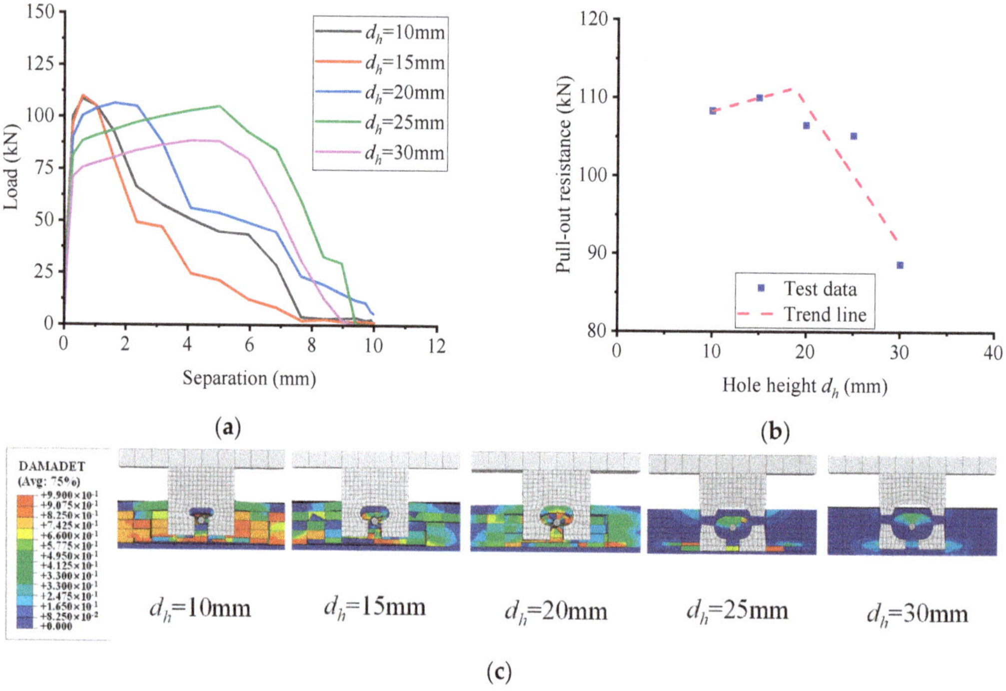

Figure 18. Influence of hole height. (**a**) Load–separation curves; (**b**) influence analysis; (**c**) failure mode.

Figure 19 demonstrates the influence of hole length on the pull-out performance of notched T-perfobond connectors. As the hole length increased from 20 mm to 30 mm, 40 mm, 50 mm, and 60 mm, the pull-out resistance of these connectors changed by 3.8%, 2.43%, 2.43%, and −36.96%, respectively. The results indicate that increasing the hole length up to 50 mm slightly augments the tensile strength of the connectors; however, a further increase to 60 mm significantly reduces the pull-out resistance. This phenomenon can be primarily attributed to the reduction in the perfobond rib's thickness at the hole's edge as the hole length expands. Consequently, the moment at the notch of the perfobond rib within the hole and at the inner edge of the hole at its maximum length increases. Although the projected contact area between the concrete dowel and the perfobond connectors may increase, the earlier yielding and damage of the perfobond rib at the inner edge of the hole at its maximum length occur, leading to a decrease in pull-out resistance.

5.4. Influence of Perfobond Size

Figure 20 illustrates the significant impact of the perfobond connector's thickness on the pull-out performance of notched T-perfobond connectors. The pull-out resistance increased dramatically as the thickness of the notched perfobond rib was incremented from 2 mm to 4 mm, 6 mm, 8 mm, and 10 mm, with respective increases of 93.22%, 146.44%, 156.71%, and 147.87%. The enhancement in pull-out resistance is primarily attributed to the robust anchoring effect of the bottom flange of the perfobond rib. For thicknesses between 2 mm and 6 mm, the perfobond rib weaker than the other structural elements; thus, the pull-out resistance increases linearly with the thickness of the perfobond rib. Once the thickness exceeds 6 mm, the perfobond rib's strength surpasses the constraints imposed by the concrete and rebars in the holes. Consequently, further increases in the perfobond rib's thickness lead to a continued rise in pull-out resistance. However, beyond a certain point, additional thickness does not yield a significant change in pull-out resistance.

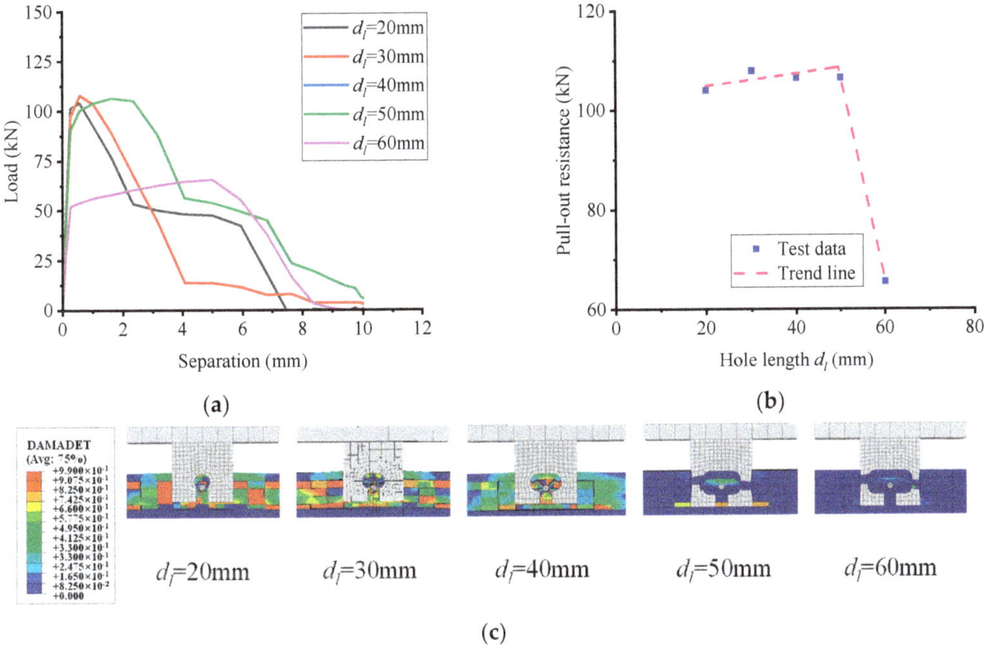

Figure 19. Influence of hole length. (**a**) Load–separation curves; (**b**) influence analysis; (**c**) failure mode.

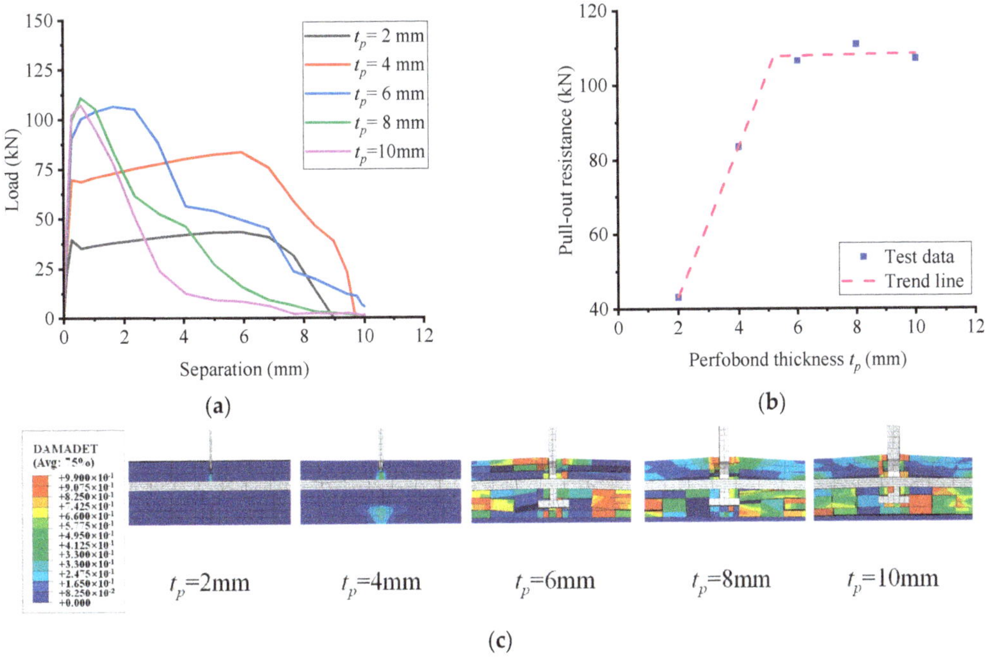

Figure 20. Influence of perfobond thickness. (**a**) Load–separation curves; (**b**) influence analysis; (**c**) failure mode.

Figure 21 demonstrates the significant impact of the notched perfobond rib's length on the pull-out performance of notched T-perfobond connectors. As the perfobond rib's length increased from 60 mm to 70 mm, 80 mm, 90 mm, and 100 mm, the pull-out resistance of the notched T-perfobond connectors correspondingly rose by 45.2%, 68.8%, 85.83%, and 83.11%, respectively. The enhancement in pull-out resistance is primarily due to the substantial anchoring effect of the bottom flange of the perfobond rib. Between lengths of 60 mm and 90 mm, the perfobond rib is weaker than the other structural constraints, and the ultimate resistance of the connectors increases linearly with the length of the perfobond rib. When the length exceeds 90 mm, the strength of the perfobond rib surpasses the constraints imposed by the concrete and rebars in the holes, suggesting that further increases in rib length may not significantly enhance pull-out resistance.

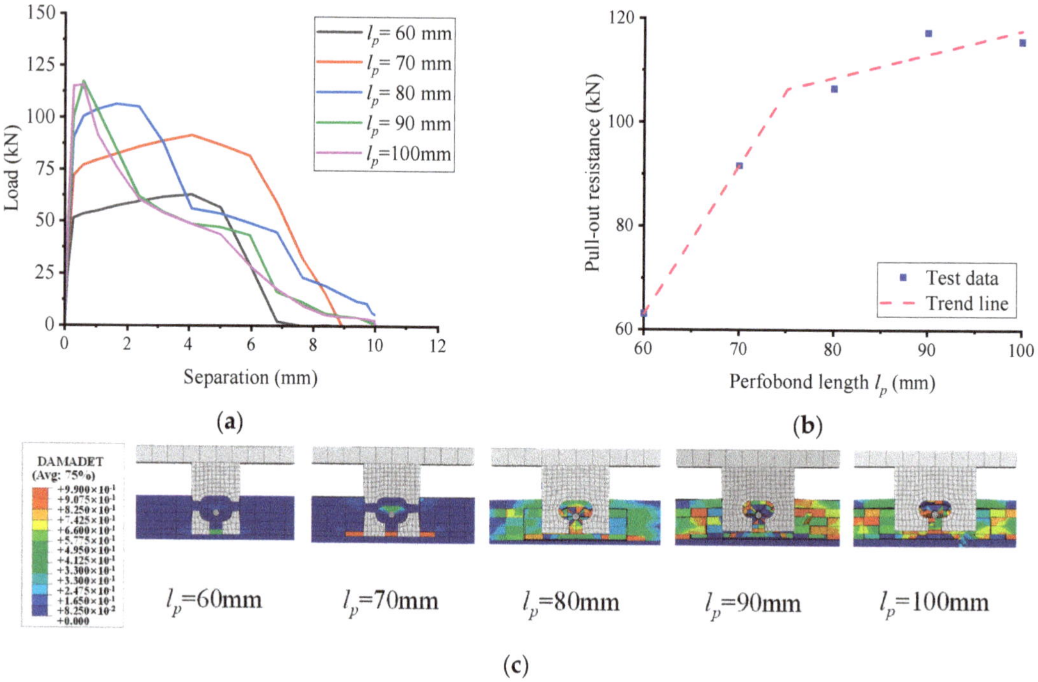

Figure 21. Influence of perfobond length. (**a**) Load–separation curves; (**b**) influence analysis; (**c**) failure mode.

5.5. *Influence of Rebar Diameter*

Figure 22 illustrates the impact of rebar diameter on the pull-out performance of notched T-perfobond connectors. Increments in rebar diameter from 4 mm to 6 mm, 8 mm, 10 mm, and 12 mm resulted in minimal changes in pull-out resistance of −0.12%, −3.08%, −1.46%, and −0.34%, respectively. The findings indicate that the rebar diameter exerts a negligible influence on the ultimate pull-out resistance of the specimens. This outcome is likely attributed to the dominant anchorage effect provided by the flange and the notches, which overshadows the impact of rebar diameter variations on the ultimate resistance to pull-out.

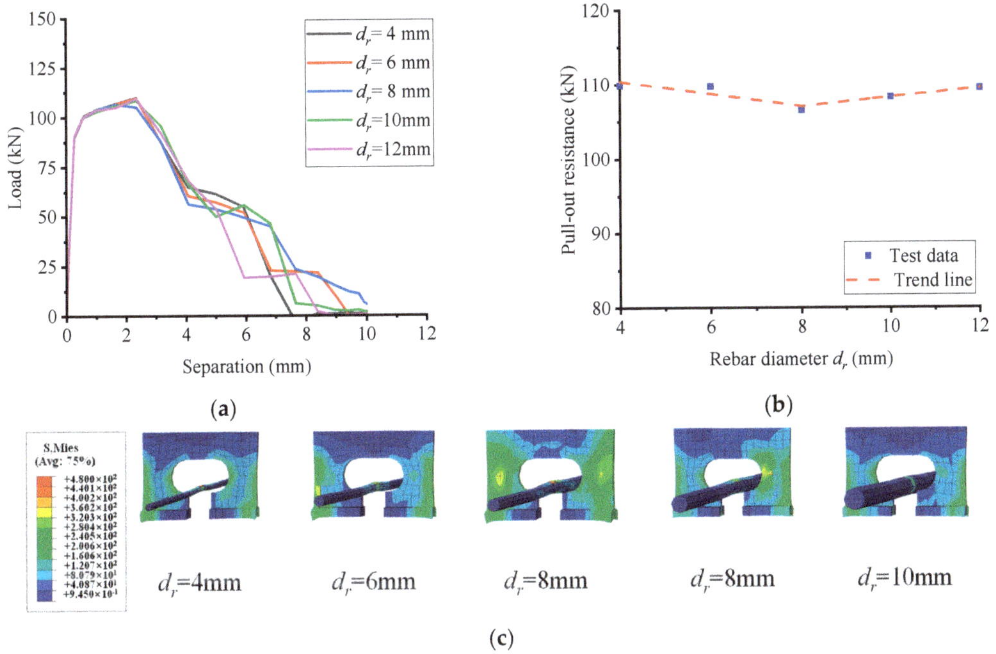

Figure 22. Influence of rebar diameter. (**a**) Load–separation curves; (**b**) influence analysis; (**c**) failure mode.

5.6. Influence of Material Strength

Figure 23 demonstrates the influence of concrete strength on the pull-out performance of notched T-perfobond connectors. As concrete strength escalated from 80 MPa to 95 MPa, 110 MPa, 130 MPa, and 140 MPa, the corresponding increases in pull-out strength were 8.87%, 22.79%, 25.31%, and 25.29%, respectively. The data indicate that the pull-out resistance of these connectors increases significantly with rising concrete strength, likely due to the strong anchorage effect of the flange. The predominant mode of specimen damage is concrete punching, and the pull-out resistance correlates linearly with concrete strength. Therefore, utilizing high-strength concrete can enhance the pull-out resistance of notched T-perfobond connectors. However, beyond a concrete strength of 110 MPa, there is a tendency for the pull-out resistance to plateau. This saturation may occur because, at certain strength levels, the concrete's anchoring effect on the flange surpasses the strength of the notched perfobond rib, making the rib itself the limiting factor.

Figure 24 illustrates the impact of rebar strength on the pull-out performance of notched T-perfobond shear connectors. Notably, the pull-out resistance remained essentially constant as rebar strength increased from 235 MPa to 335 MPa, 400 MPa, and 500 MPa. The findings indicate that variations in rebar strength have a negligible effect on the pull-out resistance of the specimens. This outcome is likely due to the overwhelming anchorage effect of the flange and the notches, which diminishes the influence of rebar strength changes on pull-out resistance.

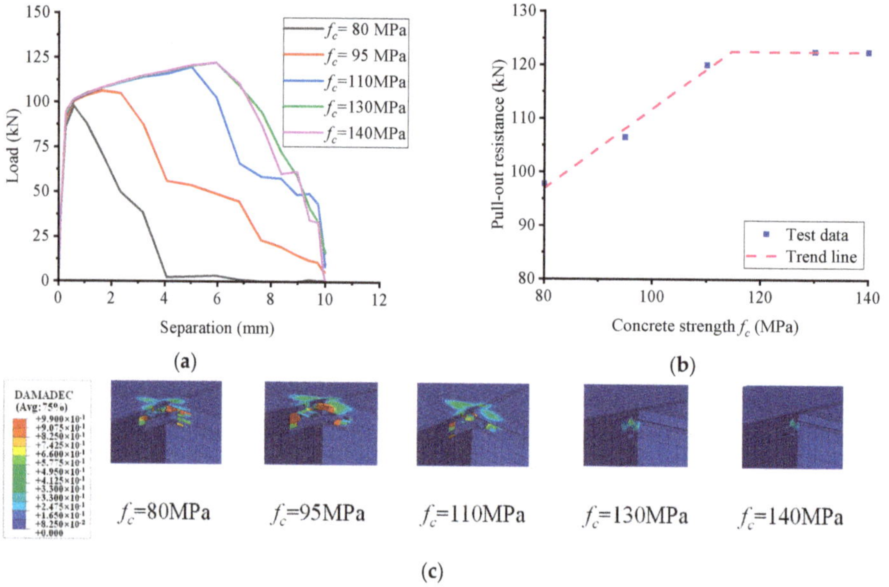

Figure 23. Influence of concrete strength. (**a**) Load–separation curves; (**b**) influence analysis; (**c**) failure mode.

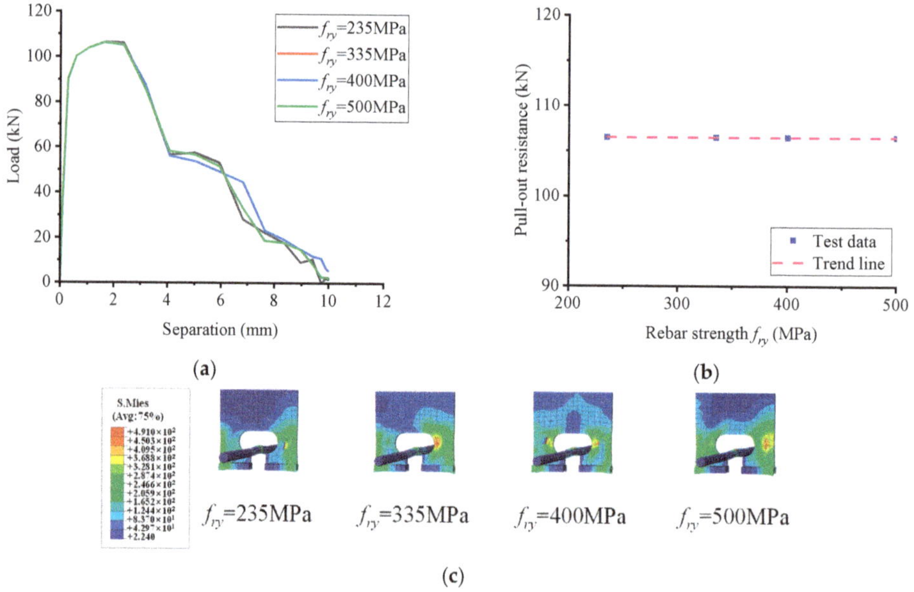

Figure 24. Influence of rebar strength. (**a**) Load–separation curves; (**b**) influence analysis; (**c**) failure mode.

Figure 25 demonstrates the impact of the perfobond rib's strength on the pull-out performance of notched T-perfobond connectors. As the strength of the perfobond rib increased from 235 MPa to 345 MPa, 390 MPa, 420 MPa, and 460 MPa, the pull-out strength correspondingly rose by 41.65%, 41.82%, 41.29%, and 52.61%, respectively. The results indicate that the pull-out resistance of the notched T-perfobond connectors increases significantly with the strength of the notched perfobond ribs. This substantial increase may be attributed to the linear relationship between pull-out resistance and the strength of the perfobond rib, facilitated by the exceptionally strong anchorage provided by the bottom flange of the notched perfobond rib, which is relatively weaker than other restraining elements.

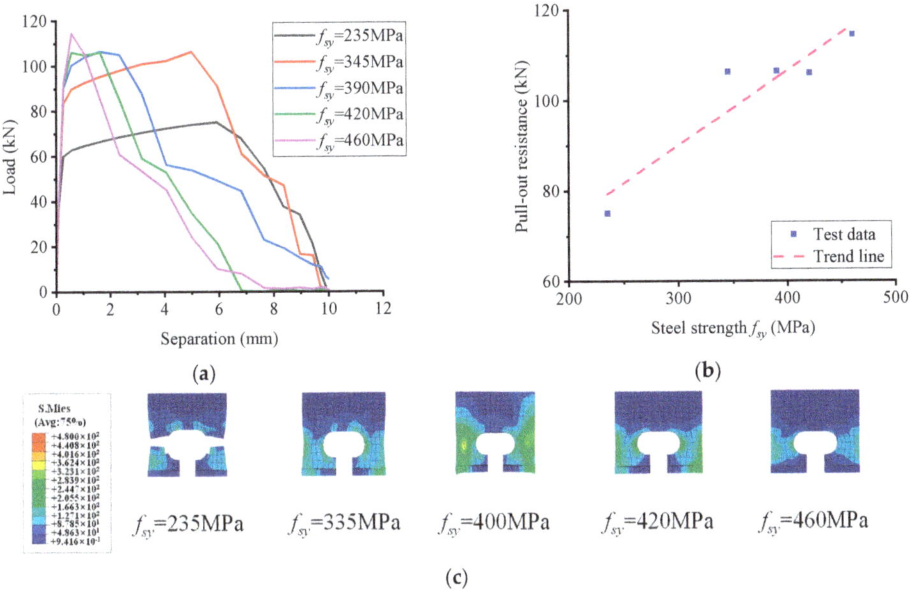

Figure 25. Influence of steel strength. (**a**) Load–separation curves; (**b**) influence analysis; (**c**) failure mode.

6. Analytical Work

The damage modes are categorized into the following two types: perfobond rib fracture and concrete punching. For each specimen, the ultimate pull-out resistance, denoted as T_u, is defined as the minimum value between the calculated resistances for perfobond rib fracture (T_s) and concrete punching (T_c).

$$T_u = \min\{T_s, T_c\} \tag{5}$$

6.1. Fracture Damage of Perfobond Ribs

When the strength of the perfobond rib is lower than the anchoring effect provided by the concrete and the perforating rebar on the flange, the perfobond rib typically experiences stress concentration at the hole's maximum length first. This concentration of stress initiates at the point where the rib is most likely to fracture. The fracture damage of the perfobond rib can be assessed using the calculation detailed in Equation (6).

$$T_s = f_s A_s = f_s(l_p - d_l)t_p \tag{6}$$

where f_s represents the tensile strength of the perforated steel plate, a measure of its capacity to withstand pulling forces before breaking. Meanwhile, A_s denotes the cross-sectional area at the point where the perforated steel plate fractures, which is considered to be the largest area involved in the breakage.

6.2. Punching Damage of Concrete

As depicted in Figure 26, the model resembles a wedge-shaped inverted pyramid with diagonal edges, each forming a 45° angle with the horizontal plane. Owing to the exceptionally strong anchorage provided by the flange, once the concrete is lifted and damaged, the concrete dowel and the rebar within the hole no longer contribute to the ultimate tensile resistance. Consequently, Equation (7) considers the concrete damage as the determining factor based on the specimens' damage patterns, and Equation (8) accounts for the effects of a reduction in the flange area of the notched specimens on the pull-out resistance, with the discounting factor being a piecewise function.

$$T_c = f_t S_h = f_t h\left[2\left(b_f + h\right) + 2\left(l_f + h\right)\right] \quad (7)$$

$$T_c = \beta f_t S_h = \beta f_t h\left[2\left(b_f + h\right) + 2\left(l_f + h\right)\right] \quad (8)$$

where T_c signifies the pull-out resistance corresponding to the concrete damage mode. The tensile strength of the concrete, represented by f_t, is considered to be one-tenth of the concrete's compressive strength (f_c). Furthermore, S_h denotes the projected area of the concrete subjected to punching damage. The discount factor for the notch width (β) is defined as $\beta = 1 - c_w/l_f \geq 0.8$, with the constraint that $0 \leq c_w/l_f \leq 0.2$.

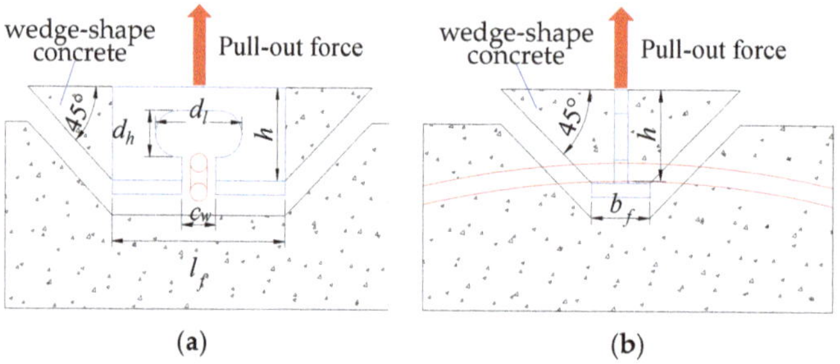

Figure 26. Punching damage of concrete. (**a**) Side view; (**b**) front view.

6.3. Comparison and Validation

Figure 27 presents a comparison of the calculated pull-out resistance with the results from pull-out tests and parametric analyses to validate the previously proposed equation. As illustrated in Figure 27a, Equation (7) tends to overestimate the pull-out resistance of notched T-perfobond connectors. This discrepancy may stem from the omission of the effect caused by discounting the notched area in Equation (7). The pull-out resistance calculated using a modified version of Equation (8) aligns more closely with both the test and numerical simulation outcomes, as shown in Figure 27b. Therefore, Equation (5), which is a combination of Equations (6) and (8), is deemed suitable for assessing the pull-out resistance of notched T-perfobond connectors in steel–concrete composite structures.

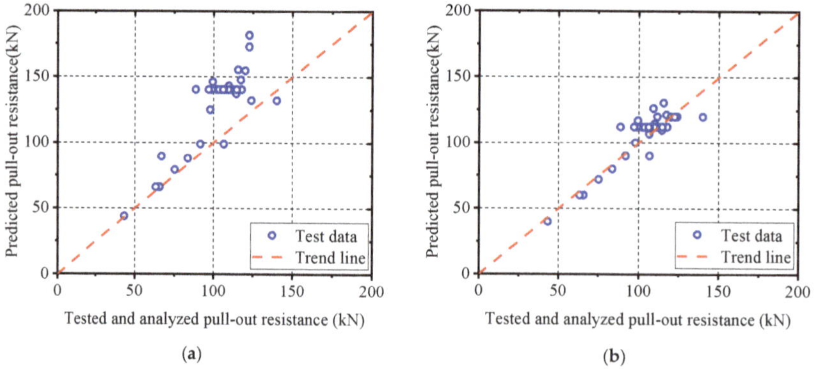

Figure 27. Comparison of calculation and tested results. (**a**) Equations (6) and (7); (**b**) Equations (6) and (8).

7. Conclusions

Based on the pull-out tests, numerical analyses, parametric studies, and analytical work on notched T-perfobond connectors, the following conclusions are drawn:

(1) The pull-out damage modes for notched T-perfobond connectors consist of perfobond rib fracture and concrete punching damage. As the pull-out force increases, the interaction between the perforating rebar and the notched T-perfobond rib diminishes, making the connectors more susceptible to concrete damage in a wedge-shaped inverted pyramid pattern. In contrast, for un-notched T-perfobond connectors, the strong interaction between the perforating rebar and the perfobond rib, along with the anchorage effect of the concrete, leads to the fracture of the perfobond connectors. Notably, these damage modes do not exhibit the extensive fine cracks on the concrete surface that are typical of conventional concrete.

(2) The proposed finite-element method is verified by comparison with the pull-out test results, demonstrating a favorable similarity to the observed pull-out damage. The analytical pull-out resistance values are 106.0% and 84.5% of the average test results for notched and un-notched T-perfobond connectors, respectively. This suggests that the finite-element model can generate rational parametric results for the pull-out performance of these connectors.

(3) A total of 54 parameter simulations indicate that pull-out performance is mainly affected by flange size, connector thickness, and perfobond rib strength. Rebar diameter and strength have little effect. Concrete strength has a notable impact up to a point. Specifically, the flange thickness increases resistance from 2 to 8 mm, then plateaus. The flange width increases resistance significantly from 6 to 12 mm, then declines. The notch width has a minor influence and stabilizes beyond 16 mm. The hole height leads to a slight increase at 15 mm, then a decline. The hole length leads to an increase up to 50 mm, then a significant decrease. The perfobond rib thickness results in a dramatic increase, then plateaus. The perfobond rib length results in a linear increase up to 90 mm. The concrete strength leads to a significant increase and plateaus beyond 110 MPa. The perfobond rib strength leads to a significant linear increase. Design should balance these parameters for the desired resistance.

(4) An analytical model is proposed to estimate the pull-out resistance of notched T-perfobond connectors. This model's formula for calculating pull-out resistance correlates well with test and numerical simulation results, demonstrating its utility for evaluating the pull-out resistance of these connectors.

Author Contributions: Z.D. conducted the finite-element simulation and wrote the article. S.Z. applied for project funding and made overall plans. L.J. completed model experiments. X.X. participated in formulating the experimental plan. Y.Y. and Z.G. participated in experiments and simulations. H.L. participated in formulating project planning schemes. All authors have read and agreed to the published version of the manuscript.

Funding: The research described in this paper was financially supported by Grants from the Natural Science Foundation of China (51808235), the Fujian Provincial Natural Science Foundation of China (2018J05083), the Promotion Program for Young and Middle-aged Teacher in Science and Technology Research of Huaqiao University (ZQN-813), the Huaqiao University Scientific Research Funding (16BS804), and the College Students' Innovation and Entrepreneurship Training Program Project (S202310385013).

Institutional Review Board Statement: Not applicable.

Informed Consent Statement: Not applicable.

Data Availability Statement: Data is contained within the article.

Conflicts of Interest: The authors declare no conflicts of interest.

Nomenclature

d_l	hole length
d_r	diameter of the rebar
t_f	flange thickness
l_f	flange length
c_w	cut width
T_u	pull-out resistance
f_{cu}	concrete cube strength
f_{ry}	yield strength of the reinforcing bar
f_{sy}	yield strength of the perfobond rib
σ_c	compressive stress of concrete
ε_{cp}	peak strain of concrete
E_c	Young's modulus of concrete
n	ratio of peak point cut-line modulus to initial modulus of elasticity
f_t	tensile strength of concrete
w_c	crack width upon the complete release of stress
c_i	constants (i = 1, 2)
d_t	uniaxial damage variables of concrete due to tensile cracking
$\varepsilon_c{}^{in}$	inelastic strain of concrete
w^{pl}	"plastic" crack width of concrete
l_0	unit length
T_u	pulling resistance
T_s	tensile resistance calculated based on the strength of the perfobond connectors
S_h	projected area of concrete punching damage surface
d_h	hole height
t_p	perfobond rib thickness
b_f	flange breadth
l_p	perfobond rib length
h	perfobond rib burial depth
$T_{u,avg}$	averaged pull-out resistance
f_c	uniaxial compressive strength
f_{ru}	tensile strength of the reinforcing bar
f_{su}	tensile strength of the perfobond rib
ε_c	compressive strain of concrete
ξ	ratio of strain to peak strain
E_{cp}	peak point cut-line modulus
σ_t	tensile stress of concrete
w	crack width of concrete
G_F	fracture energy required to create a unit area of stress-free crack
d_c	uniaxial damage variables of concrete due to compressive crushing
$\varepsilon_c{}^{pl}$	plastic strain of concrete
b_c	constant factor for concrete damage in compression
b_t	constant factor for concrete damage in tension
E_s	Young's modulus of steel
T_c	tensile bearing capacity calculated based on concrete strength
A_s	fractured cross-sectional area of perfobond connectors
β	influence factor of the notch

References

1. He, S.; Yang, G.; Jiang, Z.; Wang, Q.; Dong, Y. Effective width evaluation for HSS-UHPC composite beams with perfobond strip connectors. *Eng. Struct.* **2023**, *295*, 116828. [CrossRef]
2. Ma, Y.; Zhang, B.; Peng, A.; Wang, L. Experimental and analytical investigation on shear mechanism of steel-UHPC composite T-perfobond connectors. *Eng. Struct.* **2023**, *286*, 116061. [CrossRef]
3. Sun, L.; Liu, Y.; Wang, H.; Shi, F.; Liu, J.; Jiang, L. Tensile stiffness of perfobond rib connectors in steel–concrete composite pylon of bridges. *Eng. Struct.* **2023**, *284*, 115931. [CrossRef]
4. Zhan, Y.; Huang, W.; Li, Y.; Li, J.; Shao, J.; Tian, B. Experimental investigation on mechanical behavior of T-type perfobond rib shear connectors under combined shear and tension. *J. Build. Eng.* **2023**, *73*, 106840. [CrossRef]

5. Cândido-Martins, J.P.S.; Costa-Neves, L.F.; Vellasco, P.C.G.S. Experimental evaluation of the structural response of perfobond connectors. *Eng. Struct.* **2010**, *32*, 1976–1985. [CrossRef]
6. Vianna, J.D.C.; De Andrade, S.A.L.; Vellasco, P.D.S.; Costa-Neves, L.F. Experimental study of perfobond connectors in composite construction. *J. Constr. Steel Res.* **2013**, *81*, 62–75. [CrossRef]
7. Kim, Y.H.; Kang, J.Y.; Koo, H.B.; Kim, D.J. Pull-Out Resistance Capacity of a New perfobond connector for Steel Pile Cap Strengthening. *Adv. Mater. Sci. Eng.* **2016**, *2016*, 1374689. [CrossRef]
8. He, S.; Fang, Z.; Mosallam, A.S. Push-out tests for perfobond strip connectors with UHPC grout in the joints of steel-concrete hybrid bridge girders. *Eng. Struct.* **2017**, *135*, 177–190. [CrossRef]
9. Di, J.; Zou, Y.; Zhou, X.; Qin, F.; Peng, X. Push-out test of large perfobond connectors in steel–concrete joints of hybrid bridges. *J. Constr. Steel Res.* **2018**, *150*, 415–429. [CrossRef]
10. Zheng, S.; Liu, Y.; Liu, Y.; Zhao, C. Experimental and parametric study on the pull-out resistance of a notched perfobond connector. *Appl. Sci.* **2019**, *9*, 764. [CrossRef]
11. Oguejiofor, E.C.; Hosain, M.U. Numerical analysis of push-out specimens with perfobond rib connectors. *Comput. Struct.* **1997**, *62*, 617–624. [CrossRef]
12. Al-Darzi, S.Y.K.; Chen, A.R.; Liu, Y.Q. Finite element simulation and parametric studies of perfobond rib connector. *Am. J. Appl. Sci.* **2007**, *4*, 122–127. [CrossRef]
13. Zheng, S.; Liu, Y.; Yoda, T.; Lin, W. Parametric study on shear capacity of circular-hole and long-hole perfobond connector. *J. Constr. Steel Res.* **2016**, *117*, 64–80. [CrossRef]
14. Kim, K.S.; Han, O.; Gombosuren, M.; Kim, S.H. Numerical simulation of Y-type perfobond rib shear connectors using finite element analysis. *Steel Compos. Struct.* **2019**, *31*, 53–67.
15. Karam, M.S.; Yamamoto, Y.; Nakamura, H.; Miura, T. Numerical evaluation of the perfobond (PBL) shear connector subjected to lateral pressure using coupled rigid body spring model (RBSM) and nonlinear solid finite element method (FEM). *Crystals* **2020**, *10*, 743. [CrossRef]
16. Suzuki, A.; Suzuki, K.; Kimura, Y. Ultimate shear strength of perfobond connectors subjected to fully reversed cyclic loading. *Eng. Struct.* **2021**, *248*, 113240. [CrossRef]
17. Vianna, J.D.C.; Costa-Neves, L.F.; Vellasco, P.D.S.; De Andrade, S.A.L. Experimental assessment of Perfobond and T-perfobond connectors' structural response. *J. Constr. Steel Res.* **2009**, *65*, 408–421. [CrossRef]
18. Munemoto, S.; Sonoda, Y. An analytical study on dynamic strength evaluation of perfobond strip shear connectors. *Appl. Mech. Mater.* **2014**, *566*, 179–184. [CrossRef]
19. Zheng, S.; Liu, Y.; Yoda, T.; Lin, W. Shear behavior and analytical model of perfobond connectors. *Steel Compos. Struct.* **2016**, *20*, 71–89. [CrossRef]
20. Zheng, S.; Zhao, C.; Liu, Y. Analytical model for load–slip relationship of perfobond connector based on push-out test. *Materials* **2018**, *12*, 29. [CrossRef]
21. Zhu, Y.Y.; Nie, X.; Fan, J.S.; Cui, B.; Huang, L.J. Experimental and analytical investigation on pull-out performance of multihole thin-rib perfobond connectors. *J. Bridge Eng.* **2019**, *24*, 04019037. [CrossRef]
22. He, S.; Guan, P.; Wang, Q.; Fang, Z.; Mosallam, A. Investigation on structural performance of perfobond strip connector group in steel-concrete joints. *Eng. Struct.* **2021**, *242*, 112571. [CrossRef]
23. Zhao, Q.; Zhang, G.; Liao, S.; Wu, N.; Fang, Z.; Fang, S.; Liu, F.; Li, L.; Zhu, H.; Xiong, Z. Static behavior of small-rib-height perfobond connectors embedded in UHPC: Experimental and analytical studies. *J. Build. Eng.* **2024**, *85*, 108750. [CrossRef]
24. Liu, Y.; Yang, H.; Luan, L.; Xu, X.; Liu, Y.; Du, X. Three failure modes of high-strength steel (HSS) perfobond connector embedded in UHPC. *Eng. Struct.* **2023**, *286*, 116147. [CrossRef]
25. He, S.; Li, Q.; Yang, G.; Zhou, X.; Mosallam, A.S. Experimental study on flexural performance of HSS-UHPC composite beams with perfobond strip connectors. *J. Struct. Eng.* **2022**, *148*, 04022064. [CrossRef]
26. Dassault System. *ABAQUS Documentation, Version 6.10*; Dassault System: Waltham, WA, USA, 2010.
27. Hordijk, D.A. Tensile and tensile fatigue behaviour of concrete; experiments, modelling and analyses. *Heron* **1992**, *37*, 1.

applied sciences

MDPI

Article

Research on Mechanical Performance of In-Service Continuous Rigid-Frame Bridge Based on Vehicle-Bridge Coupling Vibration

Yixiang Liu [1], Lingbo Wang [1,2,*], Cong Jiang [1,3] and Hao Shu [1]

1 School of Highway Engineering, Chang'an University, Xi'an 710064, China; 2023121086@chd.edu.cn (Y.L.); 15608634331@163.com (C.J.); 2022221081@chd.edu.cn (H.S.)
2 Key Laboratory of Transport Industry of Bridge Detection Reinforcement Technology, Chang'an University, Xi'an 710064, China
3 Hubei Provincial Expressway Industrial Development Co., Ltd., Wuhan 430050, China
* Correspondence: wanglingbo@chd.edu.cn

Abstract: Based on a damaged continuous rigid-frame bridge in Shaanxi Province, this study deduced the crack damage simulation algorithm and the vehicle-bridge coupling numerical algorithm. Then, it established a finite element analysis model using ANSYS APDL. The Newmark-β iterative method was used to study the dynamic response of different speeds, vehicle weights, and damage degrees before and after the structural damage. In the analysis of the influence of different speeds, the results showed that the dynamic stress responses of key sections of the undamaged bridge reached the maximum when the speed was 80 km/h, indicating that the undamaged bridge was sensitive to a speed of 80 km/h. The peak response of the damaged bridge was 90 km/h. In addition, the displacement peaks and the stress peaks rose and fell together. The analysis of different vehicle weights and damage degrees showed that with the increase in them, the displacement and impact coefficients of each section increased significantly. It can be concluded that the dynamic performance of the in-service bridge decreases continuously with the aggravation of the damage. Therefore, the influence of vehicle-bridge coupling should be emphasized in maintenance, and the frequent cracking area at the midspan should be strengthened in time to prevent further damage.

Keywords: bridge engineering; vehicle-bridge coupling; impact coefficient; continuous rigid-frame bridge

1. Introduction

With the continuous development of the highway transportation industry, the number of highway bridges is steadily increasing. The continuous rigid-frame bridge has large longitudinal and transverse stiffness and remarkable mechanical performance and span capacity. It integrates the mechanical advantages of the continuous beam and T-type rigid-frame bridge and also retains certain characteristics; thus, it allows a reduction in the daily maintenance of the expansion joints and the avoidance of the replacement of large tonnage supports in the later period [1–3]. These characteristics make it one of the main bridge types built in mountainous and canyon areas in northwest China.

The existing research on the vehicle-bridge coupling vibration of high-pier and long-span continuous rigid-frame bridges is relatively well-established, and the factors involved are comparatively comprehensive. For example, Li et al. [4] developed a vehicle-bridge coupling analysis program based on Intel Visual Fortran language. Combined with the corresponding finite element model, the effects of carriageway position, vehicle-bridge mass ratio, and the bridge support form on vehicle-bridge coupling vibration were studied. Zhang et al. [5] studied the vehicle-bridge coupling response of a curved bridge using a finite element simulation method. The results showed that vehicle weight and road roughness were the most influential factors. Peng et al. [6] established a finite element model of a five-span continuous beam bridge according to the general drawings of bridges and culverts and applied the numerical simulation method to analyze the dynamic response of the bridge under vehicle load according to the two aspects of deflection and stress. Finally, it was concluded that there would be a sizable dynamic amplification factor in the middle of the side span and in the middle of the secondary side span and midspan and that the simple support structure at one end of the side span had a greater influence on this factor. Li et al. [7] took a three-span continuous beam bridge with a variable cross-section as a research object to carry out variable parameter analysis and studied the effect of jump-off on the vehicle-bridge coupling response. In addition, Lu et al. [8] further studied the coupling response of continuous rigid-frame bridges under the action of random traffic flow, fitted the critical dynamic deflection, and then provided the reliability index of the bridge under the corresponding working conditions. Lin et al. [9] studied the impact coefficient of a high-speed railway continuous rigid-frame–arch-combined bridge and obtained the fitting formula of the impact coefficient for each component. A. Camara et al. [10] established an analytical framework for the wind–vehicle-bridge interaction and analyzed driving comfort based on 2-D road roughness. Thomas J. Matarazzo et al. [11] proposed a simplified analysis method of vehicle-bridge coupling and proved that it can significantly reduce computational costs. Liu et al. [12] proposed a new method of reordering finite element model numbers using the improved Gibbs–Poole–Stockmeyer algorithm and pointed out that the method could significantly improve the iterative calculation of vehicle-bridge coupling efficiency. Sandro Carbonari et al. [13] combined the mass change method with the OMA method to obtain the modal scaling factors for the mass normalization of mode shapes. The effects of vehicle-bridge coupling effect are considered in the study, and the cases that need to be considered and can be ignored are distinguished. Soheil Sadeghi Eshkevari et al. [14] used STRIDEX algorithm to process the mobile sensor data and identify the modal characteristics of the bridge under different road roughness level. Li et al. [15] established a 2-D vehicle model and 3-D vehicle model, and the effects of the different models on the coupling vibration of the vehicle-axle are analyzed. Arturo Gonzalez et al. [16] carried out a parameterization study on the uncoupled vehicle-bridge interaction and calculated the influence of different factors on the algorithm error.

Through the above investigations, it can be seen that the theory of vehicle-bridge coupling vibration under the continuous rigid-frame bridge system is now mature after nearly 70 years of research. However, the research focus is mainly on the vehicle-bridge coupling vibration response of the undamaged continuous rigid-frame bridge, while the research on the dynamic response of the damaged continuous rigid-frame bridge under vehicle-bridge coupling vibration is still insufficient. Using modified concrete bridge damage parameters [17,18], Yin et al. [19] carried out a vehicle-bridge coupling analysis of a damaged three-span continuous concrete box beam with variable sections and pointed out that the impact coefficient of the multi-span variable sections of the damaged beam varied with the vehicle speed and was related to the crack parameters. Xie et al. [20] studied the vehicle-bridge coupling response of an old concrete truss arch bridge by combining a real bridge detection method and finite element modeling and then gave the impact coefficient of different bridge locations under different vehicle speeds. Zhou et al. [21] analyzed the vehicle-bridge coupling vibration of an in-service curved-string truss bridge based on an analysis method that considered the jump-off period and calculated the dynamic response of the space vehicle fleet crossing the bridge. However, structural diseases such as beam cracking, steel corrosion, and the local spalling of concrete are also very common in continuous rigid-frame bridges. In addition, this type of bridge has developed rapidly and in large numbers in northwest China (especially in Shaanxi Province). With the increase in their service lives, a large number of in-service bridges with cracks are in continuous operation. When cracks appear in the beam body, their damage develops under the repeated influence of vehicle-bridge coupling vibration, which has an impact on the structural durability and load-bearing safety of the concrete bridge. If it is not controlled in time, when the cracks expand to a certain extent, the collapse of the bridge structure will become a threat to people's lives and property.

Therefore, this study mainly investigates the dynamic response of the high-pier and long-span continuous rigid-frame bridge with crack damage under different vehicle parameters and the degrees of structural damage; then, it analyzes the influence of the bridge stress, displacement, and impact coefficient.

2. Crack Damage Simulation Algorithm

Based on the statistical analysis of the number, length, width, and height of the cracks in the continuous rigid-frame bridge, this study simulates the crack damage by reducing the stiffness in the areas with high concentrations of cracks and establishes a crack model accordingly. This can improve the accuracy of the results and can also establish a theoretical foundation for the subsequent influence of the coupling vibration of the vehicle bridge on the dynamic response of the bridge structure.

According to Wu Xiaoguang's [22], the investigation of 74 continuous rigid-frame bridges in Shaanxi Province, the section cracking of continuous rigid-frame bridges is serious in the L/4, L/3, and midspan closure sections. Among them, the longitudinal cracks in the bottom plate of L/4 and the assembly section, such as the midspan closure section, are particularly serious and have become the most serious disaster [23] area for crack development. Therefore, L/4, L/3, and the midspan in the following text refer to the corresponding position of the second span (middle span). The cracks are divided into four categories, and each category is shown in Figures 1–4.

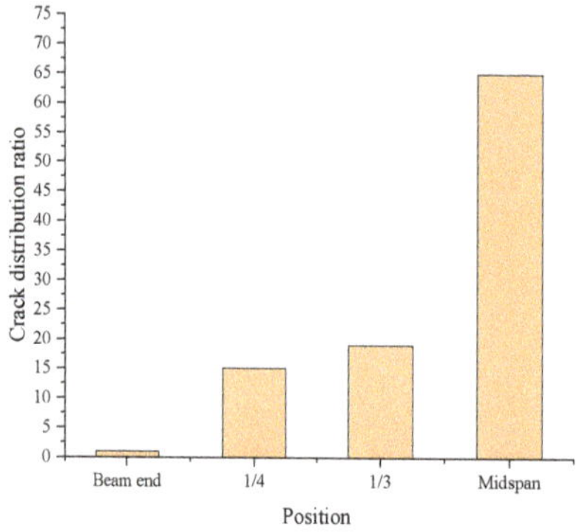

Figure 1. Distribution ratio of cracks in the first type of bridge.

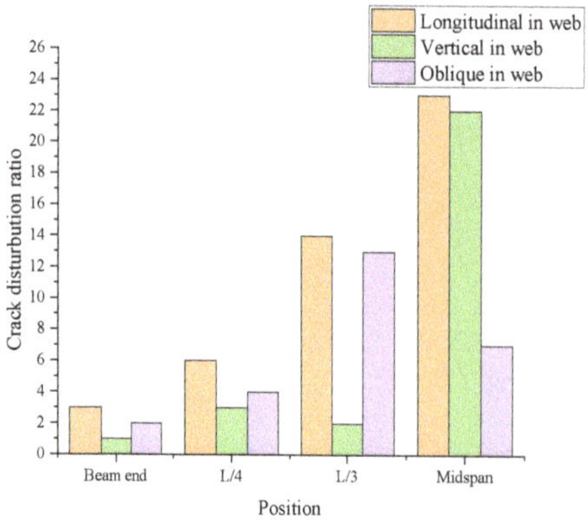

Figure 2. Distribution ratio of cracks in the second type of bridge.

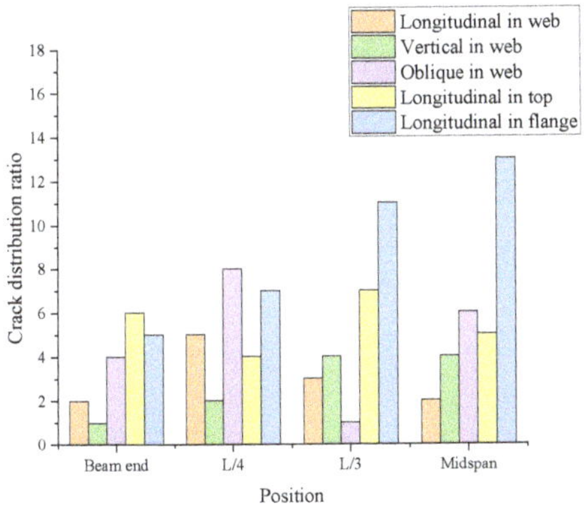

Figure 3. Distribution ratio of cracks in the third type of bridge.

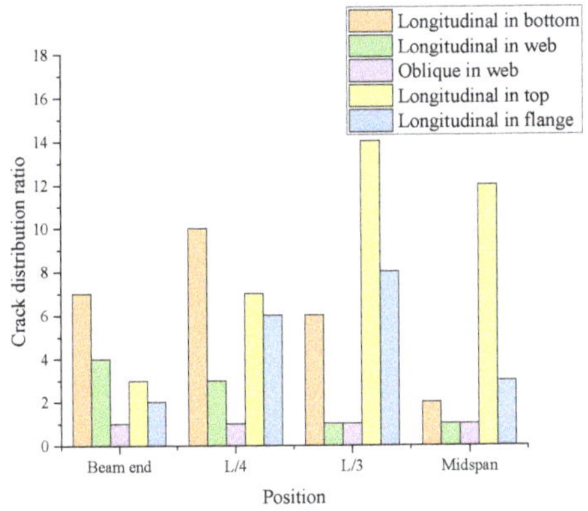

Figure 4. Distribution ratio of cracks in the fourth type of bridge.

The first type of bridge cracks is mainly distributed on the bottom plate, running longitudinally along the bridge span direction. The second type of bridge cracks is mainly distributed on the web of the beam body, and the longitudinal, vertical, and oblique cracks extend from the end of the beam to the middle of the span in an increasing trend. The third type of bridge cracks is mainly distributed in the web, roof, and flange plate. These cracks include longitudinal, vertical, and oblique cracks. Of these, the top plate and flange plate mainly have longitudinal cracks, while the web plate mostly has inclined cracks. The fourth type of bridge cracks is mainly distributed in the top plate, bottom plate, web and flange plate. The top plate, bottom plate and flange plate have longitudinal cracks, while the web has longitudinal cracks and oblique cracks.

The beam cracks mainly appear at the midspan, and the length of the cracks along the longitudinal position (near the end, one-fourth, one-third, and the midspan) typically measures around 1.5 m; in addition, a small number of cracks are longer than 2 m and concentrated in the area from one-third to the midspan. From the perspective of crack width, cracks larger than 0.2 mm are mainly concentrated in the area from one-fourth span to approximately middle span position. Therefore, this research focused on the areas around these three positions. According to the results of the crack field investigation, a cracked beam section was selected as the research object (as shown in Figure 5), where h_{cri} represents the degree of concrete withdrawal from the work in the tension area; δ_{cri} is the degree of cracking, i.e., the crack width; l_{cri} is the crack spacing, which represents the number of cracks in the curved section beam; and l_{cr} represents the range of crack distribution and is related to the position of the external load form.

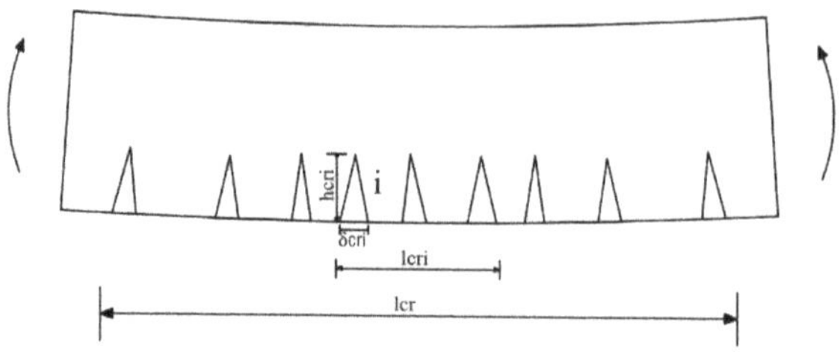

Figure 5. Schematic diagram of cracks in beam bending section.

Zhao Yu [24] pointed out that the statistical parameters of the cracks in the section are as follows: the average height of the cracks is $\overline{h_{cr}}$, the average statistical spacing of the cracks is $\overline{l_{cr}}$, the total width of the cracks is Δ_{cr}, etc.

$$\begin{cases} \overline{h}_{cr} = \dfrac{\sum\limits_{i=1}^{n} h_{cri}}{n} \\ \overline{l}_{cr} = \dfrac{\sum\limits_{i=1}^{n} l_{cri}}{n-1} = \dfrac{l_{cr}}{n-1} \\ \Delta_{cr} = \sum\limits_{i=1}^{n} \delta_{cri} \end{cases} \tag{1}$$

According to the bridge appearance survey, the cracks with similar length and width were merged to form an overall gradient beam model (see Figure 6). Specifically, the cracks with similar heights in the crack section (the difference between the crack heights should be less than one-tenth of the beam height) were merged, the section of the cracked section was reduced, and the reduction was compensated by the dead weight.

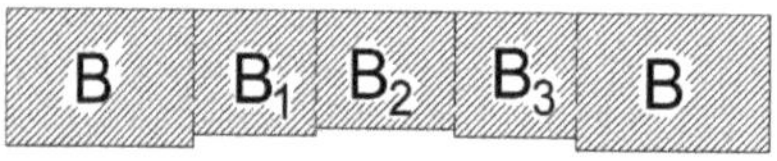

Figure 6. Overall gradient beam model.

Then, as shown in Figure 6, a mechanical analysis of the structure was carried out to solve the internal force distribution of the structure with cracks. At this point, the nominal tensile stress (strain) of the concrete on the tensile side can be expressed as:

$$\overline{\varepsilon}_{cr} = \Delta_{cr}/l_{cr} \tag{2}$$

$$\sigma_{nt} = E_c\overline{\varepsilon}_{cr} \tag{3}$$

By making the tensile stress of the concrete at the lower edge of the section equal to the nominal tensile stress, the stress state of the reinforcement of the concrete in the tensile zone can be obtained based on statistics and detection; the stress balance relation of the crack section can be established on this basis, as shown in Figure 7, where σ_{cc} is the stress at the edge of the pressure zone; σ_{nt} is the nominal stress on the edge of the tension zone; α_{Es} is expressed as the ratio of the elastic modulus between the ordinary reinforcement and the concrete; and α_{Ep} is the ratio of the elastic modulus between the prestressed reinforcement and the concrete.

Through the balance of the cross-sectional stress relationship, the actual compression height of the concrete, the position of the neutral axis, and the edge stress of the concrete in the compression zone can be solved numerically.

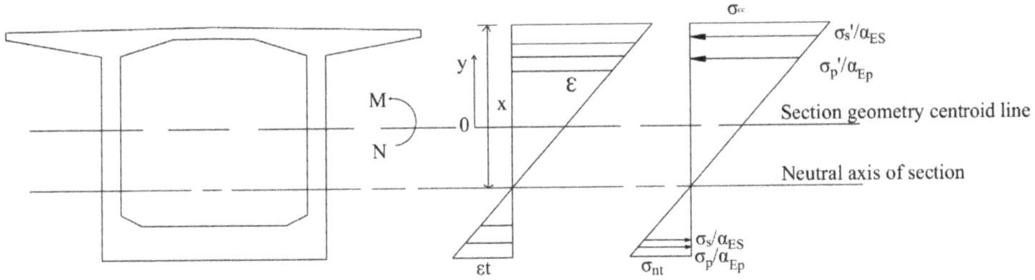

Figure 7. Stress (strain) equilibrium state of the section of the crack area.

As the height x of the compression zone and the bending curvature φ of the section are unknown, a trial algorithm can be used to solve the problem. When the calculated bending moment of the current round is close enough to that of the previous round, the result is considered final.

According to the calculated concrete crack model, the stiffness of the cracked concrete beam was reduced. It was assumed that the initial flexural rigidity of the box girder section is E_cI, and the flexural rigidity becomes αE_cI when cracks appear. α is the reduction coefficient of the flexural stiffness:

$$\alpha = \frac{M_i}{\varphi E_c I} \tag{4}$$

where M_i represents the bending moment of the section, and φ represents the bending curvature of the section.

The process used to solve the reduction coefficient α of the flexural rigidity is shown in Figure 8. Since the height x of the compression zone and the bending curvature φ of the section are unknown, a known initial value x and a changing value φ can be assumed first, making the relative error satisfy $N_t = |N_i - N_{p0}|/|N_{p0}|$ and $M_t = |M_i - M_0|/|M_0|$. When the results of the last two iterations are close enough, the obtained value can be put into Formula (4) to find α. In order to avoid the interaction of multiple factors, the single control variable method was used in this study to obtain the average stiffness damage degree of the three sections and to reduce the stiffness of the solid elements in the finite element model in this area, so as to simulate the bridge crack damage. It can be seen from the trial calculation that the reduction coefficient of the flexural stiffness α is 0.76.

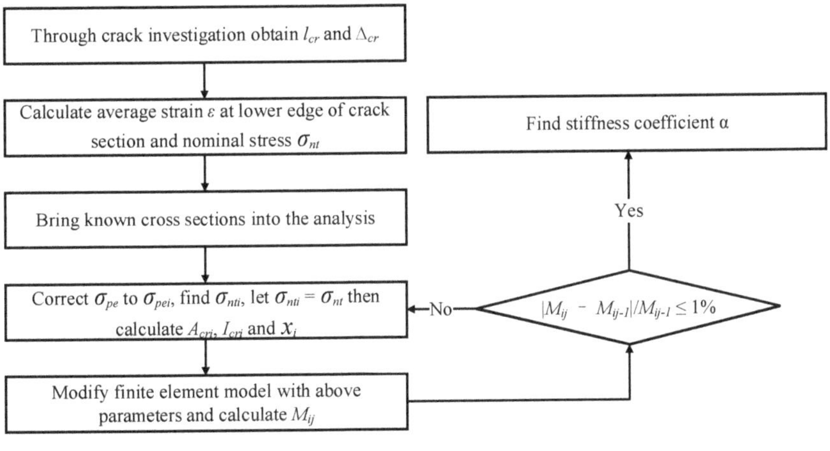

Figure 8. Process of solving the reduction coefficient α of flexural stiffness.

3. Numerical Algorithms of Vehicle-Bridge Coupling

The vehicle and bridge models were established by assuming that the wheel and bridge deck maintain favorable contact and that there would be no temporary separation due to turbulence vibration during the driving process. Using the displacement coordination at the contact point and the mechanical equation, an ANSYS model was established to achieve the vehicle-bridge coupling simulation algorithm.

The bridge model built in this study adopted the linkage and solid element model. For the existing bridge after cracking, its dynamic equation can be expressed as

$$M_b \ddot{Y}_b + C_b' \dot{Y}_b + K_b' Y_b = F_b \quad (5)$$

where M_b, C_b', and K_b' represent, respectively, the structural unit mass matrix, the modified damping matrix, and the modified stiffness matrix of the bridge; $\ddot{Y}_b$, $\dot{Y}_b$, and Y_b represent, respectively, the acceleration vector, velocity vector, and displacement vector of the bridge structural unit nodes; and F_b represents external force. The vehicle is modeled using the spring–damp–mass model, and the equation of motion can be expressed as

$$M\ddot{Y} + KY + C\dot{Y} = G + F_b \quad (6)$$

where M is the vehicle body mass matrix; K is the stiffness matrix of the car body; C is the damping matrix of the car body; G is the gravity load vector; and F_b is the instantaneous coupling load between the tire and the ground contact point.

To solve iteratively the coupled vibration equations of the beam arch VBI system, the detailed processes are summarized as follows.

Step 1: determine the working conditions to be analyzed and build the finite element model of the bridge.

Step 2: input the relevant calculation parameters according to the working condition and generate vehicle file, track irregularity file and coupling analysis macro command file.

Step 3: engage ANSYS transient dynamic analysis and obtain the coordinates of the contact points between each wheel and the bridge, node displacement D(n,ti), node velocity V(n,ti), and node acceleration A(n,ti) of node n at time ti.

Step 4: calculate the interaction force between vehicle and bridge at corresponding time based on the displacement coordination condition.

Step 5: Check whether all vehicles are off the bridge or not. If so, proceed to the next step. If not, set i = i + 1 and repeat from Step 3.

Step 6: Read the RST result file and export the results. Then check if all working conditions finished and repeat from Step 2 when negative. The analysis process is shown in Figure 9.

3.1. Build the Bridge Model

This study is based on a prestressed concrete continuous rigid-frame bridge in Shaanxi Province. The length of the main bridge is 292 m, and the span combination is 76 + 140 + 76 m. The main span is a 140 m continuous prestressed concrete rigid frame with a variable section comprising a single box and a single chamber. The roof width of the box girder is 12 m, the width of the bottom plate is 6.5 m, and the length of the flange cantilever is 2.75 m. The top plate of the box girder is provided with a bidirectional 2% transverse slope, and the transverse bridge direction of the bottom plate is horizontal. The overall layout of the main bridge type is shown in Figure 10.

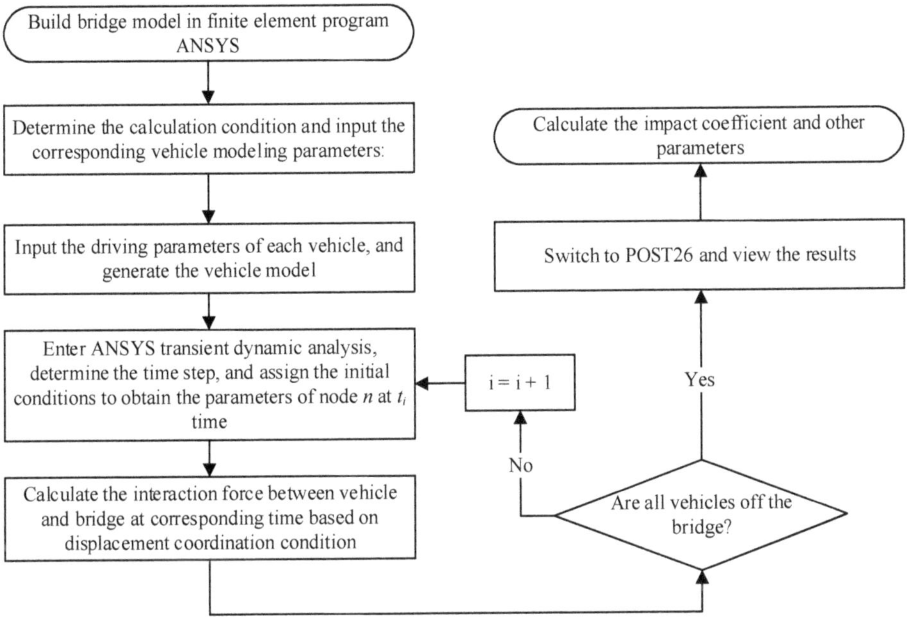

Figure 9. Process of vehicle-bridge coupling analysis.

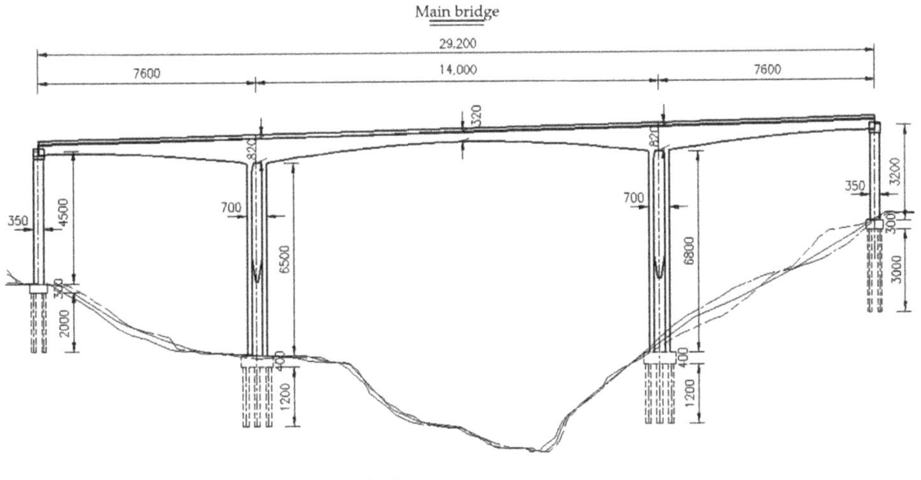

Figure 10. General layout of the main bridge type.

In this study, a multi-scale mixed-elements model was established according to the different requirements of the modeling accuracy. Solid65 (3-D-reinforced concrete solid element) element modeling was used to simulate the cracking, crushing, and stress release state of the concrete. In addition, beam44 (3-D elastic tapered asymmetric beam element) and link8 (3-D rod or truss element) were used to simulate other components. Considering that the in-service bridge could no longer meet the linear stress characteristics, nonlinear analysis was required. The constitutive relation of the concrete adopted the "miso" model, and the constitutive relation of the steel bar adopted the "biso" model. The FE model is shown in Figure 11.

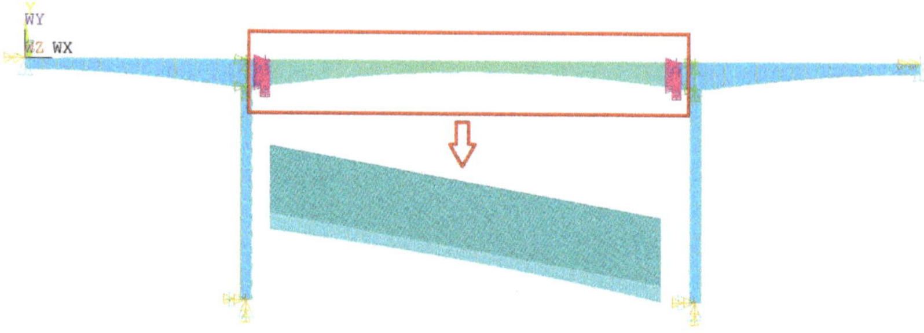

Figure 11. Multivariate hybrid finite element model.

3.2. Establishment of Vehicle Model

Through the image analysis of the surveillance video of the continuous traffic flow in a section of Shaanxi Province over a period of 24 h, the frequency of each vehicle on the highway was obtained. The types of vehicles that travel on the highway bridges in this area are mainly dual-axle vehicles. Therefore, in the analysis of the vehicle-bridge coupling vibration, spatial dual-axle (car) and spatial dual-axle (bus) were selected as the research objects, as shown in Figure 12. The higher the number of vehicle dimensions and degrees of freedom, the higher the accuracy of the vehicle model and the more realistic the vibration process [25,26]. According to Section 1, The research on the stiffness matrices of various vehicle models is very extensive; therefore, it is not necessary to go into detail here.

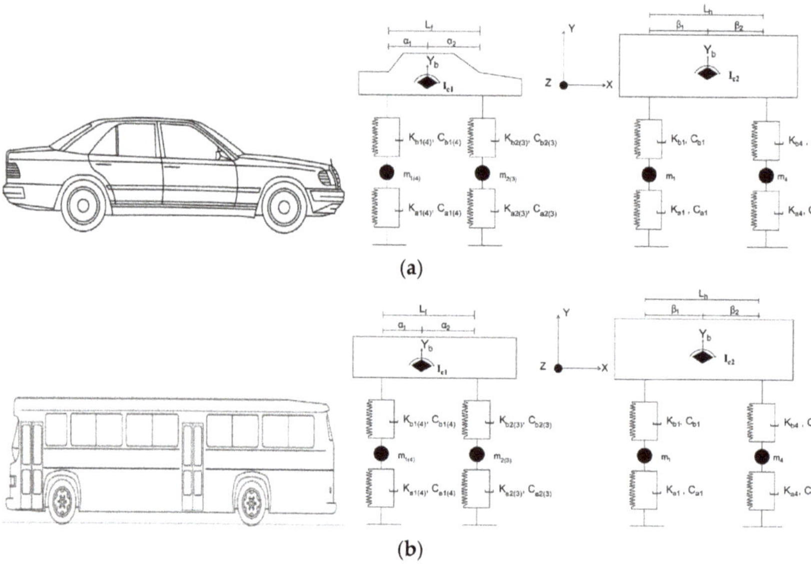

Figure 12. Schematic diagram of the bus. (**a**) The longitudinal and transverse mechanics of the car. (**b**) The longitudinal and transverse mechanics of the bus.

3.3. Bridge Deck Roughness Simulation

At present, the roughness data of a bridge (bridge deck) are usually obtained via field measurement or numerical simulation. The application cost of the measured data is high, while the numerical simulation method is lower in cost and may be more efficient. Based on the inverse Fourier transform method, the numerical calculation model of the road surface (bridge deck) smoothness sequence generated in this study is shown in Formulas (7) and (8).

$$r(x) = \frac{1}{N}\sum_{i=0}^{N-1} X_i e^{\frac{2\pi ikm}{N}} \quad (m = 0, 1, 2, \cdots, N-1) \tag{7}$$

$$X_i = |X_i| e^{k\varphi_i} \quad (i = 0, 1, 2, \cdots, \frac{N}{2}) \tag{8}$$

In accordance with the ISO standard [27], MATLAB was used to simulate and generate the first three levels of the pavement smoothness sequence obtained via Fourier inversion, as shown in Figure 13.

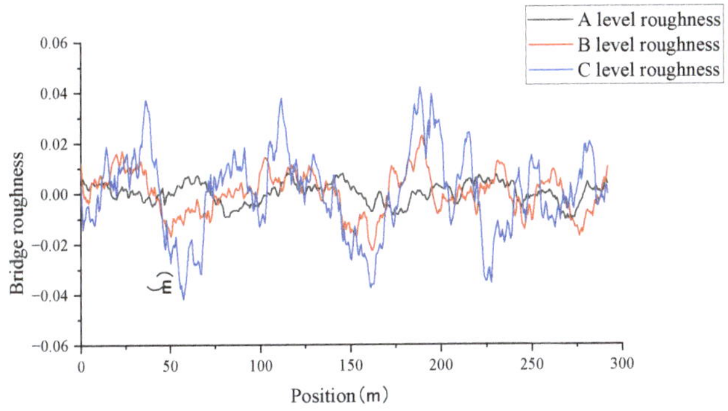

Figure 13. Roughness sequence of three levels of pavement (bridge deck).

In the investigation of continuous rigid-frame bridges in Shaanxi Province, it was found that most of the highway pavement grades had reached Class B; therefore, the road roughness examined in this study is mainly Class B.

4. Results of Vehicle-Bridge Coupling Analysis

4.1. Impact Analysis of Vehicle Parameters

4.1.1. Influence of Different Vehicle Speeds

In the study of speed parameters, the 35 t spatial dual-axis vehicle model was adopted, and the analyzed speeds were 60 km/h, 80 km/h, 90 km/h, 100 km/h, and 120 km/h. The stiffness reduction coefficients were 1 (undamaged) and 0.76 (damaged), with a total of 10 working conditions. According to the discussion in Section 2, crack damage is realized by stiffness reduction. The positions of stiffness reduction are mid-span, one-third, and one-fourth of the middle span. When analyzing these conditions, the stiffness of the beam segment at the corresponding positions in the numerical model is multiplied by the stiffness reduction factor.

Through the analysis of the dynamic response of the bridge under different vehicle speeds, the stress peak value of the concrete at the lower edge of the cross-section was extracted, and the change rule is shown in Figures 14 and 15.

By comparing the stress values of the damaged bridges and undamaged bridges, it can be seen that the vehicle-bridge coupling vibration response of each section changes greatly after the bridge is damaged. Taking the maximum stress peak values of 80 km/h and 90 km/h as examples, the midspan stress increases by 84.71% and 107.89%, respectively. The stress of the one-fourth section of the midspan increases by 49.07% and 81.64%, respectively. The stress of the one-third section of the middle span increases by 40.07% and 68.39%, respectively. From the above data, it can be seen that the vehicle-bridge coupling vibration response of the damaged bridge is more significant. From Figure 14, the stress at the speed of 80 km/h is significantly greater than that of the other three speed conditions for any of the sections. Therefore, when the bridge is undamaged, the bridge response under vehicle-bridge coupling vibration is sensitive to the speed of 80 km/h. As can be seen in Figure 15, the peak value of the vehicle-bridge coupling vibration response is the highest when the vehicle speed is 90 km/h at the midspan and one-fourth section positions. However, at the one-third section of the midspan, the structural response is greater than that at the other speeds when the vehicle speed is 80 km/h, and the peak stress value is 2.14% larger than that at the 90 km/h speed. This indicates that although the dynamic response of the bridge at the one-third section of the midspan is the largest when the vehicle speed is 80 km/h, the dynamic response is also obvious at the 90 km/h speed.

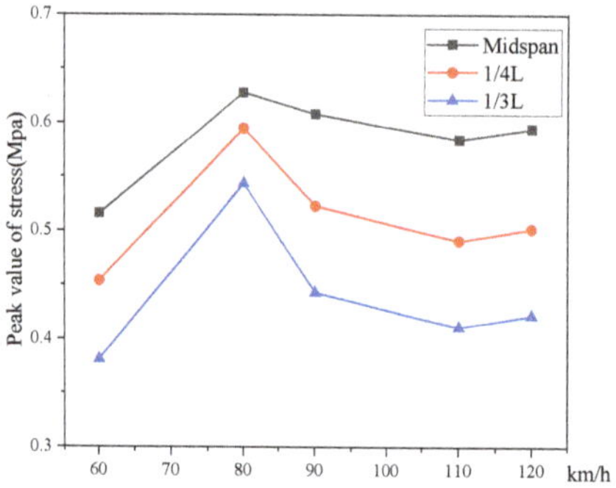

Figure 14. Stress peaks of different sections of undamaged bridges at different speeds.

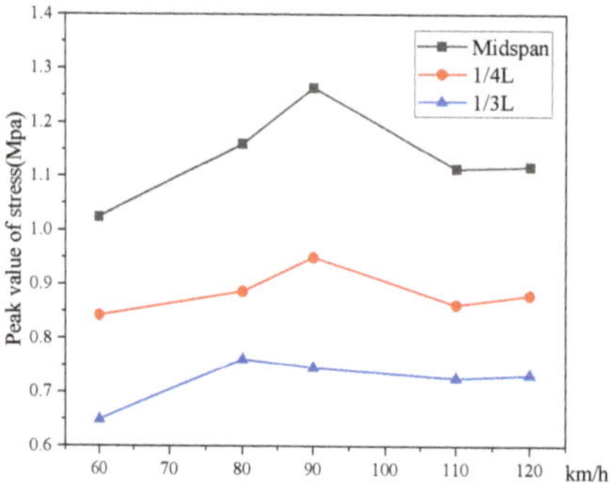

Figure 15. Stress peaks of different sections of damaged bridges at different speeds.

Taking the damaged bridge as the research object, the reduction coefficient is 0.76, and the other vehicle parameters remain unchanged. According to the above research on the relationship between the vehicle parameters and the damaged bridge, the vehicle speeds in this section are still 60 km/h, 80 km/h, 90 km/h, 100 km /h, and 120 km/h. The vehicle weight is 35 t, and the vehicle type is the spatial dual-axis model; the road roughness grade remains Class B, and the dynamic response of different vehicle speeds to the bridge is studied. Figures 16 and 17 show the variation diagrams of the peak displacement law and the impact coefficient.

As shown in Figure 16, the dynamic response of each section of the middle span to the speed is different. For the damaged bridge, the fluctuation of the displacement curve of each section corresponds to the peak stress, reflecting the law that the peak displacement and peak stress increase and decrease at the same time. In combination with the vehicle speed, the peak displacement of both the midspan section and the midspan one-fourth section reaches its maximum when the vehicle speed is 90 km/h. At the one-third cross-section of the middle span, the displacement reaches the maximum when the speed is 80 km/h. However, when the speed is from 80 km/h to 90 km/h, the displacement change rate at the cross-section is −3.46%, and the displacement peak does not change much. These data also show the correlation between the peak displacement and the peak stress.

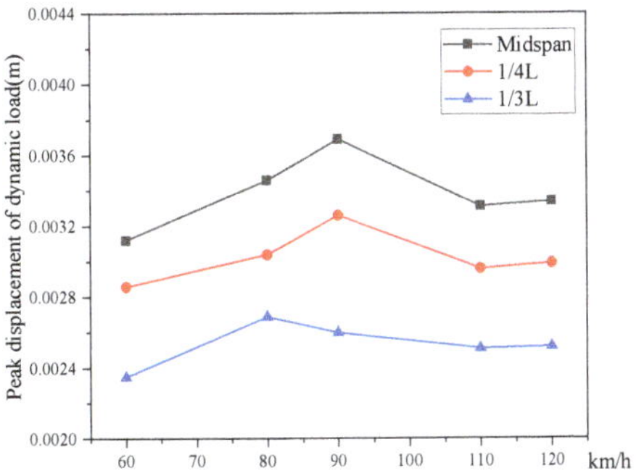

Figure 16. Peak displacement of each section at different vehicle speeds.

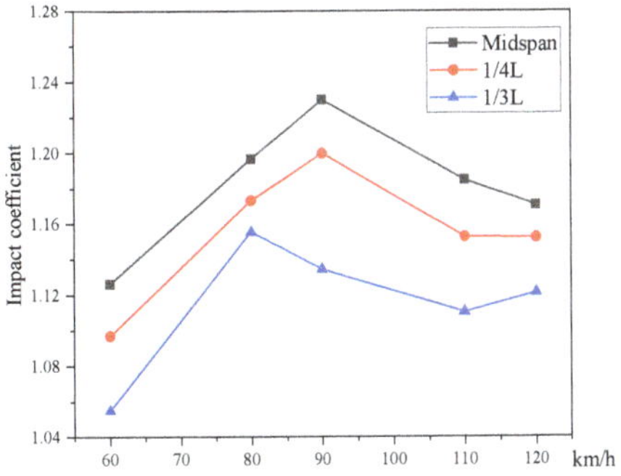

Figure 17. Dynamic amplification factor of each section at different vehicle speeds.

As shown in Figure 17, when the vehicle speed reaches 90 km/h, the impact coefficient at the one-fourth section of the midspan and the midspan is the largest. When the vehicle speed reaches 80 km/h, the impact coefficient at the one-third section of the midspan is the largest. When the speed is 60 km/h, the impact coefficient of each section is the smallest. According to the change in the peak displacement and impact coefficient, the bridge is more sensitive to the speed of 90 km/h when the damage occurs.

4.1.2. Influence of Different Vehicle Weights

With the continuous development of the highway traffic industry, the phenomena of heavy loads and excessive loads are becoming more and more frequent. The above situation will cause the static load stress of the bridge to become very large over a period of time; furthermore, it will make the damage caused by the dynamic load on the bridge more serious. Therefore, this section chooses six working conditions to study the dynamic response of the bridge. The vehicle weights are 35 t, 45 t, and 55 t, respectively. The stiffness reduction coefficients are 1 and 0.76, respectively.

Through the analysis of different vehicle weight conditions, the stress peak values of the lower edge of the concrete section at the midspan, the one-fourth section of the middle span, and the one-third section of the middle span were obtained, and the change rule is shown in Figure 18.

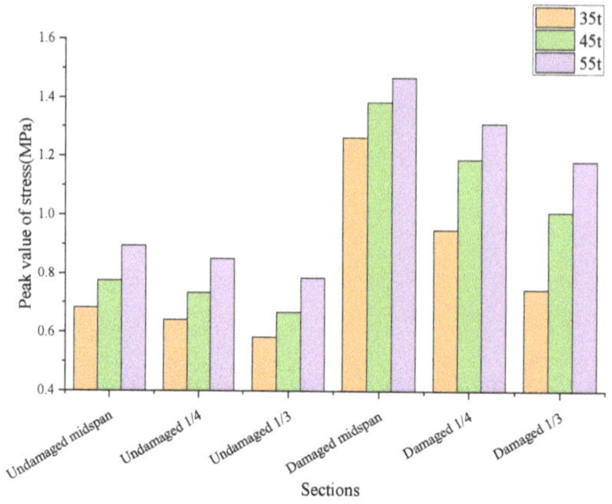

Figure 18. Stress peak values of each section under different vehicle weights.

It can be seen from Figure 18 that the peak stresses of each section of the damaged bridge under different vehicle weight conditions have obvious changes. When the vehicle weight increases from 35 t to 55 t, the increase rate of the peak stress of the middle section of the middle span is 16.21%, that of the one-fourth section of the middle span is 38.11%, and that of the one-third section of the middle span is 58.58%. Although the dynamic response of the midspan section is the largest, the one-third section has the fastest stress increase rate.

For the undamaged bridges, with the increase in vehicle weight the stress change of each section is similar to that of the damaged bridges. Under different vehicle weights, the stress at the midspan position is the largest, followed by the one-fourth section, and the one-third section is the smallest. When the vehicle weight increases from 35 t to 55 t, the stress increase rate of the middle span section is 31.03%, the stress increase rate of the one-fourth section is 32.76%, and the stress increase rate of the one-third section is 34.65%. It can be seen that the vehicle weight parameter has a significant effect on the dynamic response of the bridge under the vehicle-bridge coupling vibration, and the impact on the dynamic response of the damaged bridge is more significant.

The variation laws of the peak value and impact coefficient of the displacement at the one-fourth section and one-third section are shown in Figures 19 and 20. It can be seen that with the increase in vehicle weight, the displacement of each section also increases significantly. Among them, the increase is more obvious in the midspan position. When the vehicle weight increases from 35 t to 55 t, the peak increase rate of the midspan displacement is 22.49%. Figure 20 shows that the displacement impact coefficient of each section of the damaged bridge is different under the condition of vehicle weight, and the impact coefficient keeps increasing with the increase in vehicle weight. The increase in vehicle weight occurs at the same rate, but the impact coefficient increases in a nonlinear way. When the vehicle weight increases from 35 t to 55 t, the midspan impact coefficient increases by about 16.24%. The impact coefficient at the position of the one-fourth section increases by about 13.15%, and the impact coefficient at the position of the one-third section increases by about 11.10%. Although the increase in vehicle weight did not result in a large increment in the impact coefficient, the impact coefficient only reflected the ratio relationship between the dynamic response and the static response. The increase in the dynamic response of each section can indicate that the increase in vehicle weight has a significant impact on the dynamic response of the bridge.

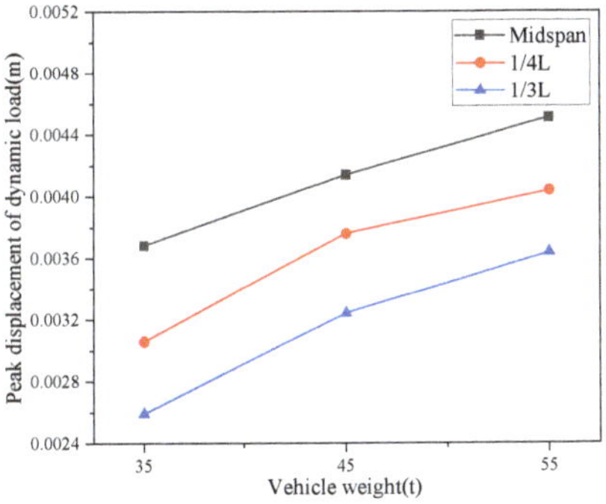

Figure 19. Peak displacement of each section under different vehicle weights.

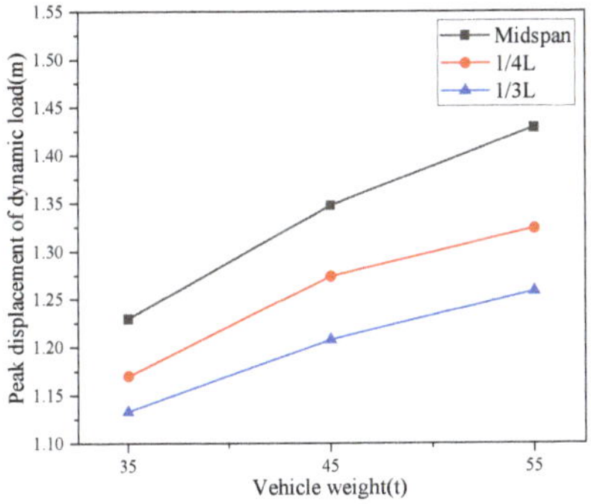

Figure 20. Impact coefficient of each section under different vehicle weights.

4.2. Impact Analysis of Damage Parameters

Based on the research above on the dynamic response of different vehicle parameters to the damaged bridge, this section studies the dynamic response of the vehicle-bridge coupling vibration to the bridge with sustained damage. The vehicle speed is 90 km/h, the vehicle weight is 35 t, and the vehicle is a spatial dual-axle model. As the bridge continues to serve after damage, the stiffness of the bridge itself will decrease with the passage of time. Assuming that the stiffness reduction coefficient starts at 0.76, six working conditions are considered: 0.76, 0.7, 0.65, 0.6, 0.55, and 0.5.

According to Figure 21, it is obvious that with the continuous damage to the bridge, the stress of each section continues to increase. The stress growth trend is different at different cross-section locations. For the midspan position, the stiffness reduction coefficient is from 0.76 to 0.5, and the change in stress value shows obvious fluctuation. When the reduction coefficient is from 0.76 to 0.7, the stress changes faster, and the growth rate is 12.5%. In the range of 0.7–0.65, the stress changes slowly, and the growth rate is 1.78%. In the range of 0.65–0.6, the stress growth rate is 7.02%, which is faster than the stiffness decay in the second stage. When the reduction coefficient is in the range from 0.6 to 0.5, the curve of the stress change is almost linear, and the average stress growth rate is 8.83%. It can be inferred that with the continuous damage to the bridge, the stress increase rate will be larger and larger. If the stress in the span continues to increase until the tensile strength is reached or even exceeded, the concrete will be damaged. For the position of the one-fourth section of the middle span, the stress value fluctuates to a certain extent. In the range of the stiffness reduction coefficient from 0.76 to 0.5, the slope of each straight line changes as follows: $k_{0.6\text{–}0.55} > k_{0.76\text{–}0.6} > k_{0.55\text{–}0.5}$. It can be seen from the slope that the stress increases slowly and linearly in the range of 0.76 to 0.60. In the range of 0.60 to 0.55, the stress increases rapidly; from 0.55 to 0.50, the rate slows down again. As the damage continues, the growth rate of the stress at the position of the one-fourth section first increases and then decreases for the damaged concrete. For the position of the one-third section of the middle span, the stress growth rate does not change rapidly in the range of 0.76 to 0.6 and always maintains a gentle rate of increase. In the range of 0.6 to 0.55, the stress growth rate has a large change and becomes smaller again after 0.55 to 0.5. It can be inferred that the stress growth rate will increase up to failure at the position of the one-third section as the stiffness continuously attenuates.

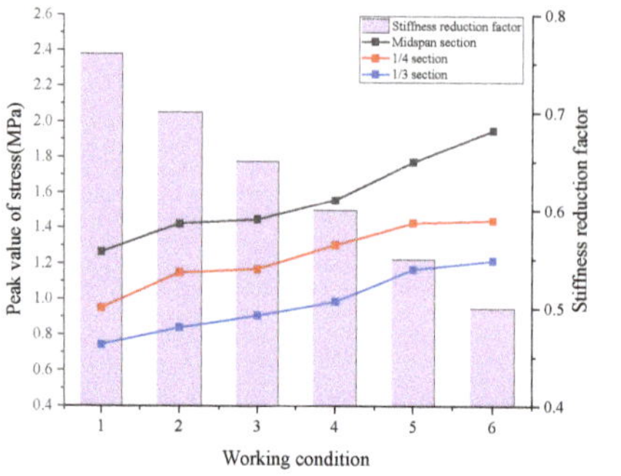

Figure 21. Stress of each section under different damage degrees.

The results of the dynamic displacement peak value and impact coefficient also show similar rules. As can be seen from Figure 22, in the process of the continuous damage to the bridge, the displacement of each section is increasing, and the displacement increment of the midspan section is the largest. When the stiffness reduction coefficient reaches 0.6, the displacement increases almost linearly. When the stiffness of the one-fourth section reaches 0.6, the displacement growth fluctuates a little. The displacement change is gentler than that in the previous stage when the reduction coefficient is 0.55 to 0.5. When the reduction coefficient is in the range from 0.76 to 0.5, the displacement is relatively gentle at the one-third section compared with the other two places. When the damage coefficient decreases from 1 to 0.76, the displacement of each section increases rapidly, with an increase of 86.36% in the midspan, 114.08% in the one-fourth section, and 192.13% in the one-third section. When the reduction coefficient increases from 0 to 0.5, the displacement growth rate of the midspan section is 335.86%, that of the one-fourth section is 316.9%, and that of the one-third section is 452.8%. It can be seen from these data that with the continuous damage to the bridge, the displacement of each section will continue to increase and that the growth trend is the fastest in the midspan.

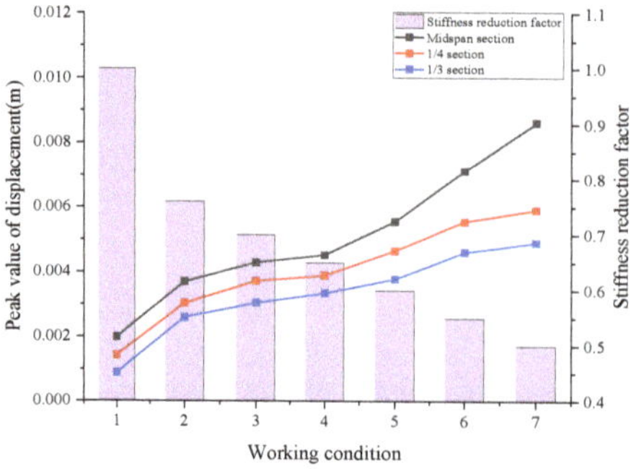

Figure 22. Peak displacement of each section under different damage degrees.

As can be seen from Figure 23, the impact coefficient of the bridge also changes greatly, and the impact coefficient at the midspan increases the fastest. By comparing the change curves of the impact coefficients at the three cross-section locations, the approximate slope of each reduction coefficient curve is ordered as follows: $k_{0.65–0.5} > k_{0.76–0.7} > k_{0.7–0.65}$. When the reduction coefficient of stiffness reaches 0.65, the growth trend of the impact coefficient of each cross-section increases significantly compared to that before 0.65, and the three curves are almost linearly distributed. However, before the reduction coefficient reaches 0.65, the impact coefficient curves of each section fluctuate to a certain extent. Therefore, the impact coefficient obviously increases after the damage occurs.

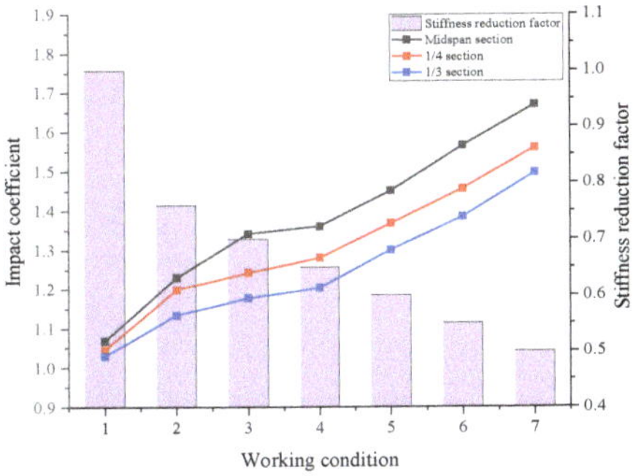

Figure 23. Impact coefficient of each section under different damage degrees.

In summary, when the stiffness reduction coefficient is in the range of 0.65 to 0.6, the stress growth rate of each section does not change much. In other ranges, the stress growth rate of the different sections is different. According to the stress growth mode shown in the figure, the stress on the lower edge of the concrete in the midspan section will first reach the tensile strength and crack, followed by the one-fourth section and, finally, the one-third section. With the continuous damage to the bridge, the displacement and impact coefficient of each section will have an increasing trend, with the fastest growth in the midspan. Therefore, when these two indexes of the bridge structure increase significantly, it indicates that the structural stiffness is decaying, and certain reinforcement measures should be taken at this time.

5. Conclusions and Prospect

This study investigated the dynamic response of an in-service continuous rigid-frame bridge under vehicle load based on the crack damage report, the finite element software ANSYS 17.0, and custom software. The main findings and conclusions are as follows:

(1) The crack damage of continuous rigid-frame bridges in Shaanxi Province was investigated, and it was found that the positive cracks of the structure floor were mostly concentrated in the midspan, the one-fourth section, and the one-third section of the middle span. The characteristic parameters and damage model of the cracks were established, and the simplified crack zone unit was obtained. The stress state of the crack element area was analyzed using a numerical method, and the total bending moment of the section and the resultant force of the reinforcement were solved. The process was iterated until convergence was achieved, and the resulting coefficient of the sectional flexural rigidity reduction for the cracked area was obtained. Based on this parameter, the in-service damage to the bridge was simulated.

(2) The investigation of traffic flow and vehicle types on highway bridges concluded that the largest proportion of the highway traffic consisted of dual-axle vehicles; therefore, the dual-axle vehicle model was adopted in this study. According to the analysis of different vehicle parameters, the results showed that under different speed conditions, the undamaged bridge was more sensitive to the speed of 80 km/h, and the damaged

bridge was more sensitive to the speed of 90 km/h at the midspan and one-fourth span sections. For the one-third section, the sensitive speed of the damaged bridge was still 80 km/h, but the difference between the stress peak and the value at 90 km/h was very small. Therefore, it is generally recommended to focus on a speed limit of 90 km/h for damaged bridges.

(3) In the investigation of the dynamic responses under different vehicle masses, the results showed that the dynamic response was obvious for the damaged bridge and that the increase in the concrete stress at the lower edge of each section was prominent. Therefore, it is recommended to strengthen the middle span of the bridge in the crack disaster area in time to prevent further cracking.

(4) The results of the analysis of the dynamic displacement and impact coefficients of the bridge showed that the dynamic response of vehicle mass to bridge structure was more obvious. With the increase in vehicle mass, the displacement response and impact coefficient response of each section continued to increase. With the continuous damage to the structure, the dynamic response and impact coefficient of each section continued to increase, and the growth rate of the impact coefficient increased. This indicates that the dynamic performance of the in-service bridge will continue to decline with the aggravation of the damage. Therefore, the influence of vehicle-bridge coupling should be emphasized in the maintenance of the in-service bridge.

In this study, only the factors of speed, vehicle weight, and damage degree were considered. In the subsequent research, more factors can be introduced for analysis, and the superposition of multiple factors can also be considered. In addition, this study investigated the dynamic response of damaged bridges based on single-vehicle crossing conditions; thus, the influence of multi-vehicle conditions and random traffic flow can be considered in the future.

Author Contributions: Conceptualization, L.W.; methodology, L.W.; software, Y.L. and C.J.; validation, L.W., Y.L. and H.S.; formal analysis, Y.L.; investigation, C.J.; resources, L.W.; writing—original draft preparation, Y.L. and C.J.; writing—review and editing, Y.L.; visualization, H.S.; supervision, L.W.; project administration, L.W.; funding acquisition, L.W. All authors have read and agreed to the published version of the manuscript.

Funding: This research was funded by National Key Research and Development Program of China (Grant No. 2021YFB1600305); the Natural Science Basic Research Program of Shaanxi (Program No. 2022JC-23); and Shaanxi Provincial Natural Science Foundation 2020JM-230.

Institutional Review Board Statement: Not applicable.

Informed Consent Statement: Not applicable.

Data Availability Statement: The original contributions presented in this study are included in the article. Further inquiries can be directed to the corresponding author.

Conflicts of Interest: Author Cong Jiang was employed by the company Hubei Provincial Expressway Industrial Development Co., Ltd. The remaining authors declare that the research was conducted in the absence of any commercial or financial relationships that could be construed as a potential conflict of interest.

References

1. Wang, D.; Tong, L.; Wang, R.; Sun, Z. Review on Advances in Seismic Research of Large-Span Prestressed-Concrete Continuous Rigid-Frame Bridges. *J. Southwest Jiaotong Univ.* **2023**, *58*, 511–526.
2. Zhu, J.; Zhang, Y.; Chen, X. Crack propagation of concrete beams under moving vehicle loads. *J. Southeast Univ. (Nat. Sci. Ed.)* **2018**, *48*, 678–686.
3. Wang, H.; Wang, L.; Zhang, Z.; He, X. Safety early warning technology of continuous rigid frame bridges based on crack width variation. *J. Jilin Univ. (Eng. Sci.)* **2023**, *53*, 1650–1657. [CrossRef]
4. Li, Z.; Li, Z.; Zhang, Y.; Chen, D.; Xu, S.; Zhang, Y. Theoretical analysis and experimental study of vehicle-bridge coupled vibration for highway bridges. *Arch. Appl. Mech.* **2024**, *94*, 21–37. [CrossRef]
5. Zhang, X.; Chen, E.; Li, L.; Si, C. Development of the Dynamic Response of Curved Bridge Deck Pavement under Vehicle-Bridge Interactions. *Int. J. Struct. Stab. Dyn.* **2022**, *22*, 2241003. [CrossRef]

6. Peng, W.; Gao, Q.; Xu, Z. Dynamic Response Analysis of Continuous Beam Bridge Based on Vehicle-bridge Coupling Vibration. *For. Eng.* **2022**, *38*, 146–155.
7. Li, X.; Zhang, Z.; Xiao, R. Parameters influence analysis of vehicle bridge coupling vibration under vehicle jumping impact. *J. Appl. Mech.* **2023**, 1–8. Available online: https://kns.cnki.net/kcms/detail//61.1112.o3.20230227.1306.014.html (accessed on 28 February 2023).
8. Lu, W.; Wang, K.; Wang, H.; Liu, Y. Dynamic Reliability of Continuous Rigid-Frame Bridges under Stochastic Moving Vehicle Loads. *Shock Vib.* **2020**, *2020*, 8811105. [CrossRef]
9. Lin, P.; Hou, S.; Wang, Y. Impact Factor of Through Continuous Rigid Frame-Arch Composite Bridge for High-Speed Railway. *China Railw. Sci.* **2023**, *44*, 73–80.
10. Camara, A.; Kavrakov, I.; Nguyen, K.; Morgenthal, G. Complete framework of wind-vehicle-bridge interaction with random road surfaces. *J. Sound Vib.* **2019**, *458*, 197–217. [CrossRef]
11. Soheil, S.E.; Thomas, J.M.; Shamim, N.P. Simplified vehicle-bridge interaction for medium to long-span bridges subject to random traffic load. *J. Civ. Struct. Health Monit.* **2020**, *10*, 693–707.
12. Liu, X.; Han, W.; Yang, Q.; Yuan, Y.; Liu, B.; Su, N. Application Research on Improving Gibbs-Poole-Stockmeyer Algorithm to Accelerate Vehicle-Bridge Coupling Vibration Analysis. *China J. Highw. Transp.* **2024**, 1–18. Available online: http://kns.cnki.net/kcms/detail/61.1313.u.20240301.1434.003.html (accessed on 5 August 2024).
13. Sandro, C.; Vanni, N.; Riccardo, M.; Fabrizio, G. Dynamics of bridges during proof load tests and determination of mass-normalized mode shapes from OMA. *J. Eng. Struct.* **2024**, *310*, 118111.
14. Soheil, S.E.; Thomas, J.M.; Shamim, N.P. Bridge modal identification using acceleration measurements within moving vehicles. *J. Mech. Syst. Signal Process.* **2020**, *141*, 106733.
15. Li, J.; Feng, D. A comparative study of vehicle-bridge interaction dynamics with 2D and 3D vehicle models. *J. Eng. Struct.* **2023**, *292*, 116493. [CrossRef]
16. Arturo, G.; Enrique, C.; Miguel, C. Verifying the suitability of uncoupled numerical methods for solving vehicle- bridge interaction problems. *Struct. Infrastruct. Eng.* **2023**, *19*, 1407–1424.
17. Wahab, M.A.; De Roeck, G.; Peeters, B. Parameterization of damage in reinforced concrete structures using model updating. *J. Sound Vib.* **1999**, *228*, 717–730. [CrossRef]
18. Law, S.S.; Zhu, X.Q. Dynamic behavior of damaged concrete bridge structures under moving vehicular loads. *J. Sound Vib.* **2004**, *26*, 1279–1293. [CrossRef]
19. Yin, X.; Fang, Z. Dynamic Response of Cracked Concrete Continuous Beam Bridge under Moving Vehicular Loads. *Eng. Mech.* **2009**, *26*, 174–180.
20. Xie, R.; Li, M.; Xu, C. Study on the Technical Status of Old Concrete Truss-Arch Bridge based on Vehicle-Bridge Interaction. *J. Adv. Civ. Eng.* **2022**, *2022*, 9946957. [CrossRef]
21. Zhou, R.; Wang, Y.; Chen, H. Analysis of vehicle-bridge coupled vibration of long-span curved truss bridge in service. *J. Vib. Eng.* **2022**, *35*, 103–112. [CrossRef]
22. Wu, X.; He, P. Investigation and analysis of cracks in longitudinal cracks of continuous rigid frame bridge in Shaanxi Province. *J. Henan Univ. Urban Constr.* **2017**, *26*, 54–60. [CrossRef]
23. Sobczyk, B.; Pyrzowski, Ł.; Miśkiewicz, M. Damage of a post-tensioned concrete bridge—Unwanted cracks of the girders. *Eng. Fail. Anal.* **2024**, *160*, 108178. [CrossRef]
24. Zhao, Y.; He, S.; Li, C.; Ge, Y. Evaluating System of Bearing Capacity for Existing Cracked Prestressed Concrete Box Girder Based on Crack Feature. *J. Tongji Univ. (Nat. Sci.)* **2010**, *38*, 1271–1275.
25. Zhou, Y.; Xue, Y.; Gao, X.; Li, R.; Wang, Y.; Zhao, Y. Research on dynamic amplification factor of highway simply supported girder bridge based on modal superposition method. *J. Traffic Transp. Eng.* **2023**, *23*, 146–155. [CrossRef]
26. Feng, D.; Li, J.; Wu, G.; Zhang, J. Fast modeling and dynamic analysis method for vehicle-bridge interaction system considering road roughness. *J. Southeast Univ. (Nat. Sci. Ed.)* **2022**, *52*, 1088–1094.
27. *ISO 8608:2016*; Mechanical Vibration—Road Surface Profiles—Reporting of Measured Data. International Organization for Standardization: Geneva, Switzerland, 2016.

applied sciences

MDPI

Article

A Novel Debonding Damage Identification Approach of Hidden Frame-Supported Glass Curtain Walls Based on UAV-LDV System

Haoyang Zheng [1], Tong Guo [2,*], Guoliang Zhi [1] and Zhiwei Hu [1]

[1] School of Civil Engineering, Southeast University, Nanjing 210096, China; haoyangzheng@seu.edu.cn (H.Z.); zhiguoliang@seu.edu.cn (G.Z.); 220231319@seu.edu.cn (Z.H.)
[2] Key Laboratory of Concrete and Prestressed Concrete Structures, Ministry of Education, Southeast University, Nanjing 210096, China
* Correspondence: guotong@seu.edu.cn

Abstract: This study introduces a novel Unmanned Aerial Vehicle-mounted (UAV-mounted) Laser Doppler Vibrometer (LDV) system for detecting debonding damage in Hidden Frame-Supported Glass Curtain Walls (HFSGCW). The established system enables UAVs to transport the LDV to high altitudes for operation. The vibration signals acquired by the UAV-LDV system are decomposed into different energy bands by wavelet packet analysis, and then the occurrence and location of the damage are identified by the Sum of Squared Differences (SSD) of the wavelet packet bands' energy. This paper investigates the potential factors affecting the performance of the Unmanned Aerial Vehicle-Laser Doppler Vibrometer (UAV-LDV) system, including the arrangement of measuring points, measuring distance, noise level, and wind speed through the first-order natural frequency, the normalized frequency response functions, and the SSD indicator. Experimental and simulation results confirm the effectiveness of the UAV-LDV system, highlighting its advantages over traditional methods by offering remote, non-contact, and efficient debonding detection. This method not only indicates the presence of the damage, as traditional indicators do, but also pinpoints the exact location of it, ensuring safety and cost-effectiveness in high-rise inspections. The proposed method and indicator offer advantages in terms of convenience, visualization, and efficiency. The study discusses the impact of measurement point arrangement, measuring distance, noise levels, and wind speed on the system's performance. The findings demonstrate that while the UAV-LDV system introduces new capabilities in rapid and reliable structural damage assessment, operational challenges such as wind and noise levels significantly influence its accuracy.

Keywords: building curtain wall; structural damage detection; laser doppler vibrometer; remote vibration measurement; unmanned aerial vehicle

1. Introduction

The glass curtain wall is an aesthetically innovative method of architectural wall decoration and a distinctive feature of modernist high-rise buildings [1]. However, the extended life of glass curtain walls may result in sudden and dangerous detachment which can cause serious injury or property damage [2]. For instance, structural adhesives can suffer from deterioration and debonding problems. Debonding damage to existing glass curtain walls is mainly identified by visual inspection [3], manual inspection [4], contact sensors [5], numerical simulation [6], health monitoring systems [7], etc. These methods rely on experienced workers or contact sensors for stationary testing, which are insufficient to cope with the increasing height of skyscrapers, the emerging variety of curtain walls and the various extreme working conditions in modern buildings. Since these conventional stationary inspection approaches are inefficient [8], hazardous [9], complicated [10], and hard to employ [11], despite being very effective at detecting debonding damage of glass curtain walls, they cannot be widely used. In light of the rapid development of Unmanned Aerial Vehicle (UAV) and new laser vibrometry technologies, mobile inspection approaches show promising advantages in application [12] and further investigations.

The increasing demand for structural inspections, focusing on safety [13], durability [14,15], and comfort [16,17], makes the intelligent inspection of UAVs an inevitable trend in health monitoring processes [18]. The development of UAV technology has provided formidable technical support for this demand, including improved stability [19], enhanced wind resistance [20], increased load capacity [21], extended endurance [22], strengthened obstacle avoidance capability [23], etc. For example, the DJI Mavic 3 is a powerful and efficient drone available to the consumer designed by SZ DJI Technology Co., Ltd. (Suzhou, China). It can bear strong winds up to 12 m/s (strong breeze) and has a vertical hovering accuracy of ±0.1 m and a horizontal hovering accuracy of ±0.3 m when the visual positioning system is working properly [24]. These outstanding properties make such UAVs reliable for inspecting structural health in windy scenarios, such as high-altitude inspections of curtain wall structural damage, seismic performance of bridges [25], blade damage of wind turbines [26], etc.

Additionally, UAVs have been involved in numerous civil engineering applications for structural health inspections. They have been used for detecting bridge cracks [27], identifying dynamic parameters of bridge components [28], measuring geometry shapes of bridge components [29], and the 3D reconstruction of bridges [30]. The utilization of UAVs under certain weather conditions can improve efficiency and ensure the safety of the inspection personnel compared to manual inspections [31]. Similarly, wind turbines are often constructed in remote locations, such as offshore [32], in deserts [33], on mountaintops [34], and other areas where wind energy is abundant but difficult for humans to access. The application of UAVs replaces manual detection and shutdown maintenance, making regular inspections of wind turbines possible [35]. UAVs are also increasingly employed in other inspection processes, such as power transmission lines [36], pipelines [37], dams [38], and water towers [39]. This technology has become a promising solution for labor-intensive, accessibility-challenging, and sometimes dangerous inspection tasks in civil engineering [40].

There are a variety of practical difficulties associated with detecting debonding damage on glass curtain walls using traditional methods. Manual inspections on the facade of an office building can be disruptive to the normal work of the building's occupants, challenging for the inspector [4], and costly for the maintenance company. Wall-climbing robots struggle with the diverse shapes of glass curtain walls and often require manual assistance [41]. Health monitoring systems are not commonly used for checking the condition of glass curtain walls due to their high cost and difficult installation [42]. Moreover, not all damage conditions result in large areas of peeling and stripping of the structural adhesive. In many cases, the structural adhesive may appear to be intact from the outside but is already failing or delaminating from the inside. This issue hinders identification using traditional vision-based methods. Conventional methods often rely on periodic inspections, and the time between inspections can lead to prolonged damage and delayed repairs. Therefore, the detection method of debonding damage on glass curtain walls should be efficient, immediate, safe, and cost-effective. These problems can be avoided by identifying damage using the vibration characteristics of the curtain walls [13]. Laser technology provides benefits of range, accuracy and efficiency. However, the power of a laser is limited, which makes ground-based damage assessment an impossible task.

The combination of UAV and laser vibration measurement technology effectively addresses the challenge of detecting debonding damage on glass curtain walls. Remote, non-contact, and low-error vibration measurements can be achieved by attaching the Laser Doppler Vibrometer (LDV) to a UAV. Some researchers have identified hidden vehicles in the forest by using LDVs placed at high altitudes to detect the vibrations of leaves [43]. Although the effect of wind turbulence on leaf vibration was considered, the interference with the LDV signals was not. Previous research also proposed the concept that LDVs could be transported in aerial platforms for testing, but their experiment still used a lift vehicle, which is no different from a stationary LDV. Other scholars claimed that they have built a UAV-LDV system to achieve aerial testing. However, the utilization of the UAVs in these systems is limited to the placement of reflective targets [44], for the enhancement of the reflected laser intensity or the carriage of acoustic exciters [45], for the replacement of manual tapping, or carrying mirrors to reflect the laser light from stationary LDVs [46,47]. Their actual tests still rely on stationary LDVs, and the purported UAV-LDV systems have not been fully realized. The UAV-LDV system is susceptible to factors that can cause interference, which in turn compromises the accuracy of the measurements. It is only used for vibration monitoring in situations with a significant degree of displacement, such as dynamic lateral displacements in railways [48]. While some studies have optimized UAV-LDV systems based on the assumption that the target remains in a steady state of vibration [49] and that the main source of error is induced by vibrations from the UAV [50], it is crucial to recognize that the UAV-LDV system is notably prone to inaccuracies resulting from the UAV's aerial instability, and more damage detection methods are based on transient vibration. Specifically, the vibration monitoring of LDVs relies on detecting differences in the optical path, which is significantly affected by the deflection of the UAV-LDV system during flight. Consequently, the most substantial errors in these systems typically stem from strong winds and difficulties in the attitude adjustment of UAVs during operation.

The dynamic response of the curtain wall can be obtained by the subsequent analysis of the filtered vibration signals, which can ultimately reveal the debonding damage of glass curtain walls. This study investigates multiple sets of test conditions that differed in the location and extent of damage to the structural adhesive on curtain walls. The obtained signals were analyzed using the first-order of natural frequency, frequency response function, and the Sum of Squared Differences (SSD) of wavelet packet band energy as indicators. The reliability of the vibration signals obtained from the UAV-LDV system is verified by comparing them with those obtained from the fixed LDV. The results of the study demonstrate that the proposed SSD indicator based on the UAV-LDV system provides a more accurate identification of debonding damage to curtain walls in transient vibration. The study also explores the factors affecting the accuracy of the UAV-LDV system and their respective degrees of influence, including distance, noise, number of measuring points, and wind strength. The research provides a useful reference for the combination of UAV and laser vibration measurement technology for debonding damage detection on Hidden Frame-Supported Glass Curtain Walls (HFSGCW).

2. UAV and LDV Selection

There are two main types of UAVs: fixed-wing and multi-rotor. Fixed-wing UAVs are mainly suitable for long-duration, long-range, and wide-area flight operations [51]. They have high-speed flight capability, longer endurance, and larger load capacity, making them suitable for agricultural and environmental missions. Multi-rotor UAVs are designed for tasks that require small-scale, aerial hovering, and minor space shuttling capabilities [52]. They consist of multiple rotor blades, enabling vertical takeoff and landing, and hovering in the air. These characteristics enable the systems to be suitable for energy surveying, electrical power inspection, etc. Consequently, they are more commonly employed in detailed image acquisition and inspection of stationary structures, such as power lines and wind turbines.

Multi-rotor UAVs are preferred in this study owing to their excellent weight-carrying capability for carrying equipment such as LDVs. Additionally, they can hover steadily and align themselves with the measuring points to obtain the acceleration time history during the test. The advanced DJI Mavic 3 was then chosen for testing due to its potential use in curtain wall inspection. Table 1 shows the specifications, and Figure 1a,b illustrate its appearance, respectively. The vibration signal of HFSGCW is measured using the LDV

(MV-H100) from OMNISENSING PHOTONICS in China. Figure 1c depicts the appearance of the LDV and Table 2 shows the specifications of it.

Table 1. DJI Mavic 3 specifications.

Specific Items	Specifications
Bare Weight	895 g
Dimensions (Folded/Unfolded)	Folded (without propellers) 221 × 96.3 × 90.3 mm (Length × Width × Height) Unfolded (without propellers) 347.5 × 283 × 107.7 mm (Length × Width × Height)
Max Flight Time (no wind)	46 min
Max Hovering Time (no wind)	40 min
Max Wind Speed Resistance	12 m/s
Operating Temperature	−10° to 40° C (14° to 104° F)
GNSS	GPS + Galileo + BeiDou
Hovering Accuracy Range	Vertical: ± 0.1 m (with Vision Positioning); ± 0.5 m (with GNSS Positioning) Horizontal: ± 0.3 m (with Vision Positioning); ± 0.5 m (with High-Precision Positioning System)
Maximum Take-off Weight with Payload	1200 g

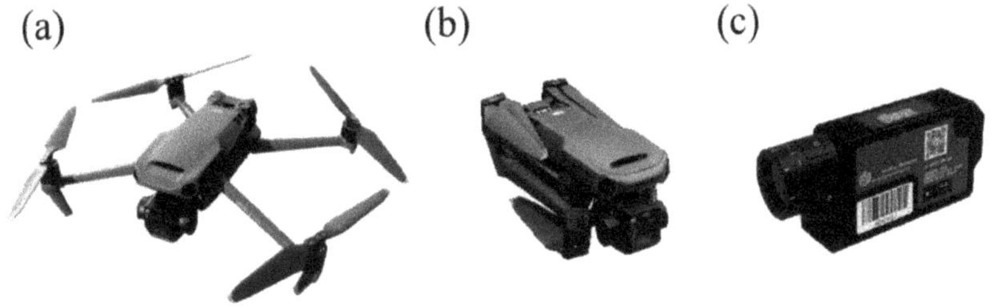

Figure 1. Instruments. (**a**) Unfolded form. (**b**) Folded form. (**c**) LDV.

Table 2. LDV specifications and weight of the system.

Frequency Range	Laser Wavelength	Displacement Resolution	Maximum Speed Range	Displacement Repeatability	Output Signal	Weight
DC~2.5 MHz	1310 mm	1.28 nm	1500 mm/s	10 nm	Digital	213.7 g
	Weight of the bracket			Weight of the system		
	90.2 g			1198.9 g		

Figure 2 shows the constructed testing system. The LDV is first attached to the bracket using a carbon fiber plate and stainless-steel screws. Then, the bracket is fitted onto the UAV's fuselage. A system for acquiring acceleration time history is constructed by combining the selected UAV and LDV. The LDV directly captures and transmits signals to a PC for analysis, enabling efficient detection of acceleration time history at high altitudes without an additional data acquisition system, as it is an integrated device.

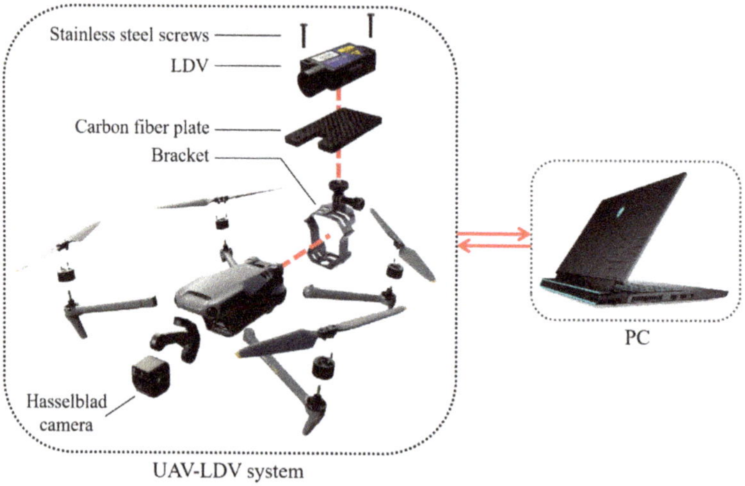

Figure 2. Testing system for examining vibration signals.

3. Influence Factors of the UAV-LDV System

The number and distribution of the measuring points, as well as the testing distance, are first considered when using the LDV to acquire the acceleration of the HSFGCW. It is important to compare the results of the mobile test with the stationary test to verify its accuracy, as the stationary test has shown great efficiency and reliability [53]. However, since the mobile test is susceptible to noise and wind disturbances, it is essential to assess their impact.

3.1. Selection of Assessment Indicators

Modal analysis can be used to detect the safety state of glass panels due to the significant impact of sealant failure on modal parameters. The first-order modal frequency is considered as the safety condition evaluation indicator because of its stability and accessibility [54]; therefore, this indicator is used in this paper to evaluate the impact of each disturbance. The inspection method based on the accumulated change rate of the frequency response function (FRF) can efficiently, accurately, and expeditiously identify the sealant failure of HFSGCW. This study employs the normalized FRF at different locations to identify the number and distribution of measuring points, according to the literature [55].

Assume that the natural frequency and mode vector of the glass panel system with n degrees of freedom are θ_r and $\{\varphi\}_r (r = 1, 2, \cdots, n)$, where the $\{\varphi\}_r$ can be expressed as $\{\varphi_1^r\ \varphi_2^r \cdots \varphi_n^r\}^T$. Thus, in the p-th degree of freedom, the acceleration FRF $H_{\mathrm{pp}}(\omega)$ is [56]

$$H_{\mathrm{pp}}(\omega) = \frac{A_{pp}(\omega)}{Q_{pp}(\omega)} = \sum_{r=1}^{n} \frac{-\omega^2 \varphi_p^{r2}}{-\omega^2 m_r + i\omega c_r + k_r} \tag{1}$$

where $A_{pp}(\omega)$ and $Q_{pp}(\omega)$ are the Fourier transforms of the displacement and force, respectively, at frequency ω; the modal mass is $m_r = \{\varphi\}_r^T [M] \{\varphi\}_r$, the stiffness coefficient is $k_r = \{\varphi\}_r^T [K] \{\varphi\}_r$, and the modal damping coefficient is $c_r = \{\varphi\}_r^T [C] \{\varphi\}_r$.

The wavelet packet of the structural dynamic response under vibration excitation can be decomposed into a number of independent frequency bands. Then the changes in the energy spectrum of the wavelet packet bands can indicate the occurrence of structural failure. This energy change can be identified by the Sum of Squared Differences (SSD) of the wavelet packet band energy, which can be defined as [57]

$$SSD = \sum_{i=1}^{m} [(E_i)_b - (E_i)_a]^2 \tag{2}$$

where m is the number of the main frequency bands, $(E_i)_a$ and $(E_i)_b$ represent the signal energy from the intact and damaged structure, respectively. The SSD indicator has been demonstrated to be effective in detecting structural sealant failure [13]. In this paper, the reliability of the UAV-LDV system and the factors affecting its operation are also verified by this indicator.

3.2. Similarity Relationship

The number and distribution of the measuring points are related to the location of the excitation and structural sealant failure during the test. Theoretically, the number of measuring points should be as high as possible to accurately localize the occurrence of structural adhesive damage. However, a large number of measuring points reduce detection efficiency and increase the difficulty of analyzing the acquired data. This acts against the original purpose of developing a UAV-LDV system for rapid and mobile inspection. Due to the difference in the frequency response function at different locations of the glass curtain wall panel system [55], this section examines the effect of the measuring point location on the identification.

As depicted in Figure 3, twelve measuring points were arranged diagonally across the panel. The method of excitation at the measuring points is employed to obtain a driving point Frequency Response Function (FRF). Point A is the hammering excitation point which is located at the geometric center of the glass panel to obtain the average first-order natural frequency. In the area where measurement points 2, 4, and 6 are located, a simulated excitation experiment was performed according to the literature [13]. The first- to fifth-order modal frequencies of the glass panel obtained by modal analysis and LDV tests are shown in Table 3.

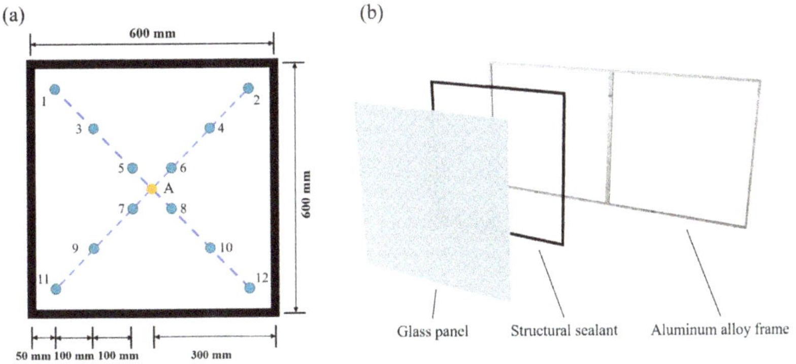

Figure 3. Diagram. (**a**) The distribution of measuring points. (**b**) Composition of HFSGCW.

Table 3. Glass panel frequency of the first five-step mode.

Order	Frequency/Hz			
	Numerical Simulation [13]	Point 2 Test	Point 4 Test	Point 6 Test
1st	101.0	99.2	95.7	93.1
2nd	234.3	/	227.1	228.9
3rd	354.2	329.6	330.8	/
4th	364.3	378.4	375.9	/
5th	449.8	436.4	435.2	433.3

After testing and analysis, the normalized frequency response functions of the acceleration at points 2, 4, and 6 are shown in Figure 4. It shows that point 4 has five resonance peaks between 0 and 500 Hz, while point 2 has four resonance peaks, and point 6 only has three resonance peaks. Table 3 shows that in the first fifth-order frequency range, the resonance peak of the frequency response function does not include the second-order modal frequency at point 2, and only includes the first-, second- and fifth-order modal frequencies at point 6. As the acceleration time history should be recorded at a location where more modal frequencies can be obtained [55], point 4 is selected as a relatively preferred location.

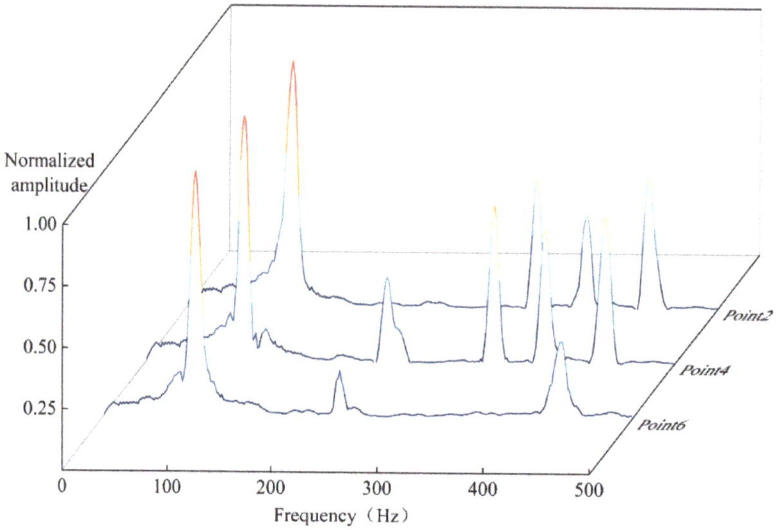

Figure 4. Normalized FRF at different locations.

Similarly, the optimal locations for the measuring points' arrangement are selected at points 3, 4, 9, and 10. In order to complete the damage assessment mapping based on the SSD indicator, the number of measuring points was finalized to 9, arranged in a square (similar to the geometry of the glass curtain wall). Points 3, 4, 9, and 10 are the corners of the square as shown in Figure 5.

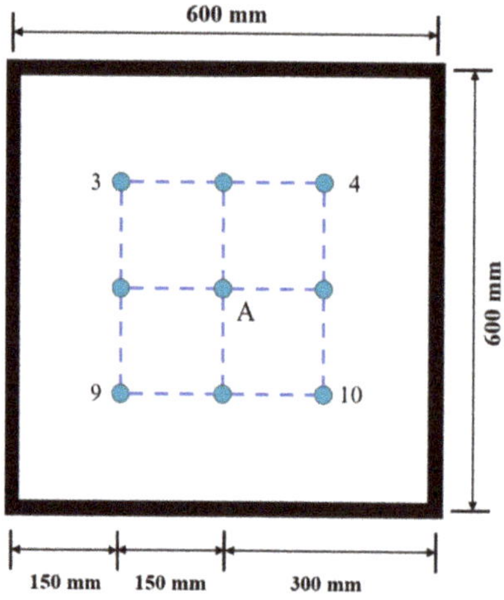

Figure 5. Arrangement of measuring points on HFSGCW.

3.3. Testing Distance

Remote testing allows for a wider field of view and more comprehensive coverage, which enables effective detection of a wide range of injuries. However, proximity testing is also indispensable for specific applications, such as damage detection based on mobile platforms such as UAVs or wall-climbing robots. Therefore, it is necessary to investigate the effect of testing distance on the system.

Comparative experiments were conducted for typical conditions of intact, unilateral, bilateral, and multilateral debonding damage, as illustrated in Figure 6. The diagram shows the glass panel (dotted shadow), the border of the specimen (blue area), and the length of the structural adhesive where damage occurred (yellow area). The four types of structural adhesive damage conditions are as follows: intact condition, complete damage on one side, half of the damage on each of the two sides, and complete damage on all three sides. The distances tested were 1 m and 10 m from the specimen, respectively.

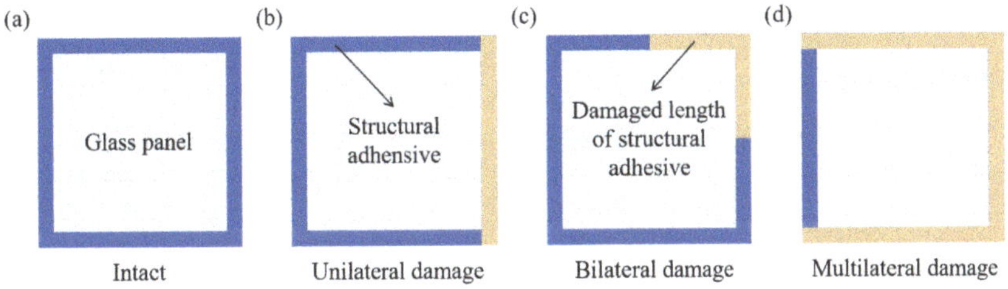

Figure 6. Testing conditions. (**a**) Condition 1. (**b**) Condition 2. (**c**) Condition 3. (**d**) Condition 4.

The first-order natural frequencies of samples under each condition were measured using the nine-point method identified in Section 3.2. Photographs of the laboratory tests are shown in Figure 7. Manual tapping is employed to create an incentive for the curtain wall. As the natural frequency is an inherent property of the structure which is independent of the magnitude of the excitation force, there is no need to deliberately ensure the same force's magnitude.

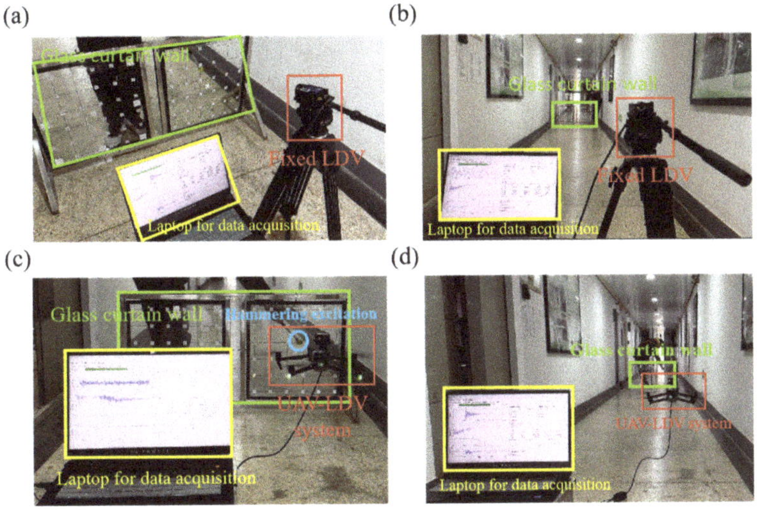

Figure 7. Testing photographs. (**a**) 1 m ranging fixed test. (**b**) 10 m ranging fixed test. (**c**) 1 m ranging mobile test. (**d**) 10 m ranging mobile test. The hammering excitation is labeled as an example in (**c**).

Table 4 compares the average first-order natural frequencies obtained from the numerical simulations with the experimental results of 1 m and 10 m ranging. The four identical panels were subjected to the same experimental procedure. The findings demonstrate that the fixed test conducted at a closer range exhibits a smaller error than the one conducted at a longer range under greater damage conditions. This is attributed to the shorter laser path, which results in a reduction in the power loss of the laser. Meanwhile, under all conditions, the mobile test consistently has a greater error than the fixed test. However, the maximum measuring error for the mobile tests was 12.91% (10 m ranging test under Condition 4). Since most curtain wall boundaries have failed, the sample is easily disrupted by external factors, resulting in unstable results. The remaining error is no more than 10%. This indicates the reliability of the UAV-LDV system for first-order natural frequency testing. The system has acceptable testing errors at various distances, but tests should be performed as closely as possible for the stability of results.

Table 4. Comparison of first-order natural frequency results.

Conditions	First-Order Natural Frequency (Hz)								
	Numerical Simulation	Fixed Test				Mobile Test			
		1 m	Error	10 m	Error	1 m	Error	10 m	Error
Condition 1	101.00	101.43	0.43%	101.27	0.27%	104.34	3.30%	106.15	5.10%
Condition 2	66.11	65.13	−1.48%	64.47	−2.48%	69.64	5.34%	70.48	6.61%
Condition 3	75.37	72.04	−4.42%	71.86	−4.66%	74.07	−1.72%	80.67	7.03%
Condition 4	15.34	15.10	−1.56%	14.09	−8.15%	15.92	3.78%	13.36	−12.91%

The effectiveness of the UAV-LDV system in identifying the debonding damage on HFSCGCW also requires verification. The acceleration time history under four conditions is tested and analyzed using the nine-point method identified in Section 3.2. In Figure 8, the results obtained by the UAV-LDV system for detecting debonding damage on HFSGCW closely align with the results predicted by numerical simulation and those obtained through fixed LDV testing. This alignment mainly refers to the intuitive indication of where the damage occurred and the clear reflection of the extent of the damage. It demonstrates the system's capability to effectively identify the occurrence and location of debonding damage. The testing distance did not significantly affect the system when using the SSD indicator to detect debonding damage.

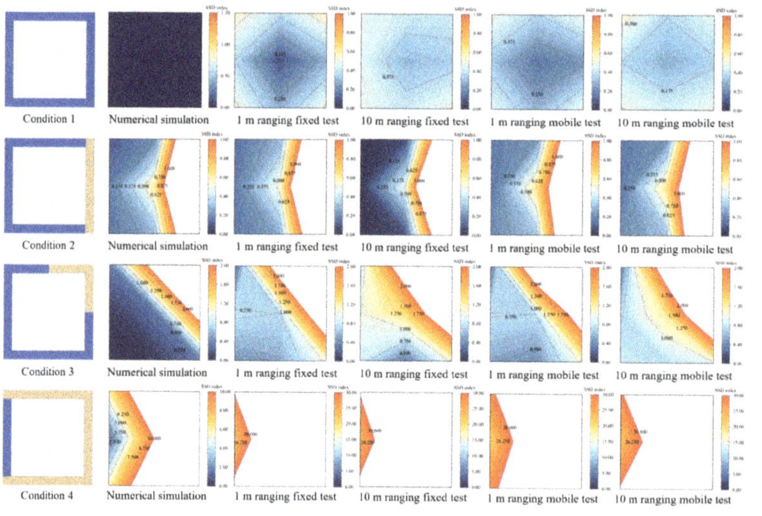

Figure 8. Comparison diagrams of SSD. Note: The results of each test under the same condition are listed in the same row.

3.4. Noise Level

The LDV exhibits self-vibration caused by machinery, transportation, passersby, wind, etc. These natural noises can be mitigated by incorporating vibration sensors on the LDV and analyzing the combined data from them, or alternatively by using two LDVs together [54]. However, random noise from the environment affects the measuring accuracy, which necessitates further investigation. Four unilateral damaged conditions are used as examples to test the performance of the UAV-LDV system under different noise levels, as illustrated in Figure 9. The damage all occurs unilaterally, which distinguishes it from the four conditions depicted in Figure 6. In the case of unilateral damage, most of the structural sealants work properly, and the overall structure remains relatively stable. If noise levels significantly interfere with structural damage detection in this situation, more severely damaged conditions will further deteriorate. Therefore, the unilateral damaged conditions are chosen for testing. The damage lengths of the four conditions are intact, 1/4 length damaged, 1/2 length damaged, and completely damaged on one side. The measuring points are arranged according to the nine-point method presented in Section 3.2. Based on the results of the study in Section 3.3, the measuring distance was set to 1 m.

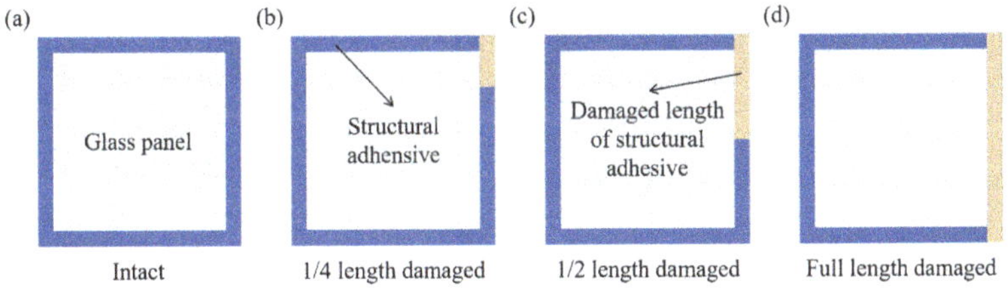

Figure 9. Testing conditions. (**a**) Condition 1. (**b**) Condition 2. (**c**) Condition 3. (**d**) Condition 4.

During the test, to simulate the actual environment, a stereo system is established using two Apple HomePod minis to play the recorded construction noises with a frequency distribution range of 0–500 Hz, as shown in Figure 10. They are located 2 m apart and symmetrically distributed along the central axis of the specimen, 0.5 m away from it. This allows them to interfere with both the curtain wall vibration and the LDV test results, resulting in a more realistic simulation of ambient noise. A decibel meter is used to measure the ambient noise level, and the test is performed when the noise level reaches 20 dB, 30 dB, 40 dB, 50 dB, 70 dB, and 100 dB, respectively.

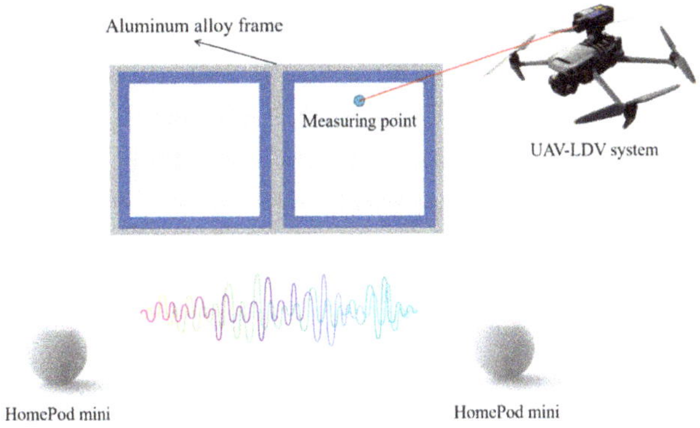

Figure 10. Test site setup.

Table 5 compares the average first-order natural frequencies obtained from the numerical simulations with the tests. The same panel was tested under different conditions with the same technique. The results indicate that the noise level has a significant impact on the reliability of the UAV-LDV system. Even in the case of unilateral damage, the error in the first-order natural frequency always exceeds 5% or even 10%, and it will be larger for the cases with bilateral and multilateral damage where boundary failure is more severe. Consequently, to acquire first-order natural frequencies, it is recommended to perform the test during night-time or at low noise levels to minimize the effect of noises on accuracy. Photos of the tests under different noise levels are shown in Figure 11.

Table 5. Comparison of first-order natural frequency results.

Conditions	Noise Level (dB)	First-Order Natural Frequency (Hz)				
		Numerical Simulation	Fixed Test	Error	Mobile Test	Error
Condition 1	20	101.00	101.73	0.72%	105.97	4.17%
	30		101.92	0.91%	105.92	3.92%
	40		102.1	1.09%	92.48	−9.42%
	50		103.21	2.19%	93.28	−9.62%
	70		98.43	−2.54%	92.65	−5.87%
	100		96.62	−4.34%	86.26	−10.72%
Condition 2	20	91.09	91.61	0.57%	96.37	5.80%
	30		90.63	−0.50%	88.67	−2.66%
	40		92.83	1.91%	92.48	1.53%
	50		91.25	0.18%	94.34	3.57%
	70		94.12	3.33%	84.05	−7.73%
	100		88.24	−3.13%	101.04	10.92%
Condition 3	20	88.52	88.81	0.33%	80.32	−9.26%
	30		88.1	−0.47%	86.58	−2.19%
	40		89.29	0.87%	92.72	4.74%
	50		90.79	2.56%	95.31	7.67%
	70		86.02	−2.82%	94.68	6.96%
	100		85.31	−3.63%	96.95	9.52%
Condition 4	20	66.11	66.28	0.26%	64.1	−3.04%
	30		66.9	1.19%	64.47	−2.48%
	40		66.1	−0.02%	61.94	−6.31%
	50		63.04	−4.64%	62.88	−4.89%
	70		68.77	4.02%	71.94	8.82%
	100		70.12	6.07%	73.67	11.44%

(a) (b)

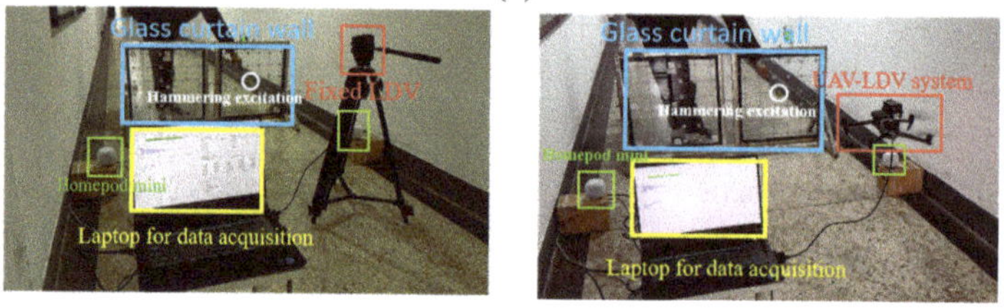

Figure 11. Testing photographs under noises. (**a**) 1 m ranging fixed test under noises. (**b**) 1 m ranging mobile test under noises.

The study also verified the effectiveness of the UAV-LDV system in identifying debonding damage to the HFSCGCW at noise exposure, and the results are shown in Figure 12. The research findings indicate that noise levels do affect the test results, as the outcomes at different noise levels vary significantly from those predicted by the numerical simulation. The results show that the damage detection capability of the UAV-LDV system is significantly affected when the noise levels are 20, 30, and 40 dB. The assessment diagrams do not accurately reflect the location and severity of the damage. This is attributed to the excitation energy level at this time, which is similar to that of the ambient noise, causing the energy level of the structure to fluctuate more and have a larger absolute value. As the background noise level continues to rise, it eventually reaches the threshold for disturbing the curtain wall vibration. At this stage, the SSD index is more sensitive to the vibration characteristics of the structure, and the system's recognition ability is gradually recovering. However, if the noise level is smaller than 20 dB or larger than 40 dB, either excitation or noise will dominate while others attenuate this tendency. As a result, the absolute value of the energy level will be smaller, leading to a smaller range of overall energy variations. Therefore, it is advisable to perform HFSGCW damage detection in a quiet environment.

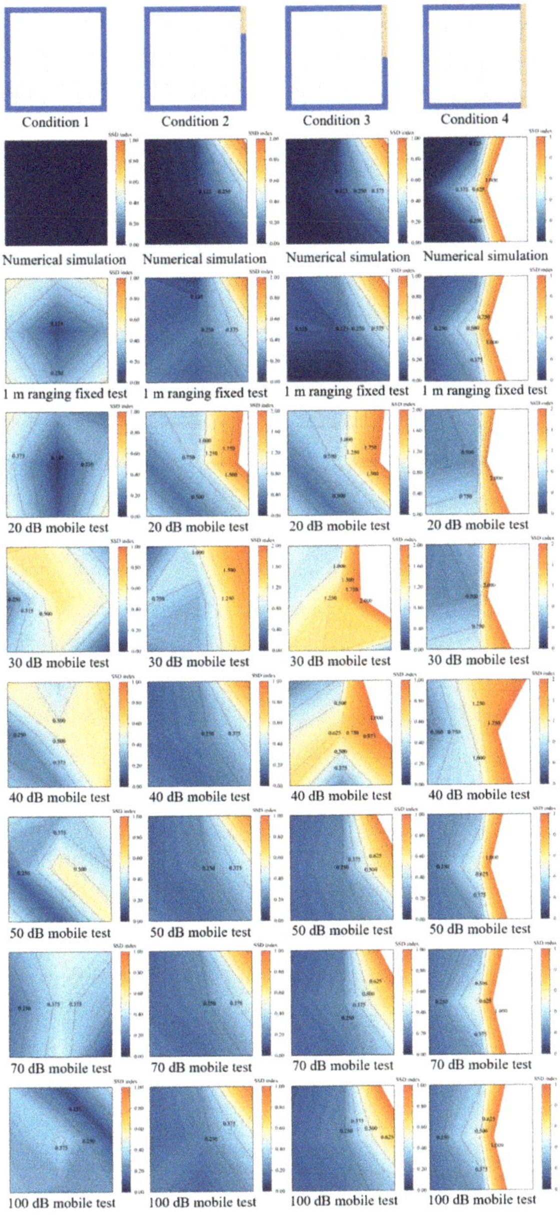

Figure 12. Comparison diagrams of SSD. Note: The results of each test under the same condition are listed in the same column.

3.5. Wind Speed

Controlling the attitude and stability of UAVs in windy conditions has been a significant technical challenge [58]. The UAV-LDV system is designed to inspect curtain walls, which requires long-term operation at high altitudes, making it subject to wind interference. This section conducts a wind resistance test to investigate the impact of wind speed on the system's performance.

The tests of the first-order natural frequency are conducted for Condition 3 in Section 3.4 under the influence of wind speed. This condition is primarily selected due to the shorter damaged length and the fact that most of the structural adhesive is still in use. The glass panel will perform better in high winds during this time, while the wind speed mainly affects the performance of the UAV-LDV system. If there were significant errors in the results under this condition, the conditions of greater severity of damage would have larger errors. The measuring points were arranged and the distances were measured in accordance with the provisions of Section 3.4 (1 m). The working UAV-LDV system was subjected to lateral wind at three different speeds: 1.5 m/s, 3.0 m/s, and 5.0 m/s (produced by an electric fan). Figure 13 shows the arrangement of the tests. Photos of the tests in the wind are shown in Figure 14.

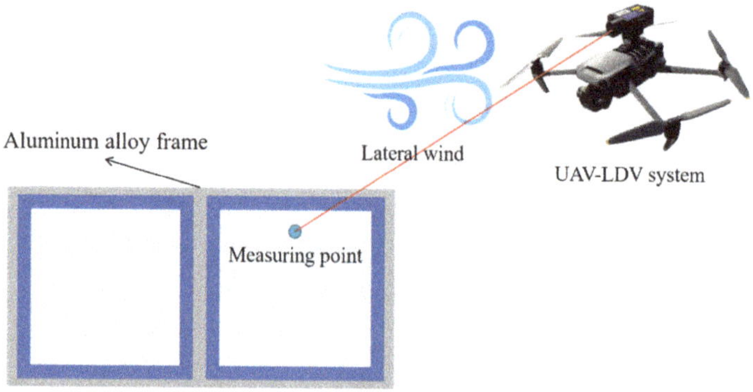

Figure 13. Test site setup.

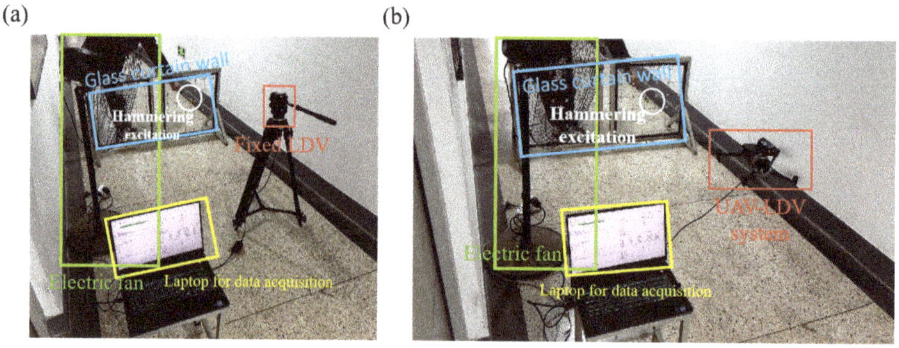

Figure 14. Testing photographs in the wind. (**a**) 1 m ranging fixed test in the wind. (**b**) 1 m ranging mobile test in the wind.

Table 6 presents the average first-order natural frequencies obtained from the nine measuring points. The results indicate that the wind speed has a significant impact on the reliability of the UAV-LDV system. Therefore, it is recommended to acquire the first-order natural frequency with this system in a windless environment or use an error correction algorithm to minimize the effect of wind speed on accuracy.

Table 6. Comparison of first-order natural frequency results of Condition 3.

Average of 1st Order Natural Frequency (Hz)					
Numerical Simulation	**Wind Speed (m/s)**	**Fixed Test**	**Error**	**Mobile Test**	**Error**
88.52	1.5	88.68	0.18%	85.41	−3.51%
	3.0	88.97	0.51%	83.48	−5.69%
	5.0	88.13	−0.44%	96.13	8.60%

The study verifies the effectiveness of the UAV-LDV system in identifying debonding damage to the HFSCGCW under the influence of wind. Figure 15 validates the results of different wind speeds through numerical simulation and fixed tests. The assessment chart can give a general indication of the locations of damaged structural adhesives, but it is not very effective in pinpointing the exact location. When the wind speed reaches 5.0 m/s, it is difficult to provide useful information about the damaged side. The results indicate that the wind greatly affects the detection capability of the UAV-LDV system. The SSD indicator has a large numerical error in windy conditions and does not clearly indicate the exact location of the damage, even in the case of lighter damage. During the test, high winds interfered with the hovering and stabilization of the UAV. Even aligning the measuring points was difficult at times. To maintain alignment in strong winds, the drone continuously maneuvers through turns and side shifts. In contrast to the drone's camera, which is stabilized by a gimbal, the LDV is fixed directly to the fuselage. This configuration leads to the laser beam deviating from the target as the orientation of the fuselage changes, and passing through transparent glass. Consequently, the laser intensity that reflects the LDV is insufficient for stable data acquisition, resulting in failed assessments. As demonstrated in Figure 15, at a wind speed of 5 m/s, the processed results suggest extensive and severe debonding damage along both boundaries, even when the actual damage is minor. This contradicts the actual conditions. In subsequent research, it is recommended that a gimbal structure be developed to stabilize the LDV device, in order to minimize the errors in optical path differences caused by attitude adjustments. This approach is simpler and more efficient than developing complex algorithms. As a result, it is not recommended to perform tests in windy weather now.

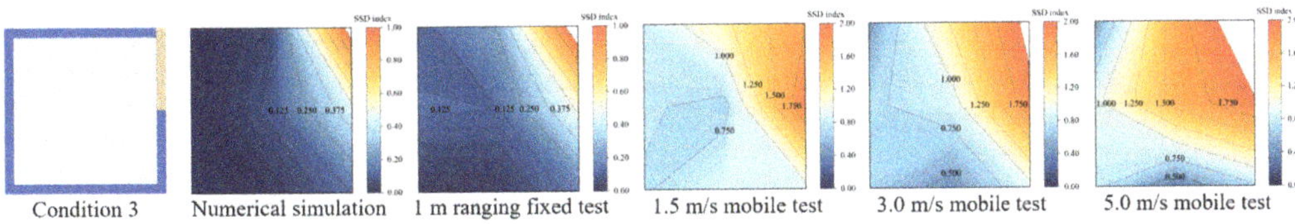

Figure 15. Comparison diagrams of SSD.

4. Conclusions

In this paper, based on previous research, an identification method of debonding damage of HFSGCW based on a UAV-LDV system is proposed. The established system enables UAVs to transport the LDV to high altitudes for operation rather than utilizing a fixed LDV which works with the assistance of a UAV. It solves the previous problem of relying on manual inspection for debonding damage, which was inefficient, costly, and risky. The new assessment indicator (SSD) based on wavelet packet band energy decomposition has the additional advantage of providing a visual representation of the precise location of the damage, which is a superior feature compared to traditional indicators such as first-order natural frequency. In addition, potential factors affecting the performance of the UAV-LDV system were investigated, including the arrangement of measuring points, measuring distance, noise level, and wind speed. The study demonstrated that a wider distribution of measurement points and closer proximity of the UAV to the curtain wall enhance the detection efficiency by increasing data density in assessment diagrams and reducing laser energy loss. The results demonstrate that despite the influence of ambient noise and wind speed on the system performance, the system is still able to provide accurate and reliable detection results under reasonable operating conditions. Compared with traditional methods, the UAV-LDV system has significant advantages. Firstly, the system is able to complete large-area inspection tasks in a shorter time, which greatly improves inspection efficiency. Secondly, the system realizes non-contact detection, avoiding the safety hazards in the traditional method. Finally, the SSD indicator is able to accurately identify the damage location, providing more intuitive damage detection results.

The main innovations of the study include the introduction of a novel UAV-LDV system for damage detection in Hidden Frame-Supported Glass Curtain Walls, the development of a new SSD indicator that offers visualization benefits and outperforms conventional indicators, the investigation of potential factors affecting the UAV-LDV system's performance, the conduction of an indoor full-scale experiment to examine the efficiency of the novel approach and the provision of valuable guidelines for utilizing the UAV-LDV system to inspect glass curtain walls.

The establishment of the UAV-LDV system provides a low-cost, remote method for the early warning of HFSGCW, similar to the health monitoring of bridges and building structures. The results of our study indicate that the proposed nine-point measuring arrangement method should be adopted. It is inadvisable to utilize the UAV-LDV system for the identification of damage in windy or noisy environments, given the potential for a reduction in accuracy. In the future, we will further optimize the wind resistance of the UAV-LDV system and methods to reduce noise interference and will integrate the system with other technologies, such as incorporating machine learning methods to further improve inspection accuracy and efficiency. Finally, more field tests will be conducted to verify the stability and reliability of the system under different environmental conditions.

Author Contributions: Conceptualization, H.Z. and T.G.; methodology, H.Z.; software, H.Z. and G.Z.; validation, H.Z.; investigation, H.Z. and G.Z.; resources, T.G.; data curation, H.Z., Z.H. and G.Z.; writing—original draft preparation, H.Z. and Z.H.; writing—review and editing, T.G. and G.Z.; visualization, H.Z. and Z.H.; supervision, T.G and G.Z.; project administration, T.G; funding acquisition, T.G. All authors have read and agreed to the published version of the manuscript.

Funding: The work described in this paper was supported by National Natural Science Funds for Distinguished Young Scholar of China (No. 52125802, Tong Guo) and SEU Innovation Capability Enhancement Plan for Doctoral Students (CXJH_SEU 24110, Guoliang Zhi).

Data Availability Statement: All data used are available from the corresponding author upon reasonable request.

Conflicts of Interest: The authors declare no conflicts of interest.

References

1. Górka, M. Use of aluminium and glass facades in urban architecture. *Bud. Arch.* **2020**, *18*, 29–40. [CrossRef]
2. Zhang, L.; Lu, H.; Li, H. Multi-scale damage detection of glass curtain wall by acoustic emission and vibrational modal analysis. *Sens. Smart Struct. Technol. Civ. Mech. Aerosp. Syst.* **2023**, *12486*, 141–150. [CrossRef]
3. China Academy of Building Research. *JGJ 102-2003 Technical Code for Glass Curtain Wall Engineering*; China Architecture & Building Press: Beijing, China, 2003.
4. Huang, B.; Lu, W.; Cao, W. Discussion on safety assessment of existingarchitectural curtain walls. *Struct. Eng.* **2006**, *3*, 76–79. (In Chinese) [CrossRef]
5. Liu, X.; Bao, Y.; Song, Y.; Qiu, Y. Safety evaluation of glass curtain walls by using dynamic method. *China Civ. Eng. J.* **2009**, *42*, 11–15. (In Chinese)
6. Fang, Z.; Luo, W. Study on the damage detection of full-scale frame-concealed glass curtain-walls based on modal curvature. *Value Eng.* **2017**, *20*, 89–93. (In Chinese) [CrossRef]
7. Wang, J. Research on Online Signal Collection and System for Glass Curtain Wall Detection. Master's Thesis, Guizhou University, Guizhou, China, 2021. (In Chinese)
8. Bedon, C.; Amadio, C. Numerical assessment of vibration control systems for multi-hazard design and mitigation of glass curtain walls. *J. Build. Eng.* **2018**, *15*, 1–13. [CrossRef]

9. Wang, P.; Xiao, J.; Duan, Z.; Li, C. Intelligent development trend of building enclosure damage detection. *JACE* **2022**, *39*, 24–37. (In Chinese) [CrossRef]
10. Huang, T.; Zhang, D.; Zhao, Y.; Liu, J.; Li, J. Comprehensive appraisal of the safety of hidden frame glass curtain wall based on fuzzy theory. *J. Build. Eng.* **2019**, *26*, 100863. [CrossRef]
11. Efstathiades, C.; Baniotopoulos, C.C.; Nazarko, P.; Ziemianski, L.; Stavroulakis, G.E. Application of neural networks for the structural health monitoring in curtain-wall systems. *Eng. Struct.* **2007**, *29*, 3475–3484. [CrossRef]
12. Duque, L.; Seo, J. Wacker, Synthesis of unmanned aerial vehicle applications for infrastructures. *J. Perform. Constr. Fac.* **2018**, *32*, 04018046. [CrossRef]
13. He, D. Research in Damage Detection Techniques of Curtain Wall Structure Based on Movable LDV. Ph.D. Dissertation, School of Civil Enginering, Southeast University, Nanjing, China, 2023. (In Chinese)
14. Liu, Z.; Guo, T.; Hebdon, M.H.; Han, W. Measurement and comparative study on movements of suspenders in long-span suspension bridges. *J. Bridge Eng.* **2019**, *24*, 04019026. [CrossRef]
15. Liu, Z.; Guo, T.; Correia, J.; Wang, L. Reliability-based maintenance strategy for gusset plate connections in steel bridges based on life-cost optimization. *J. Perform. Constr. Fac.* **2020**, *34*, 04020088. [CrossRef]
16. Zhi, G.; Xu, X.; Guo, T.; Chen, Z.; Zhang, M. Experimental and numerical investigation of vibrations in over-track scale model buildings. *J. Build. Eng.* **2023**, *77*, 107538. [CrossRef]
17. Zhi, G.; Chen, Z.; Guo, T.; Zhang, M. Investigation of structure-borne noise propagation characteristics in a novel double-story high-speed railway station. *J. Vib. Eng. Technol.* **2023**, *12*, 5325–5344. [CrossRef]
18. Zeng, J.; Wu, Z.; Todd, M.D.; Hu, Z. Bayes risk-based mission planning of Unmanned Aerial Vehicles for autonomous damage inspection. *Mech. Syst. Signal Proc.* **2023**, *187*, 109958. [CrossRef]
19. Hu, C. Attitude stability control of UAV gyroscope based on neutral statistics for smart cities. *Int. J. Syst. Assur. Eng. Manag.* **2022**, *13*, 281–290. [CrossRef]
20. Lei, Y.; Li, Y.; Wang, J. Aerodynamic analysis of an orthogonal octorotor UAV considering horizontal wind disturbance. *Aerospace* **2023**, *10*, 525. [CrossRef]
21. Brandão, A.S.; Smrcka, D.; Pairet, É.; Nascimento; Saska, M. Side-pull maneuver: A novel control strategy for dragging a cable-tethered load of unknown weight using an UAV. *IEEE Robot. Autom. Lett.* **2022**, *7*, 9159–9166. [CrossRef]
22. Hassan, M.A.A.; Phang, S.K. Optimized autonomous UAV design for duration enhancement. *AIP Conf. Proc.* **2020**, *2233*, 030004. [CrossRef]
23. Fraga-Lamas, P.; Ramos, L.; Mondéjar-Guerra, V.; Fernández-Caramés, T.M. A review on IoT deep learning UAV systems for autonomous obstacle detection and collision avoidance. *Remote Sens.* **2019**, *11*, 2144. [CrossRef]
24. DJI, DJI Mavic 3 Technical Support. Available online: https://www.dji.com/cn/mavic-3/specs (accessed on 15 May 2014).
25. Yoon, S.; Jr, B.F.S.; Lee, S.; Jung, H.J.; Kim, I.H. A novel approach to assess the seismic performance of deteriorated bridge structures by employing UAV-based damage detection. *Struct. Control. Health Monit.* **2022**, *29*, e2964. [CrossRef]
26. Car, M.; Markovic, L.; Ivanovic, A. Autonomous wind-turbine blade inspection using LiDAR-equipped unmanned aerial vehicle. *IEEE Access* **2020**, *8*, 131380–131387. [CrossRef]
27. Ding, W.; Yang, H.; Yu, K.; Shu, J. Crack detection and quantification for concrete structures using UAV and transformer. *Automat. Constr.* **2023**, *152*, 104929. [CrossRef]
28. Goricanec, J.; Ereiz, S.; Orsag, M.; Duvnjak, I. Identification of the dynamic parameters of bridge elements using unmanned aerial vehicle. *J. Sound Vib.* **2023**, *556*, 117901. [CrossRef]
29. Han, Y.; Feng, D.; Wu, W.; Yu, X.; Wu, G.; Liu, J. Geometric shape measurement and its application in bridge construction based on UAV and terrestrial laser scanner. *Automat. Constr.* **2023**, *151*, 104880. [CrossRef]
30. Li, J.; Peng, Y.; Tang, Z.; Li, Z. Three-dimensional reconstruction of railway bridges based on unmanned aerial vehicle-terrestrial laser scanner point cloud fusion. *Buildings* **2023**, *13*, 2841. [CrossRef]
31. Wang, Y.; Duan, C.; Huang, X.; Zhao, J.; Zheng, R.; Li, H. Task-driven path planning for unmanned aerial vehicle-based bridge inspection in wind fields. *Fluids* **2023**, *8*, 321. [CrossRef]
32. Christiansen, N.; Carpenter, J.R.; Daewel, U. The large-scale impact of anthropogenic mixing by offshore wind turbine foundations in the shallow North Sea. *Front. Mar. Sci.* **2023**, *10*, 1178330. [CrossRef]
33. Liang, E.P.; Ma, G.S.; Li, Y.; Zheng, X.; Wu, F.; Li, S.; Li, D. Summary of the impact of aeolian sand environment on key parts of wind turbine. *Sci. Sin-Phys. Mech. As.* **2023**, *53*, 234701. [CrossRef]
34. Ma, B.; Yang, J.; Chen, X. Revealing the ecological impact of low-speed mountain wind power on vegetation and soil erosion in South China: A case study of a typical wind farm in Yunnan. *J. Clean Prod.* **2023**, *419*, 138020. [CrossRef]
35. Sun, X.; Wu, W.; Wang, J.; Xu, L.; Jiang, R.; Sun, Y.; Fang, L. Optimization design of negative pressure adsorption car for internal defect detection of wind turbine blades on UAV. *AIP Adv.* **2023**, *13*, 025133. [CrossRef]
36. Liu, Z.; Wang, X.; Liu, Y. Application of unmanned aerial vehicle hangar in transmission tower inspection considering the risk probabilities of steel towers. *IEEE Access* **2019**, *7*, 159048–159057. [CrossRef]
37. Yu, C.; Yang, Y.; Cheng, Y.; Wang, Z.; Shi, M.; Yao, Z. UAV-based pipeline inspection system with Swin Transformer for the EAST. *Fusion Eng. Des.* **2022**, *184*, 113277. [CrossRef]
38. Yin, H.; Tan, C.; Zhang, W.; Cao, C.; Xu, X.; Wang, J.; Chen, J. Rapid compaction monitoring and quality control of embankment dam construction based on UAV photogrammetry technology: A case study. *Remote Sens.* **2023**, *15*, 1083. [CrossRef]

39. Lenda, G.; Marmol, U. Integration of high-precision UAV laser scanning and terrestrial scanning measurements for determining the shape of a water tower. *Measurement* **2023**, *218*, 113178. [CrossRef]
40. Wu, Z.; Zeng, J.; Hu, Z.; Todd, M.D. Optimization of unmanned aerial vehicle inspection strategy for infrastructure based on model-enabled diagnostics and prognostics. *Mech. Syst. Signal Proc.* **2023**, *204*, 110841. [CrossRef]
41. Xu, F.; Wang, B.; Shen, J.; Hu, J.; Jiang, G. Design and realization of the claw gripper system of a climbing robot. *J. Intell. Robot Syst.* **2018**, *89*, 301–317. [CrossRef]
42. Li, S.; Chen, S. Field monitoring and prediction on temperature distribution of glass curtain walls of a super high-rise building. *Eng. Struct.* **2022**, *250*, 113405. [CrossRef]
43. Sabatier, J.; Ekimov, A.; Aranchuk, V.; Mack, R. Remote detection of vehicle obscured by forest canopy using laser Doppler vibrometer. *J. Acoust. Soc. Am.* **2008**, *124*, 2508. [CrossRef]
44. Gioffre, M.; Cavalagli, N.; Pepi, C.; Trequattrini, M. Laser Doppler and radar interferometer for contactless measurements on unaccessible tie-rods on monumental buildings: Santa maria della consolazione temple in todi. *JPCS* **2017**, *778*, 012008. [CrossRef]
45. Sugimoto, T.; Sugimoto, K.; Uechi, I.; Utagawa, N.; Kuroda, C. Outer wall inspection using acoustic irradiation induced vibration from UAV for noncontact acoustic inspection method. In Proceedings of the IEEE International Ultrasonics Symposium Conference 2018, Kobe, Japan, 22–25 October 2018; pp. 1–9. [CrossRef]
46. Schewe, M.; Ismail, M.A.; Rembe, C. Towards airborne laser Doppler vibrometry for structural health monitoring of large and curved structures. *Insight Non-Destr. Test. Cond. Monit.* **2021**, *63*, 280–282. [CrossRef]
47. Schewe, M.; Ismail, M.A.A.; Zimmermann, R.; Durak, U.; Rembe, C. Flyable mirror: Airborne laser Doppler vibrometer for large engineering structures. *JPCS* **2024**, *2698*, 012007. [CrossRef]
48. Garg, P.; Nasimi, R.; Ozdagli, A.; Zhang, S.; Mascarenas, D.D.L.; Taha, M.R.; Moreu, F. Measuring transverse displacements using Unmanned Aerial Systems Laser Doppler Vibrometer (UAS-LDV): Development and field validation. *Sensors* **2020**, *20*, 6051. [CrossRef] [PubMed]
49. Komgom, C.N.; Mureithi, N.W.; Lakis, A.A. Application of time synchronous averaging, spectral kurtosis and support vector machines for bearing fault identification. In Proceedings of the ASME 2008 Pressure Vessels and Piping Conference, Chicago, IL, USA, 27–31 July2008; Volume 61601, pp. 137–146. [CrossRef]
50. Halkon, B.J.; Rothberg, S.J. Towards laser Doppler vibrometry from unmanned aerial vehicles. *JPCS* **2018**, *1149*, 012022. [CrossRef]
51. Ma, T.; Yang, C.; Gan, W.; Xue, Z.; Zhang, Q.; Zhang, X. Analysis of technical characteristics of fixed-wing VTOL UAV. In Proceedings of the 2017 IEEE International Conference on Unmanned Systems, Beijing, China, 27–29 October 2017; pp. 293–297. [CrossRef]
52. Sabour, M.H.; Jafary, P.; Nematiyan, S. Applications and classifications of unmanned aerial vehicles: A literature review with focus on multi-rotors. *Aeronaut. J.* **2023**, *127*, 466–490. [CrossRef]
53. Rothberg, S.J.; Allen, M.S.; Castellini, P.; Di Maio, D.; Dirckx, J.J.J.; Ewins, D.J.; Halkon, B.J.; Muyshondt, P.; Paone, N.; Ryan, T.; et al. An international review of laser Doppler vibrometry: Making light work of vibration measurement. *Opt. Laser. Eng.* **2017**, *99*, 11–22. [CrossRef]
54. Huang, Z.; Xie, M.; Chen, C.; Du, Y.; Zhao, J. Engineering application of a safety-state evaluation model for hidden frame-supported glass curtain walls based on remote vibration. *J. Build. Eng.* **2019**, *26*, 100915. [CrossRef]
55. Zheng, H.; Zhang, X.; Wang, H.; Pan, D.; Jian, K. Damage detection of bonded structure of building glass curtain wall based on origin FRP. *J. Vib. Shock* **2021**, *40*, 289–298. (In Chinese) [CrossRef]
56. Zhang, Y.; Wang, S. Damage detection for abeam based on imaginary part of its FRF. *J. Vib. Shock* **2018**, *37*, 38–42. (In Chinese) [CrossRef]
57. Tang, A. Research on Grid Structure Damage Identification Combined with Modal Strain Energy Decomposition Method and Wavelet Packet Energy Index. Master's Thesis, Chongqing University, Chongqing, China, 2021. (In Chinese)
58. Wang, B.; Wang, D.; Ali, Z.A.; Bai, T.; Wang, H. An overview of various kinds of wind effects on unmanned aerial vehicle. *Meas. Control* **2019**, *52*, 731–739. [CrossRef]

applied sciences

MDPI

Article

Bridge Assessment under Earthquake and Flood-Induced Scour

Taulant Karriqi [1], José C. Matos [1,*], Ngoc-Son Dang [1] and Ye Xia [2,3]

[1] Department of Civil Engineering, Institute for Sustainability and Innovation in Structural Engineering (ISISE), University of Minho, 4800-058 Guimarães, Portugal; id10785@alunos.uminho.pt (T.K.); sondn@civil.uminho.pt (N.-S.D.)
[2] Department of Bridge Engineering, Tongji University, Shanghai 200092, China; yxia@tongji.edu.cn
[3] Shanghai QiZhi Institute, Shanghai 200232, China
* Correspondence: jmatos@civil.uminho.pt

Abstract: Earthquakes and floods in Albania are devastating, but combining these two different hazards in terms of action on bridge structures may lead the bridge to collapse. This article presents a seismic risk assessment of a code-conforming precast reinforced concrete bridge located in a region prone to earthquakes and where local scour induced by floods is a significant concern. The seismic action is considered using a group of ground motion accelerograms generated by matching the accelerogram of the 29 November 2019 earthquake in Durres (M = 6.4), Albania, to the target response spectrum. The scouring effects on the bents of the bridge are characterized by the scour depths. A set of non-linear time-history analyses of the bridge are performed to assess the bridge's performance. The bridge fragility curves are generated and analyzed for multi-hazard scenarios at both element and system levels for different flow discharge values and PGA levels. The result shows a low seismic risk of the bridge with bearings when considering the local scour induced by flood events in the seismic analysis due to their flexibility to adapt to changes in structure geometry and significant foundation stiffness. This research also emphasizes the significance of choosing the right foundation type and depth for bridges located in areas prone to local scour induced by floods.

Keywords: seismic and flood hazards; local scour; failure probabilities; fragility curves; damage limit states

1. Introduction

Every nation relies on its infrastructure as a cornerstone, with bridges playing vital roles in highway and railway transportation networks. Bridges are critical in supporting first responders during the aftermath of destructive earthquakes and facilitating post-disaster rescue operations, particularly in remote areas. Bridges are susceptible to various natural hazards, including earthquakes, high winds, landslides, floods, and other hazards like explosions and vehicle collisions at bridge bents and abutments. Studying the causes that have led to damage up to the collapse of bridges throughout history, mainly under strong seismic actions and flood-induced scours, has significantly improved design codes and structural health monitoring techniques.

Combining two different natural hazards of earthquakes and floods in terms of action on bridge structures may lead a bridge to collapse. Albania is identified as a region with significant seismic activity resulting from the interaction of the Adriatic and Aegean Seas' multiple tectonic plates and complex tectonic processes in the Carpathian region [1]. The damage potential of active seismic faults has been evident in numerous earthquakes, such as the 30 November 1967 earthquake measuring M = 6.6 in Dibra and Librazhd, the 15 April 1979 earthquake measuring M = 6.9 in Montenegro, and the earthquakes on 21 September and 26 November 2019, registering magnitudes of M = 5.8 and M = 6.4, respectively, in Durres, Albania. Floods in Albania happen more frequently, and the sum of the impact of every single one of them may cause a substantial loss. Bridge flood-induced scour is a significant concern in areas prone to earthquakes and can be compounded by climate change. Hydraulic events are the primary cause of bridge failures globally, with flood-induced scour representing a significant factor in hydraulic-related failures [2]. Scouring results from the erosive force exerted by flowing water, which removes and transports sediment from the streambed and banks and the surrounding areas near the bents and abutments of the bridge [3]. As a result, the restraining conditions of bents at the foundation may be modified. Consequently, the lateral capacity of bents might decrease. This additional flexibility leads to varying dynamic characteristics of the structure, which can exacerbate the detrimental impact of seismic ground motions on structure performance [4].

Avşar (2009) [5] meticulously formulated a fragility analysis for road bridges in Turkey built and constructed post-1990s, a crucial step in understanding their vulnerability. These fragility curves address three levels of damage states (DS), primarily influenced by the relative movement of the deck. In 2015, Yilmaz [6] conducted a comprehensive investigation on two actual river-crossing highway bridges and two representative examples, providing invaluable insights into regional multi-hazard scenarios, including earthquakes and floods, across diverse regions in California. These bridges were built at varying points in time, making the study's findings even more significant. The study involved the generation of fragility curves and surfaces in predicting the probability of bridge failures under different levels of earthquakes and floods, a crucial tool for risk assessment.

European design codes currently lack guidelines to guarantee adequate bridge foundations that are resistant to scouring and subjected to earthquake lateral loading. This gap in our knowledge and practices is a call to action for all of us, as Eurocodes do not cover the scour and flood effects approach in bridge foundation design. The Italian Code for Construction (NTC 2018, 17 January 2018, 5.1.2.3) for new bridges suggests avoiding, where possible, bents placed in the riverbed. If necessary, the distance between consecutive bents or between a bent and bridge abutment must be at least 40 m measured perpendicular to the mainstream [7]. Much research revolves around analyzing the risk and reliability of bridges during specific extreme occurrences like earthquakes, storm surges, winds, floods, and collisions, typically relying on proprietary computational models. Recent findings underscore the reality that bridges are exposed to extreme events over their operational duration. Albania faces multiple sources of hazards. Hence, it is imperative to reevaluate the design criteria that have traditionally been formulated using a single-hazard approach. Clarke and Obrin (2016) [8] implemented a stress testing approach focused on assessing the combined risk posed to critical infrastructure networks by various severe hazard events, including earthquakes, landslides, and floods, and to assess how effectively the entire network can endure highly unlikely yet severe natural hazard events. This framework offers a valuable tool for transport infrastructure owners and managers to address the challenges posed by extreme natural hazards and enhance the resilience of transportation networks. Crespi et al. (2020) [9] and Crespi et al. (2022) [10] emphasized the importance of evaluating the impact of the corrosion effects caused by the carbonation phenomenon (for three different levels) on the seismic performance of existing concrete and precast bridges. The corrosion effects are applied by reducing the cross-section of steel rebars. The proposed procedure identifies the first structural component reaching failure providing valuable insights into retrofitting strategies.

Compared with current research on continuous concrete bridges, the study work carried out by Wang et al. (2014) [11] shows that the multi-span, simply supported bridge is more sensitive to flood hazards than a continuous concrete bridge. Also, both bridges benefit from scouring due to longer modal periods. This paper suggests enhancing the stiffness of foundations to mitigate the seismic risk of bridges with a scoured foundation, which leads to the same conclusion as the results achieved by this study.

The aim of this work is to assess the bridge's performance using a probabilistic-based approach under multi-hazard scenarios involving seismic action and local scour induced by floods to the bridge by utilizing fragility functions. This study examines how the bridge responds when facing the sequential and combined effect of local scour and an earthquake, assessing its performance under these circumstances. Moreover, the study area is situated in a region with a notably elevated annual likelihood of experiencing floods. Fragility analyses are developed, and the probabilities of bridge failure considering different earthquake levels and local scour depths are calculated. These tools are valuable for determining the likelihood of bridge failure across various potential damage scenarios, correlating ground motion intensity with local scour depths [12–14]. This article emphasizes the importance of making bridges more resilient by adapting their structural capacities to the combined effect of natural hazards compounded by climate change.

2. Methodology

A comprehensive multi-hazard framework is developed to evaluate the reliability of a bridge situated in a region in Albania prone to seismic activity and flooding. This framework aims to measure the risk confronting the bridge when affected by both earthquake activity and local scour simultaneously. At the outset, a case study is initiated, and information is compiled from existing research sources. Subsequently, an FE model is built using the collected information. In parallel, earthquake accelerograms are generated and scour depths are calculated. Four damage limit states (DLS) are delineated, and fragility curves for its components and the system are developed. The outcomes are depicted through curves and surfaces of fragility function and risk evaluation, illustrating different multi-hazard scenarios.

The methodology outlined below consists of six steps and is visually depicted in a flowchart (Figure 1).

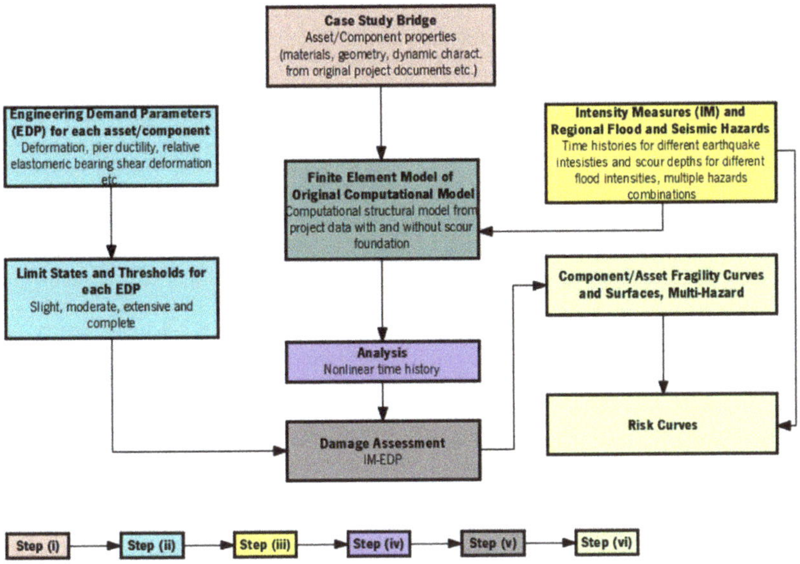

Figure 1. Framework for assessment of bridge under earthquake and flood hazards.

2.1. Case Study Bridge and Numerical Model

As a case study, a road bridge is considered for the bridge assessment under seismic action and local scour. The bridge is situated in Urake, Diber and spans the Kurvajt River. All the information is gathered from the original project and other studies, and an FEM is built on CSIBridge v25.0.0. software considering time history accelerograms and scour depths for different probabilities of exceedance.

2.2. Engineering Demand Parameters

Defining damage limit states and aligning them with meaningful parameters for evaluating bridge conditions is vital to ensure the reliability of analytical fragility curves. These definitions, specific to various bridge components, are determined through the input of experts, empirical data, or analytical models. The meticulous definition of damage limit states plays a pivotal role in shaping the resulting fragility curves, making it a critical element in the overall process of developing analytical fragility curves and surfaces. No particular method establishes a direct correlation between the damage to the bridge and the different damage states. Threshold limits for different component damage states have been derived from earlier studies and are succinctly outlined in Table 1 for three Engineering Demand Parameters (EDP).

Table 1. Damage threshold limits.

Component	EDP	Slight	Moderate	Extensive	Complete
Bents	Curvature ductility, μ_{Φ}	≥1.0 and <4.0	≥4.0 and <8.0	≥8.0 and <12.0	≥12.0
Abutment	Longitudinal deformation, $\Delta_{a,long}$ [mm]	≥35 and <55	≥55 and <79	≥79 and <276	≥276
Elastomeric bearings	Longitudinal deformation, $\Delta_{bearing}$ [mm]	≥30 and <60	≥60 and <90	≥90 and <120	≥120

There is no indication of shear failure in any bents under the specific ground motions. The evaluation of bridge bents is based on their capacity to withstand bending, with curvature ductility μ_{Φ}, to assess flexural damage. The threshold values for curvature ductility are sourced from Ramanthan's research (2012) [15] for post-1990 ductile columns. For abutment damage, the longitudinal deformation is considered ($\Delta_{a,long}$). Threshold limits are taken from Stefanidou (2016) [16]. The condition of elastomeric bearings is assessed according to their horizontal longitudinal deformation. For moderate damage and complete failure, the threshold limits are established based on longitudinal deformation. In particular, a moderate damage state is characterized by the initiation of sliding at the interface between concrete and elastomer. At the same time, complete failure occurs when the bridge superstructure becomes detached from the pads. The threshold limit for extensive DS is computed by arithmetic mean between the thresholds associated with moderate damage and complete state [6].

When any bridge element reaches or exceeds a DLS, the entire bridge system is categorized into the same DS, regardless of the conditions of other bridge elements.

2.3. Intensity Measures (IM) and Flood and Earthquake Hazards

The expected strong and ground motion characteristics are delineated to generate fragility curves derived from real accelerograms and spectrum-matching techniques. A group of real earthquake accelerograms matched to the target response spectrum employing the wavelets algorithm introduced by Al Atik and Abrahamson [17] are generated and reflect seismic properties specific to the area of the bridge ensuring alignment with the soil conditions prevalent in the locality. Spectral matching adjusts the selected time history record to align with the target response spectrum. The spectrum functions as a reference point as the produced accelerograms will undergo refinement to align with its response spectrum. Eurocode 8 target response spectrum will be defined to produce the necessary adjusted records for each seismic level (a_{gR} = 0.05 g, 0.10 g, 0.20 g, and up to 0.5 g).

The impact of scouring on the bridge will be assessed based on the depths of the scour. Various equations have been proposed to forecast local scour depth under different riverbed conditions. Drawing from model-scale experiments, two commonly utilized equations are tailored for sand and clay bed streams. Since sand is the most susceptible to erosion, this study will concentrate on a bridge in a sand-bed stream.

This study addresses scour, which arises from the erosive impact of the river, by assessing the maximum depth of local scour induced by floods at the bent and abutment foundations. Scour can manifest in three forms: long-term riverbed degradation, local scour at bents and abutments foundations, and contraction scour at the bridge [3]. This study primarily examines the scouring of bents and abutment foundations caused by flowing water. The scour depths (y_s) at bents and abutments are determined with the expression of HEC-18 [3]:

$$\frac{y_s}{a} = 2.0 \cdot K_1 \cdot K_2 \cdot K_3 \cdot \left(\frac{y_1}{a}\right)^{0.35} \cdot Fr_1^{0.43} \quad (1)$$

Correction factor K_1 accounts for the nose shape of the bents, K_2 accounts for the angle of the flow direction, and K_3 accounts for the riverbed condition, y_1 is the flow depth, a is the width of the bent, and Fr_1 is the number of Froude. The equation provided is suitable for assessing local bent scour for a live bed and clear water. However, for this particular case study, the focus is on evaluating live-bed bent scour. Bridge scour depths calculated from Expression (1) will be incorporated into finite element analyses.

2.4. *Analysis of Numerical Model*

A 3D FEM was developed on CSIBridge v25.0.0 software using data related to bridge dimensions and materials gathered from the original project.

2.5. *Damage Assessment*

The damage assessment for the selected bridge components is conducted for each earthquake and scour scenario.

2.6. *Fragility Curves and Surfaces Development*

In the case of the studied bridge, each point along the fragility curve indicates the probability of achieving/surpassing a specific damage threshold due to seismic action at various scour depths. IM represents this motion on the fragility curve.

2.6.1. Fragility Curves at the Component Level

A two-parameter log-normal distribution (log-ND) characterizes the seismic fragility curve and is defined as follows [18]:

$$F(x_j; \theta_k, \beta_k) = \Phi\left(\frac{\ln(x_j/\theta_k)}{\beta_k}\right) \quad (2)$$

The function of fragility function $F(x_j;\theta_k,\beta_k)$ gives the probability of a seismic action with IM (PGA) = x_j will lead to bridge element failure at DS k (slight, moderate, major, and complete). The function $\Phi(\ )$ represents the cumulative distribution (CDF) with fragility parameters, θ_k (median) and β_k (standard deviation), of the log-ND at damage state k, respectively. Given the logical presumption that x_j for each seismic level operates independently, the fragility parameters are evaluated with the maximum likelihood method for truncated incremental dynamic analysis (IDA) [12]:

$$L = \left(\prod_{j=1}^{m} \Phi\left(\frac{\ln(x_j/\theta_k)}{\beta_k}\right)\right)\left(1 - \Phi\left(\frac{\ln(x_{max}/\theta_k)}{\beta_k}\right)\right)^{n-m} \quad (3)$$

For n seismic levels used, there will be m seismic levels that cause failure at x_j levels less than x_{max}, and $n-m$ seismic levels that result in failure prior to the termination of the analyses.

2.6.2. System-Level Fragility Curves

The overall damage condition of the bridge, under a particular seismic level and scour depth induced by flood events, is established by pinpointing the most severe damage across all critical bridge elements. The parameters that characterize the fragility curve at the system level are computed utilizing the maximum likelihood approach.

2.6.3. Fragility Surfaces

Fragility surfaces illustrate the cumulative impact of seismic action in the presence of local scour on the likelihood of bridge failure across different damage states. Statistical independence is assumed among random variables of the intensity measures, which depict flood and earthquake hazards. The joint CDF of seismic action and local scour induced by flood yields the structural failure probability in a combined-effect scenario. Fragility surfaces are derived using the maximum likelihood method of bivariate log-normal distribution.

$$L = \left(\prod_{j=1}^{m} F\left(x_{1,j}, x_{2,j}; \theta_{1,k}, \theta_{2,k}, \beta_{1,k}, \beta_{2,k}\right)\right)\left(1 - F\left(x_{1,j}, x_{2,j}; \theta_{1,k}, \theta_{2,k}, \beta_{1,k}, \beta_{2,k}\right)\right)^{n-m} \quad (4)$$

In this expression, the random variables of $x_{1,j}$ and $x_{2,j}$ denote peak ground acceleration and the annual peak flow discharge, respectively. The values of medians $\theta_{1,k}$ and $\theta_{2,k}$ and log-standard deviations $\beta_{1,k}$ and $\beta_{2,k}$ are given for damage state k. The bivariate log-normal distribution's probability density function is derived using the Jacobian of the transformation and can be articulated as follows:

$$f\left(x_{1,j}, x_{2,j}; \theta_{1,k}, \theta_{2,k}, \beta_{1,k}, \beta_{2,k}\right) = \frac{1}{2\pi x_{1,j} x_{2,j} \beta_{1,k} \beta_{2,k}} \exp\left(-\frac{\left(\frac{\ln\left(x_{1,j}/\theta_{1,k}\right)}{\beta_{1,k}}\right)^2 + \left(\frac{\ln\left(x_{2,j}/\theta_{2,k}\right)}{\beta_{2,k}}\right)^2}{2}\right) \quad (5)$$

2.7. Risk Evaluation

The expected risk associated with earthquake and flood hazards concerning the bridge is delineated through risk curves. These curves present the annual probabilities of exceedance across different types of bridge rehabilitation expenses within the multiple impacts of seismic action and local scour induced by flood context. In the multi-hazard scenario denoted as m, the bridge rehabilitation cost for repair or replacement, designated as C_{RPm}, is explicitly defined as follows [19]:

$$C_{RPm} = \sum_{1}^{4} p_m(DS = k|a_m, d_m) C_n r_k \quad (6)$$

The probability $p_m(DS{=}k\,|\,a_m, d_m)$ signifies the likelihood for which the examined structure can endure DS k given seismic level a_m (PGA), derived from the bridge's fragility data, occurring alongside local scour d_m. C_n denotes the replacement cost of the bridge. The values of damage ratios, r_k, associated with damage k, as outlined in Table 2 and recommended in HAZUS (2022) [20], are employed.

Table 2. Different damage ratios of bridge element.

Definition of Damage State	Optimal Damage Ratio
DS1 (Slight)	0.03
DS2 (Moderate)	0.08
DS3 (Extensive)	0.25
DS4 (Complete)	0.33 *

* When the number of spans surpasses two, the optimal estimate for complete damage ratio is determined as [2 divided by the total number of spans]. For the specific bridge under examination, this equates to 2/6 spans, resulting in 0.33.

In this research, the cost of repairing the bridge is determined in relation to its replacement value, with the damage ratio serving as the proportionality factor.

3. Case Study

Most Albanian bridges have short spans consisting of simply supported single-span or multiple-span with reinforced concrete or precast pre-stressed girders. This bridge class is of interest to assess the vulnerability and estimate the performance. Most of them are supported by pile foundations, offering protection against additional seismic action stemming from local scour induced by floods. Nevertheless, local scour can amplify the fragility of bridges, particularly at slight and moderate DS, as scouring adversely affects elements of the bridge superstructure. These insights will streamline bridge design considerations within the specified effect of combining earthquake and flood hazards.

3.1. Study Bridge

The bridge is on the segment of the "Urake—Ura e Shoshajve" road, crosses the Kurvajt River, and has six simply supported spans. The bridge was built in 2013, and drawings and details are given in Figure 2. The length of the bridge is 178.65 m. The foundation of the bridge is on a fine sandy layer. The bridge's superstructure is constructed from precast post-stressed concrete I-girders and reinforced concrete slab connected monolithically with girders. The bridge's superstructure relies on elastomeric bearings to support its I-girders. Each bridge bent consists of two reinforced concrete columns, each with a diameter of 1.50 m, interconnected with a monolithic bent cap. All bents exhibit identical cross-sectional dimensions and material characteristics. Each group of bents is supported by a set of reinforced concrete piles, each with a diameter of 1.2 m, capped with reinforced concrete pile caps. Expansion joints are installed at abutments and bents numbers 2 and 4 of the bridge. These joints create a separation of 50 mm within the bridge structure.

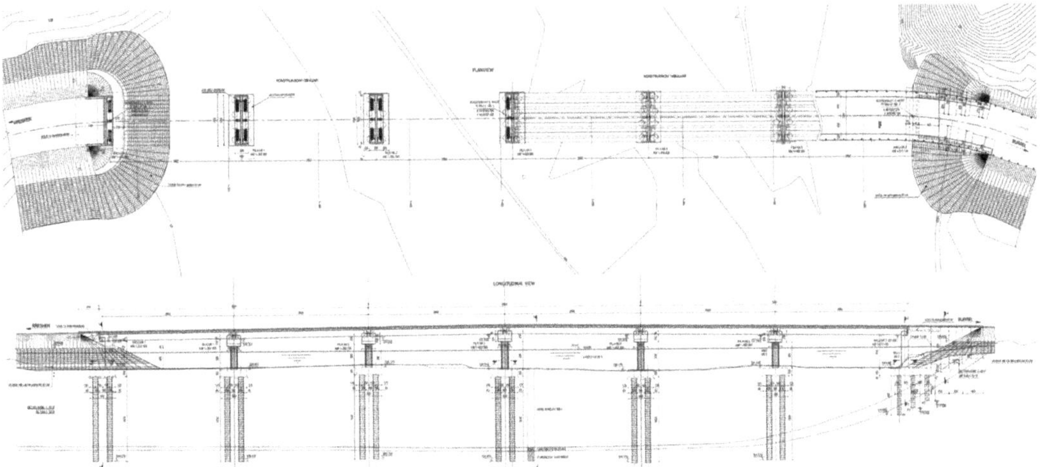

(I)

Figure 2. *Cont.*

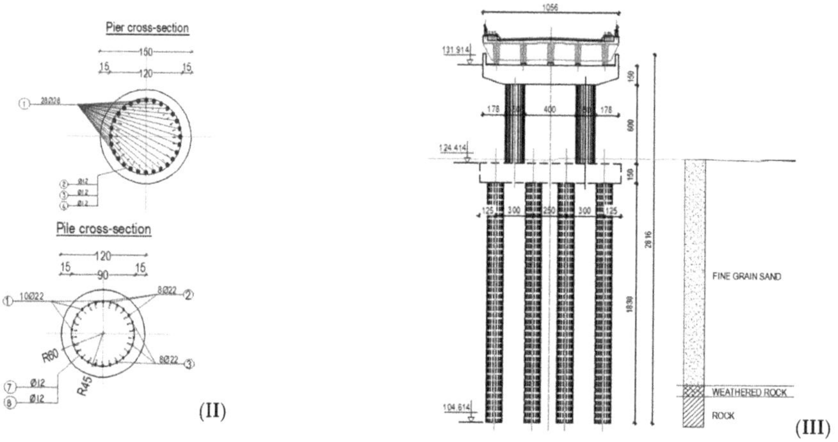

Figure 2. Bridge details: (I) plan and longitudinal view; (II) cross-section of bents and piles; (III) cross-section of bridge at bent.

3.2. *Flood and Earthquake Hazards*

3.2.1. Earthquake Hazard

In conducting this study, the earthquake hazard is evaluated by referring to the seismic hazard map of Albania (PGA, 10% probability of exceedance in 50 years, and V_{S30} site condition of 800 m/s) [21], which provides information about peak ground acceleration for a 475-year RP and ground motion of the 26 November 2019 earthquake in Durres, Albania. The time history analysis focuses on the horizontal orthogonal directions of the earthquake. The bents are not subjected to high bending moments due to vertical permanent loads of superstructure, and so the vertical component of seismic action is not considered [22]. The time history acceleration chosen from a location near the bridge site is adjusted to align with the response spectrum outlined in Eurocode 8-1 [23], and the matched records for each seismic level (a_{gR} = 0.05 g, 0.10 g, 0.20 g, and up to 0.5 g) are generated. The defined target spectrum is Type 1 of the horizontal elastic response spectrum with a damping ratio of 5% and C-ground type. The importance factor for the bridge is equal to $\gamma_I = 1.0$ (importance class II, average importance).

Seven monitoring stations (www.geo.edu.al (accessed on 8 January 2024)) detected a significant earthquake on 26 November 2019 (M = 6.4), spanning an epicentral distance range of 15 to 130 km [24]. Figure 3 illustrates the recorded ground motions in Tirana, displaying both the (N-S) and (E-W) directions (accelerometric station TIR1, 33.7 km from the epicenter). As shown in Figure 3, the horizontal peak ground acceleration is approximately 0.12 g, and in Durres, near the epicenter, this value was approximately 0.2 g [25].

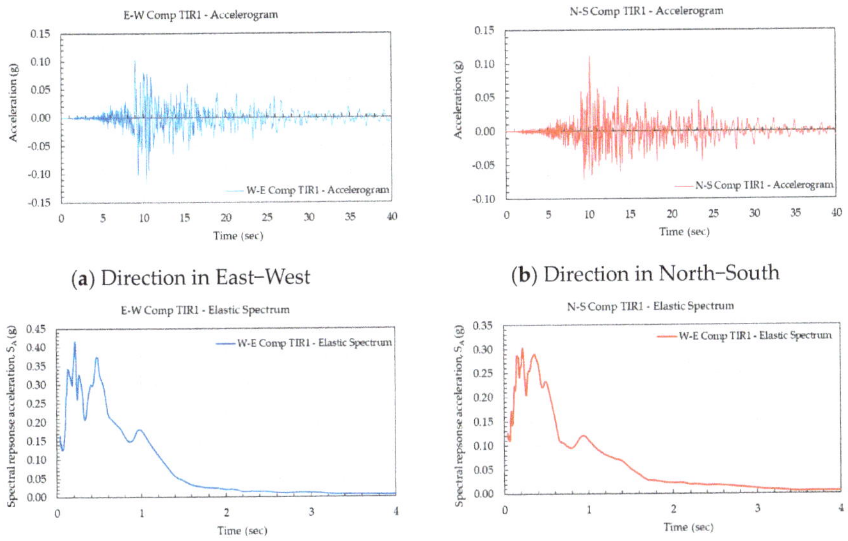

(**a**) Direction in East–West (**b**) Direction in North–South

(**c**) Elastic spectrum in East–West (**d**) Elastic spectrum in North–South

Figure 3. Accelerograms and elastic spectra obtained from the earthquake that took place on November 26 (M = 6.4), 2019, as observed by the accelerometric station located in Tirana.

3.2.2. Flood Hazard

Table 3 presents the peak flow discharges observed at the bridge site for flood events characterized by different annual exceedance probabilities [26]. The annual peak discharge data are obtained from original studies of the project.

Table 3. Local scour depths at bent foundations.

Flood Probability of Exceedance		1%	2%	5%	10%	20%
Q [m^3/s]		**661**	**595**	**528**	**463**	**397**
Scour depth (y_s), [m]	Bent No. 1	3.48	2.94	2.59	2.32	2.12
	Bent No. 2	3.40	2.86	2.51	2.24	2.03
	Bent No. 3	3.54	2.99	2.64	2.37	2.18
	Bent No. 4	3.62	3.07	2.72	2.45	2.27
	Bent No. 5	3.53	2.99	2.42	2.12	1.91

3.2.3. Local Scour at Bents Foundations Induced by Flood

In conducting an assessment of the multiple impacts of the case study bridge, flood occurrences for annual probabilities of exceedance of 20%, 10%, 5%, 2%, and 1% are studied. Bridge bent foundation scour depths, denoted as y_s, are estimated using Equation (1). Table 3 gives the resulting values at all bent foundations. The slopes on both abutments have river protection, and no local scour is expected.

3.3. FE Modeling of Elements

The structure of the bridge is analyzed using CSIBridge v25.0.0 software. The effectiveness of the analytical model for the bridge relies on its elements' functionality and interconnectivity. The FE model of the bridge shown in Figure 4 includes some assumptions and idealization, as described below. Characteristic values of materials are used as given in the project details. Concrete class C30/37 is used for reinforced concrete components and C45/55 for precast girders. The increment of existing concrete strength due to the aging of concrete is not considered. The steel class of B500C is used for reinforcing rebars. In the seismic analysis, the permanent masses of self-weight of structural elements, wearing surfaces, sidewalks, and barriers are considered, and the moving loads are excluded (EN1998-2) [22]. Also, referring to the research works carried out by Pietro et al. (2020, 2022) [9,10], the bridge is visually inspected, and it is concluded that the reinforced concrete elements and precast girders have adequate concrete cover and no exposed steel rebars are observed. So, the influence of corrosion induced by the carbonation phenomenon is not considered in the analysis.

Figure 4. 3D FE model of bridge built in CSIBridge v25.0.0 Software.

3.3.1. Superstructure

The superstructure comprises a composite section incorporating a cast-in-place reinforced concrete slab and precast post-tensioned girders, synergistically operating as a unified entity. The superstructure is anticipated to maintain its elasticity and remain undamaged during seismic events. The girders are linked to bent caps or abutments through spring elements. The modulus of elasticity of concrete is estimated according to Section 3.1 of EN 1992-1-1:2004 [27]. For the composite section of the superstructure, no stiffness reduction is applied. An ideal bond between steel rebars and concrete, ensuring that the structural components retain their planar configuration following the application of loads, is considered.

3.3.2. Bents

The bridge bents are made up of columns and cap beams. The orientation of the bents' strong axes aligns with the bent's transverse direction, attributed to the stiffness contributed by the cap beam with multiple column bents. Finite element discretization-based displacement is employed to perform nonlinear modeling of both bents and cap beams. The concrete material model proposed by Mander et al. (1988) [28] is utilized to define the stress–strain relations for unconfined and confined sections, shown in Figure 5.

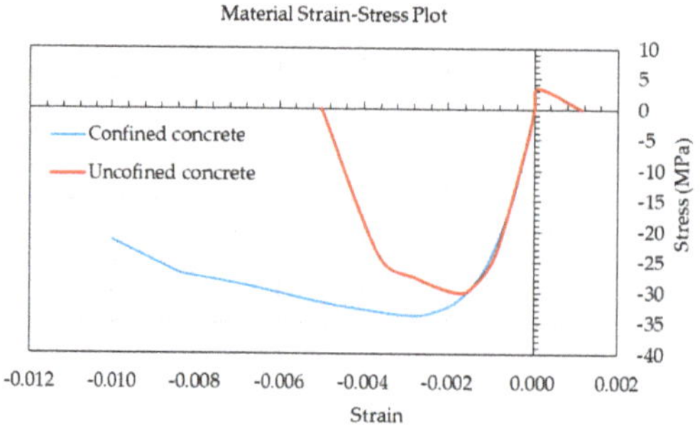

Figure 5. Material stress–strain plot for unconfined and confined concrete.

3.3.3. Abutments

The abutments interact with the earthfill behind the back wall. Therefore, soil structure is considered in abutment modeling. The space between the abutment wall and girders is represented utilizing elastic-perfectly plastic gap components. The backfill soil on the abutment wall is modeled as suggested by Caltrans (2013) [29], and for dynamic loading, the passive pressure will be raised by 50%. The abutment resistance in the bridge axis direction does not incorporate the influence of active soil pressure. Therefore, in the active direction, piles are the only resisting components. Spring elements are arranged in parallel along the bridge axis direction. The stiffness of the abutment wall is calculated using Equation (7).

$$K_{\text{abut}} = K_{\text{i}} \cdot w \cdot \left(\frac{h}{1.7} \right) \tag{7}$$

Given the parameters, w represents the width and h signifies the height of the backwall. The backfill soil stiffness is taken as K_{i} = 14.35 kN/mm/m as proposed by Caltrans. The maximum passive pressure of 239 kPa given by Caltrans is increased by 50%, and 368 kPa of soil pressure is utilized for dynamic and earthquake loads. This pressure is utilized to calculate the passive pressure force opposing movement at the abutment:

$$F_{\text{abut}} = \left(\frac{h^2 \cdot w \cdot 360}{1.7} \right) \tag{8}$$

3.3.4. Elastomeric Bearings

Under each precast post-tensioned girder of the superstructure, elastomeric bearings have been positioned and are considered free to move. The schematic models of the bridge components are shown in Figure 6. The horizontal stiffness of the elastomeric bearings is significantly lower when compared to the adjacent substructure and superstructure. In both horizontal directions, the elastomeric bearings demonstrate a behavior that can be accurately described using non-linear elements (elastic-perfectly plastic model). At the same time, the yield force aligns with the frictional force that arises at the interface between the concrete surface and the bearings. The bridge elastomeric bearings have a shear modulus of G_{bearing} = 0.90 N/mm^2 with a friction coefficient set at 0.4 [30]. For this type of bridge, bearings are deemed sacrificial components, requiring inspection for damage and subsequent replacement in the aftermath of a significant seismic event.

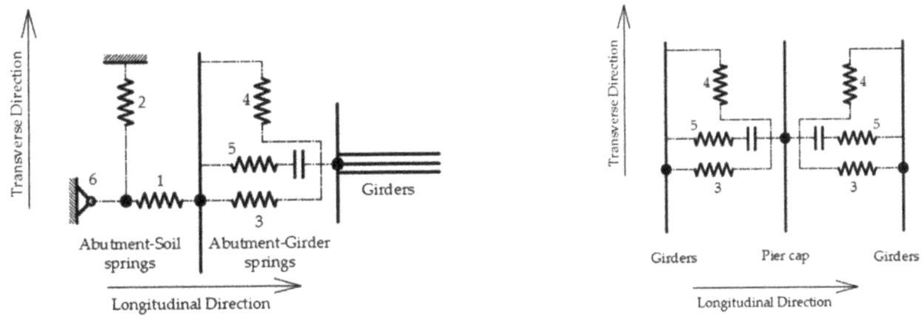

1 and 2. Longitudinal and transverse directions of abutment springs.
3 and 4. Longitudinal and transverse directions of elastomeric bearing.
5. Abutment back wall or bent cap with a gap property.
6. Complete restraint except for unrestricted movement in the transverse direction.

(**a**) Abutment (**b**) Bent cap

Figure 6. Analytical model of the bridge components: (**a**) modeling of abutment; (**b**) modeling of bent cap (schematic model referring to details in original drawings).

3.3.5. Pounding Elements

Pounded elements are utilized to simulate the collision among neighboring bridge superstructures or the bridge superstructure and the wall of abutments and piers in the bridge axis direction. For the lateral direction, pounding arises as the superstructure interacts with the concrete shear keys. The space distances for the pounding elements in both directions are set at 50 mm. Figure 7 illustrates the possible pounding location considered in the FE model analytically, taking into account the bridge's amplified response due to the potential impact of structural elements (e.g., slab and back wall of abutment).

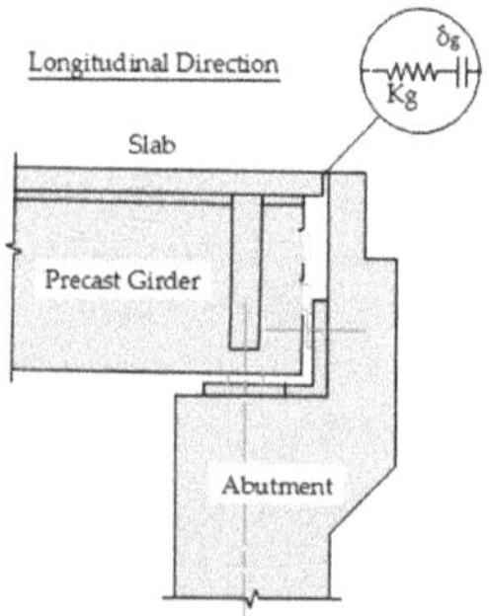

Figure 7. Pounding location at the abutment under seismic actions (schematic model referring to the original drawings used in the FE model).

3.3.6. Foundations

The connection points between the soil and structure are depicted using a collection of non-linear soil springs distributed along the foundation piles and modeled using *p-y* spring elements for both horizontal directions. In contrast, the pile's bottoms are fixed.

3.4. Non-Linear Time History Analysis

The bridge undergoes nonlinear time history analyses to assess its behavior under the ground motions generated as described above for different seismic levels by matching the real accelerogram of the Durres earthquake of Mw 6.4 to the target response spectrum with different seismic levels of Eurocode 8, as presented in Appendix A.

4. Results

In this paper, a multi-span, simply supported concrete girder bridge, which is the most common bridge type in Albania, is analyzed. The bridge is modeled with finite elements in CSIBridge v25.0.0 software and considers soil–structure interactions for different local scour depths and seismic levels. Then, the fragility curves are generated by utilizing the gathered analysis output data for the key elements.

4.1. Modal Analysis of Bridge

The bridge is analyzed in its original state. Figure 8 illustrates the fundamental mode shapes in the actual condition. Table 4 gives the fundamental modal periods for all levels of flood probabilities of exceedance. The permanent masses of self-weight of structural elements, wearing surfaces, sidewalks, and barriers are considered. The bridge is assessed with normal traffic, and the quasi-permanent values of variable actions are taken to be equal to zero.

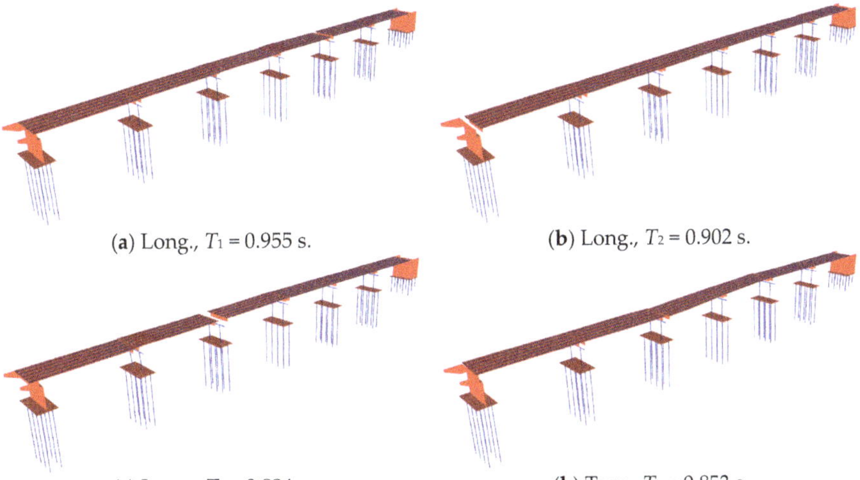

(**a**) Long., $T_1 = 0.955$ s. (**b**) Long., $T_2 = 0.902$ s.

(**c**) Long., $T_3 = 0.894$ s. (**b**) Tors., $T_4 = 0.853$ s.

Figure 8. Fundamental periods of bridge.

Table 4. Fundamental modal periods.

Flood Probability of Exceedance	**No Flood**	**1%**	**2%**	**5%**	**10%**	**20%**
Maximum Bent Scour [m]	-	**3.62**	**3.07**	**2.72**	**2.45**	**2.27**
First mode (long.) [s.]	0.955	0.975	0.970	0.967	0.966	0.966
Second mode (long.) [s.]	0.902	0.915	0.912	0.910	0.909	0.908
Third mode (long.) [s.]	0.894	0.904	0.901	0.900	0.899	0.898
Fourth mode (transv.) [s.]	0.853	0.854	0.854	0.854	0.854	0.853

Figure 8 shows that the bridge is dominated by a fundamental mode of vibration in the longitudinal direction due to the smaller stiffness of bent columns in this axis.

As the local scour depths increase, the modal periods also exhibit a corresponding slight increase. The pile foundations of the bridge have larger stiffness values than the bent columns, and the vibration modes show low sensitivity to increasing the local scour depths.

4.2. Generation of Fragility

4.2.1. Element-Level Fragility

Critical elements of the structure are subjected to seismic assessments at various depths of local scour to develop fragility curves, utilizing specific engineering demand parameters. Bents, abutments, and bearings are recognized as elements with a high damage potential. The threshold limits specified in Table 1 are utilized to compare the structural response of elements, aiding in determining the damage states of the bridge. The results of non-linear response history analyses, conducted under the influence of generated accelerograms at various seismic levels, are utilized to develop fragility curves at the element level. The maximum structural response for each ground motion of the bridge element is determined by identifying the absolute maximum from the response non-linear time history analysis for each ground motion, considering the relevant engineering demand parameters. In this research work, truncated incremental dynamic analysis is performed only up to the maximum intensity measure $IM_{max} = 0.5$ g, above which no further analyses are performed [12]. In the analysis of the bridge, $n = 10$ ground motions were applied, among which m motions led to collapse at IM (x_j) values less than IM_{max}

($x_{\max}$). Meanwhile, the remaining $n-m$ ground motions did not result in collapse before the end of the analysis. The given data are assessed using the maximum likelihood method. Equation (9) provides the mathematical expression for estimating the fragility function parameters for a specific damage limit state.

$$\{\hat{\theta}_k, \hat{\beta}_k\} = \text{argmax}_{\theta,\beta} \sum_{j=1}^{m} \left\{ \ln \Phi\left(\frac{\ln(x_j/\theta_k)}{\beta_k} \right) \right\} + (n-m)\ln\left\{ 1 - \Phi\left(\frac{\ln(x_{\max}/\theta_k)}{\beta_k} \right) \right\} \tag{9}$$

Figure 9 illustrates the fragility curves of structural elements at different local scour depths. The parameters, including median and log-normal values, are provided in these figures.

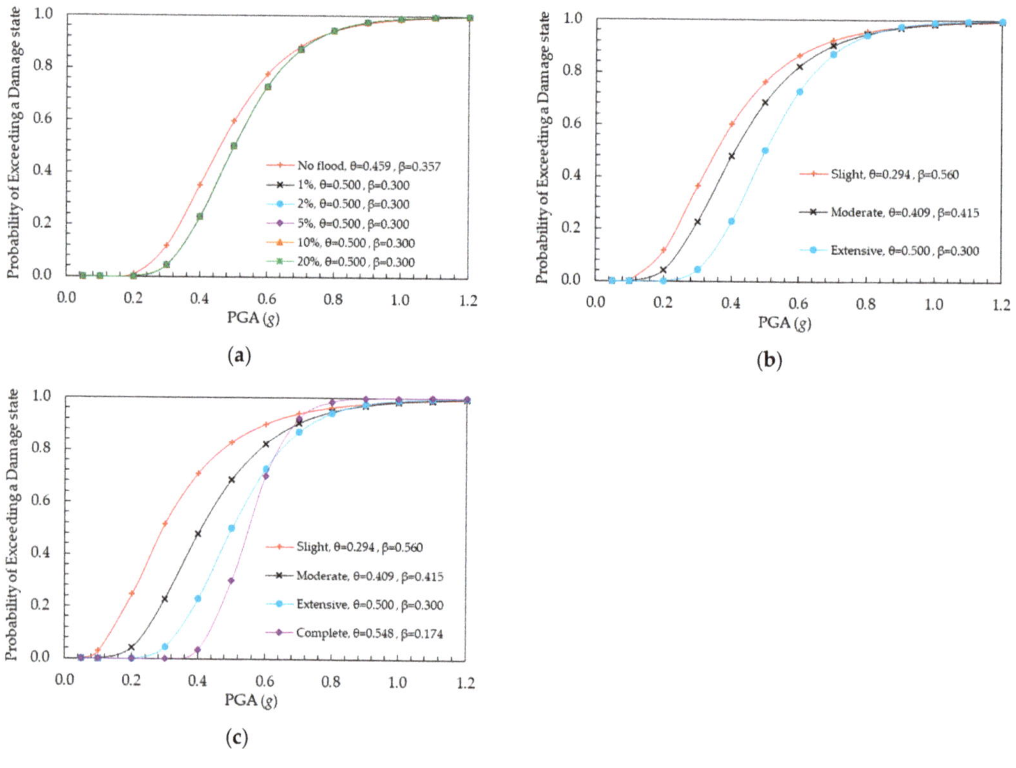

Figure 9. Element-level fragility curve: (**a**) flexural damage of bent at complete state, (**b**) deformation of abutment wall, and (**c**) deformation of elastomeric bearing at longitudinal direction.

This figure highlights that bents are the sole elements of the bridge affected by seismic activity, regardless of the local scour depths, which could potentially lead to complete collapse. Variations in the fragility of the element are noted in the scenario without local scour compared to bents with varying scour depths at different flood hazard levels. The heightened bents with scour depths enhance the flexibility of bridge bents, consequently decreasing vulnerability to seismic actions in this bridge's case.

4.2.2. System-Level Fragility Curves

If a damage threshold is reached or exceeded by any individual part of the bridge, the entire structure is deemed to be in the same DS, irrespective of the conditions of other bridge elements. Figure 10 illustrates the bridge's system-level fragility curves. Bent damage determines the complete damage state, while elastomeric bearings and abutments influence the extensive, moderate, and slight states. Additionally, increased local scour depth induced by flood events does not alter bridge seismic fragility.

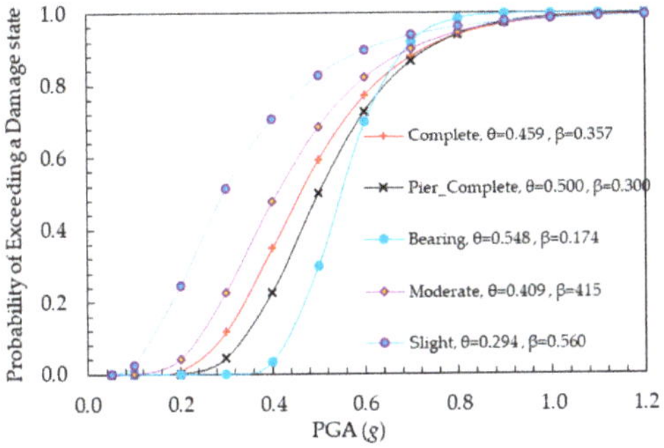

Figure 10. Fragility curves of bridge at the system level.

This paper emphasizes the importance of providing proper foundation stiffness and depth to reduce the possibility of bridge failure due to local scour induced by flood events in areas prone to earthquakes.

4.2.3. Fragility Surfaces

Figure 11 shows the fragility surfaces of the multiple impacts of seismic actions and floods treated as statistically independent random variables. The fragility surfaces are generated using the parameters outlined in Table 5 and mathematically estimated according to Equation (4). The formation of the fragility surface occurs exclusively when there is slight damage. Conversely, in the remaining three DS, no changes in the θ values of fragility curves are noted with increasing local scour depths induced by flood events. Consequently, the surfaces at these three DS extend linearly and represent the intensity measure of floods, following the respective fragility curves.

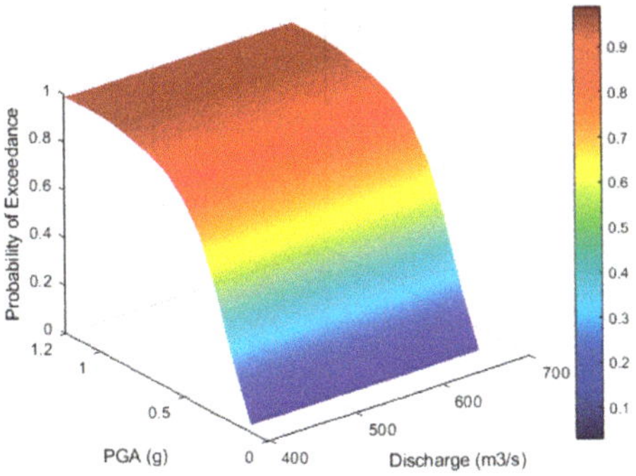

Figure 11. Fragility surface of bridge at slight DS.

Table 5. Values of θ and β at DS k.

Damage State	$\theta_{1,k}$	$\theta_{2,k}$	$\beta_{1,k}$	$\beta_{2,k}$
Slight	0.295	0.025	0.56	1.15

The slight DS has notably small values of θ of flood hazard. This suggests a minimal impact of the multiple impacts of seismic actions and local scour induced by floods.

The fragility analysis developed at the same damage conditions further supports this observation, demonstrating only small variations in bridge fragility characteristics.

4.3. Risk Evaluation

The average cost per unit area for construction of a new bridge of this type is 1150 Euro/m^2. This is the average cost of constructing bridges at current prices for the same type of structure. Also, the real cost of the bridge built in 2013 is used as a reference. Figure 12 shows the risk curve for multi-hazard scenarios. The risk curve is generated using the information gathered from the fragility surface and curves and seismic hazard curves.

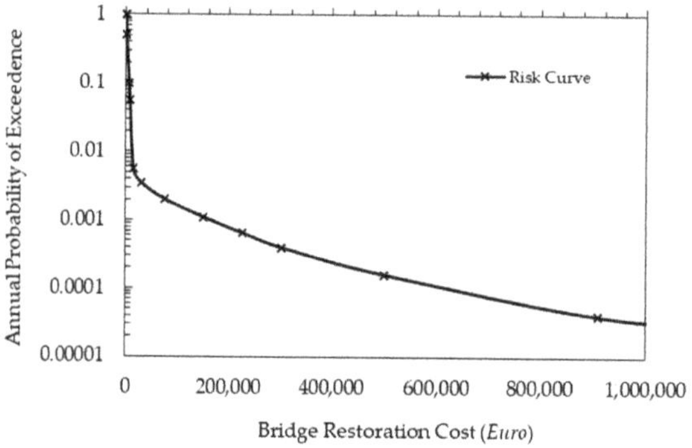

Figure 12. Seismic risk curves of bridge.

In the case of the study bridge, there is no noticeable shift in the bridge risk curve as the local scour depth induced by floods increases. This lack of change is expected, as the bridge's seismic fragility displays minimal sensitivity to scour induced by floods. In the case of slight DS, minor changes in the fragility curves of the bridge are noted for different scour depths. However, these changes are deemed insufficient to significantly alter the overall bridge risk curve.

5. Discussion

The presence of local scour induced by flood events can heighten the vulnerability of bridges imposed by later loads, such as that caused by an earthquake. Conversely, bridge flexibility might increase and help reduce earthquake load. However, the design philosophy and criteria used in the project could influence the overall safety of the bridge in scenarios involving local scour caused by flood hazards. In the last decade, several bridges in Albania, similar to the bridge under study, have had their foundations with piles uncovered by several meters due to the increase in flood events and, therefore, the increase in the risk of local scour. This situation significantly affects the structural stability of the bridge by causing a loss of lateral support at the foundations. Therefore, assessing the performance of the bridge for different local scour depths resulting from various annual peak flow discharges is crucial.

The results show that the fundamental periods of the bridge do not increase significantly with the scour depth, and this is due to the considerable stiffness of piles. The fragility curves of the study bridge for element and overall system levels, developed for multi-hazard scenarios, exhibit minor sensitivity to regional flood hazards. The bridge collapse is governed by the flexural failure of the bent (comprised of two columns each) due to seismic actions regardless of the local scour depths. A similar observation is also made by analyzing the graph of surface fragility and seismic curves. Also, the bridge is outfitted with elastomeric bearings, demonstrating reduced vulnerability to the harmful consequences of local scour. This is due to the damping characteristics and flexibility inherent in elastomeric materials. Additionally, the elastomeric bearings allow for greater adaptability to variations in load and movement. However, the increase in the local scour depth increases the probability of unseating the bridge superstructure in strong seismic activity.

The results are summarized as follows:

- The multi-span simply supported concrete bridge comprising elastomeric bearings and stiff foundations has low sensitivity to flood hazards.
- The seismic risk of a scoured bridge can be reduced by increasing the stiffness and depth of the foundations.

This study increases the attention of research communities and government agencies for the importance of studying bridges in areas prone to local scour induced by floods and earthquakes, which can be compounded by climate change. It will also serve as a basis for future research on implementing comprehensive bridge design that accounts for the potential impacts of earthquakes and changing climate conditions in multi-hazard scenarios, including the increased frequency and intensity of extreme weather events. Further research should focus on the impact of local scour induced by flood events on the seismic risk of these bridges (i.e., supported single or multi-span concrete bridges) and compare them with integral or continuous bridges. For the same type of bridge, the impact of foundation types on the bridge fragility will be of interest. It should also compare the components of the bridge that affect their functionality.

Author Contributions: Conceptualization, T.K. and J.C.M.; data curation, T.K., J.C.M. and N.-S.D.; formal analysis, T.K.; funding acquisition, T.K.; investigation, T.K.; methodology, T.K., J.C.M. and N.-S.D.; project administration, J.C.M.; resources, T.K.; software, T.K.; supervision, J.C.M.; validation, T.K.; visualization, T.K.; writing—original draft, T.K.; writing—review and editing, J.C.M., N.-S.D. and Y.X. All authors have read and agreed to the published version of the manuscript.

Funding: The paper is supported by the Project to Attract Foreign Experts (G2023133018L) from the Ministry of Science and Technology of China.

Institutional Review Board Statement: Not applicable.

Informed Consent Statement: Not applicable.

Data Availability Statement: The data presented in this study are available upon request from the corresponding author.

Conflicts of Interest: The authors declare no conflicts of interest.

Appendix A

Generated accelerograms used in time history analysis.

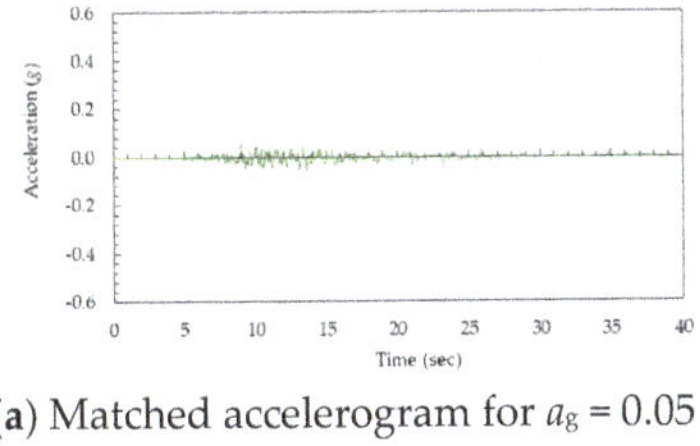

(**a**) Matched accelerogram for $a_g = 0.05$ g

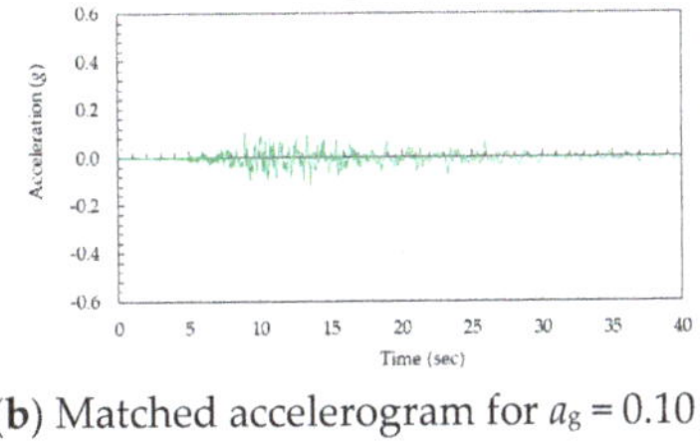

(**b**) Matched accelerogram for $a_g = 0.10$ g

Figure A1. *Cont.*

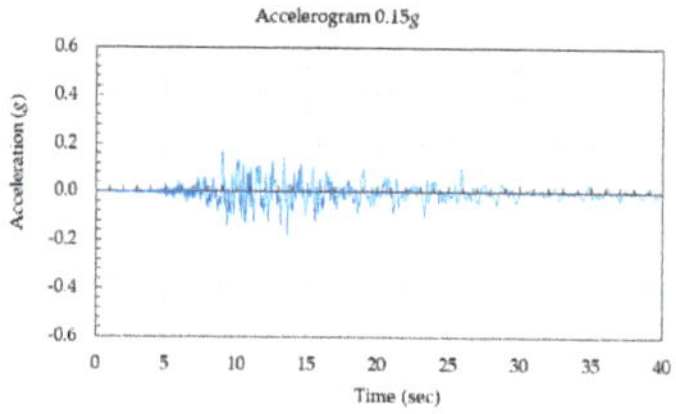

(**c**) Matched accelerogram for a_g = 0.15 g

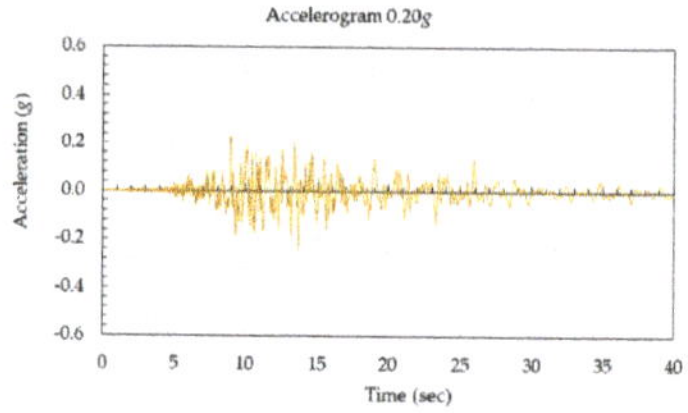

(**d**) Matched accelerogram for a_g = 0.20 g

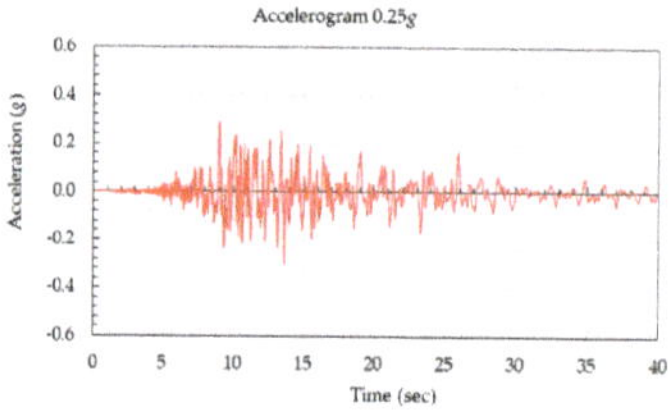

(**e**) Matched accelerogram for a_g = 0.25 g

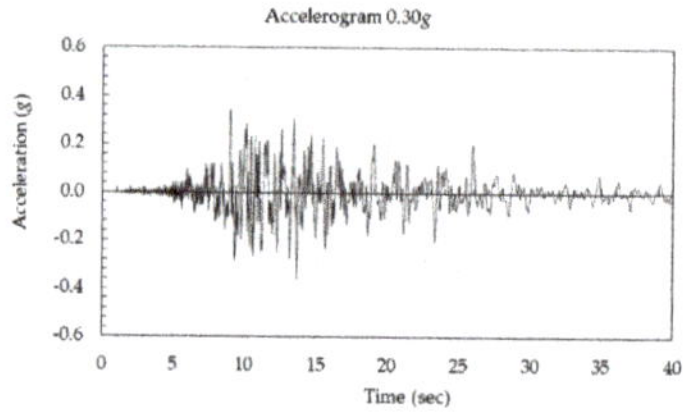

(**f**) Matched accelerogram for a_g = 0.30 g

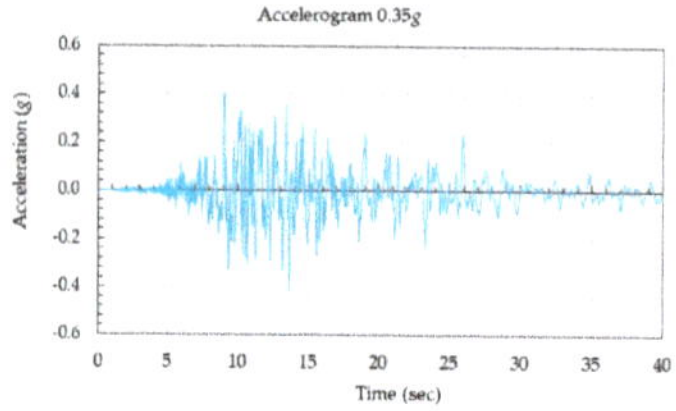

(**g**) Matched accelerogram for a_g = 0.35 g

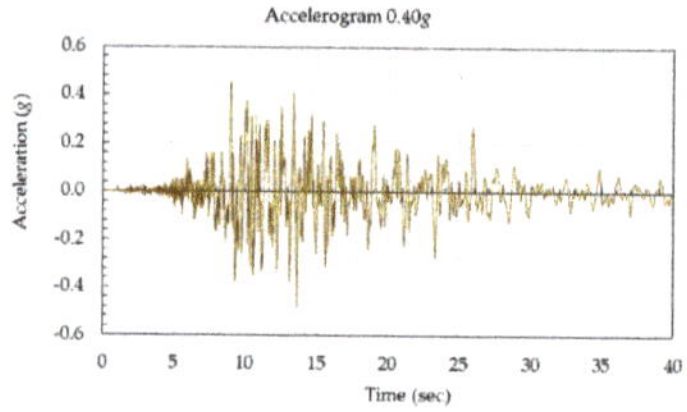

(**h**) Matched accelerogram for a_g = 0.40 g

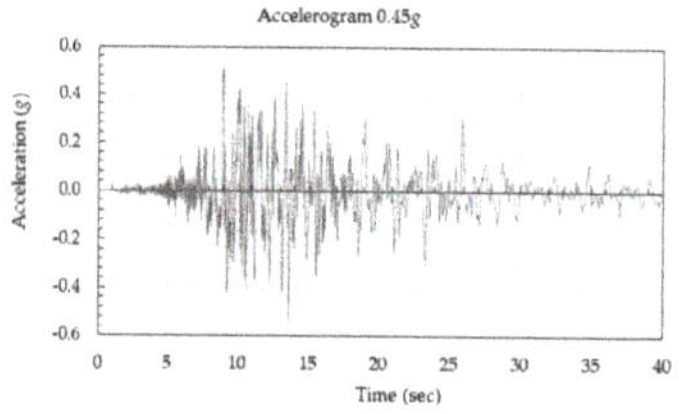

(**i**) Matched accelerogram for a_g = 0.45 g

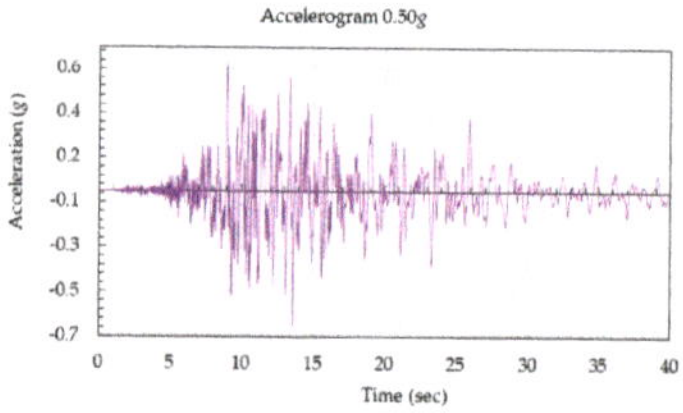

(**j**) Matched accelerogram for a_g = 0.50 g

Figure A1. Matched acceleration time histories.

References

1. Aliaj, S.; Koçiu, S.; Muço, B.; Sulstarova, E. *Seismicity, Seismotectonics and Seismic Hazard Assessment in Albania*; Academy of Sciences: Tirana, Albania, 2010.
2. Nemry, F.; Demirel, H. Transport and Climate Change: A Focus on Road and Rail Transport Infrastructures. JRC72217 2012, EUR 25553 EN. Available online: https://publications.jrc.ec.europa.eu/repository/handle/JRC72217 (accessed on 27 January 2024).
3. Arneson, L.A.M.; Zevenbergen, L.W.; Lagasse, P.F.; Clopper, P.E. *Evaluating Scour at Bridges. Hydraulic Engineering Circular (HEC) No.18*; Publication No. FHWA-HIF-12-003; Federal Highway Administration, U.S. Department of Transportation: Washington, DC, USA, 2012. Available online: https://www.fhwa.dot.gov/engineering/hydraulics/pubs/hif12003.pdf (accessed on 17 December 2023).
4. Bennett, C.R.; Lin, C.; Parsons, R.; Han, J. Evaluation of behavior of a laterally loaded bridge pile group under scour conditions. In Proceedings of the SEI 2009 Structures Congress, Austin, TX, USA, 30 April–2 May 2009; pp. 290–299. [CrossRef]
5. Avşar, Ö. Fragility Based Seismic Vulnerability Assessment of Ordinary Highway Bridges in Turkey. Ph.D. Thesis, Middle East Technical University, Ankara, Turkey, 2009. Available online: https://open.metu.edu.tr/handle/11511/18754 (accessed on 26 November 2023).
6. Yilmaz, T. Risk Assessment of Higway Bridges under Multi-Hazard Effect of Flood-Induced Scour and Earthquake. Ph.D. Thesis, The Pennsylvania State University, University Park, PA, USA, 2015. Available online: https://etda.libraries.psu.edu/files/final_submisssions/10677 (accessed on 22 November 2023).

7. NTC, Norme Tecniche per le Costruzioni, DM 17/01/2018 (In Italian). 2018. Available online: https://www.gazzettaufficciale.it/eli/gu/2018/02/20/42/so/8/sg/pdf (accessed on 1 December 2023).
8. Clarke, J.; O'Brien, E. A multi-hazard risk assessment methodology, stress test framework and decision support tool for resilient critical infrastructure. *Transp. Res. Procedia* **2019**, *14*, 1355–1363. [CrossRef]
9. Crespi, P.; Zucca, M.; Valente, M. On the collapse evaluation of existing RC bridges exposed to corrosion under horizontal loads. *Eng. Fail. Anal.* **2020**, *116*, 104727. [CrossRef]
10. Crespi, P.; Zucca, M.; Valente, M.; Longarini, N. Influence of corrosion effect on the seismic capacity of existing RC bridges. *Eng. Fail. Anal.* **2022**, *140*, 106546. [CrossRef]
11. Wang, Z.; Dueñas-Osorio, L.; Padgett, J.E. Influence of scour effects on the seismic response of reinforced concrete bridges. *Eng. Struct.* **2014**, *76*, 202–214. [CrossRef]
12. Baker, J.W. Efficient analytical fragility function fitting using dynamic structural analysis. *Earthq. Spectra* **2015**, *31*, 579–599. [CrossRef]
13. Nielson, B.; DesRoches, R. Analytical seismic fragility curves for typical bridges in the central and southeastern united states. *Earthq. Spectra* **2007**, *23*, 615–633. [CrossRef]
14. Argyroudis, S.; Mitoulis, S.; Winter, M.; Kaynia, A. Fragility of transport assets exposed to multiple hazards: State-of-the-art review toward infrastructural resilience. *Reliab. Eng. Syst. Saf.* **2019**, *191*, 106567. [CrossRef]
15. Ramanathan, K.N. Next Generation Seismic Fragility Curves for California Bridges Incorporating the Evolution in Seismic Design Philosophy. Ph.D. Thesis, Georgia Institute of Technology, Atlanta, GA, USA, 2012. Available online: https://repository.gatech.edu/server/api/core/bitstreams/1ec64ecb-b5e2-435d-ace3-e5ff0d3c0560/content (accessed on 6 January 2024).
16. Stefanidou, S.; Kappos, A. Methodology for the development of bridge-specific fragility curves. *Earthq. Eng. Struct. Dyn.* **2017**, *46*, 73–93. [CrossRef]
17. Atik, L.; Abrahamson, N. An improved method for nonstationary spectral matching. *Earthq. Spectra* **2010**, *26*, 601–617. [CrossRef]
18. Shinozuka, M.; Feng, M.Q.; Lee, J.; Naganuma, T. Statistical analysis of fragility curves. *J. Eng. Mech.* **2000**, *126*, 1287–1296. [CrossRef]
19. Zhou, Y.; Banerjee, S.; Shinozuka, M. Socio-economic effect of seismic retrofit of bridges for highway transportation networks: A pilot study. *Struct. Infrastruct. Eng.* **2010**, *6*, 145–157. [CrossRef]
20. Federal Emergency Management Agency (FEMA). Hazus Earthquake Model Technical Manual: HAZUS 5.1. Washington, DC, USA. 2022. Available online: https://www.fema.gov/sites/default/files/documents/fema_hazus-earthquake-model-technicalmanual-5-1.pdf (accessed on 15 December 2023).
21. Kuka, N.; Duni, L.; Koçi, R.; Dushi, E.; Xhahysa, A. *Probabilistic Seismic Hazard Assessment of Albania*; Department of Seismology, Institute of Geosciences: Tirana, Albania, 2020. Available online: https://www.geo.edu.al/rc/doc/New_PSHA_for_Albania_5460.pdf (accessed on 15 December 2023).
22. *EN 1998-2:2005*; Eurocode 8—Design of Structures for Earthquake Resistance—Part 2: Bridges. European Committee for Standardization: Brussels, Belgium, 2005; pp. 44, 50.
23. *EN 1998-1:2004*; Eurocode 8—Design of Structures for Earthquake Resistance—Part 1: General Rules, Seismic Actions and Rules for Buildings. European Committee for Standardization: Brussels, Belgium, 2004; p. 37.
24. Duni, L.; Theodoulidis, N. Short Note on the November 26, 2019, Durres (Albania) M6. 4 Earthquake: Strong Ground Motion with Emphasis in Durrës City. *EMSC Online Rep.* 2019. Available online: https://www.ssis.org.al/wp-content/uploads/2020/01/Short-Note_EMSC_Duni-Theodoulidis-1.pdf (accessed on 2 December 2023).
25. Freddi, F.; Novelli, V.; Gentrile, R.; Veliu, E.; Andreev, S.; Andonov, A.; Greco, F.; Zhuleku, E. Observations from the 26th November 2019 Albania earthquake: The earthquake engineering field investigation team (EEFIT) mission. *Bull. Earthq. Eng.* **2021**, *19*, 2013–2044. [CrossRef]
26. Kolaneci, M.; Jegeni, T. Climatic and hidrological conditions of road "Urake to Ura e Shoshajve". Tirana, Albania, 2009.
27. *EN 1992-1-1:2004*; Eurocode 2—Design of Concrete Structures—Part 1-1: General rules and Rules for Buildings. European Committee for Standardization: Brussels, Belgium, 2004; p. 29.
28. Mander, J.B.; Priestley, M.J.N.; Park, R. Theoretical stress-strain model for confined concrete. *J. Struct. Eng.* **1988**, *114*, 1804–1826. [CrossRef]
29. Caltrans, California Department of Transportation. Seismic Design Criteria Version 1.7, Sacramento, CA, 2013. Available online: https://dot.ca.gov/-/media/dot-media/programs/engineering/documents/seismicdesigncriteria-sdc/f0007585seismicdesigncriteriasdc17fullversionoeereleasea11y.pdf (accessed on 26 November 2023).
30. Caltrans, California Department of Transportation. Seismic Design Criteria Version 2.0, Sacramento, CA, 2019. Available online: https://dot.ca.gov/-/media/dot-media/programs/engineering/documents/seismicdesigncriteria-sdc/202007-seismicdesigncriteria-v2-a11y.pdf (accessed on 26 November 2023).

MDPI AG
Grosspeteranlage 5
4052 Basel
Switzerland
Tel.: +41 61 683 77 34

Applied Sciences Editorial Office
E-mail: applsci@mdpi.com
www.mdpi.com/journal/applsci

www.ingramcontent.com/pod-product-compliance
Lightning Source LLC
LaVergne TN
LVHW070452061126
851858LV00011B/221